HOLLOWED
GROUND

Great Lakes Books

A complete listing of the books in this series can be
found online at wsupress.wayne.edu

EDITOR

Charles K. Hyde
Wayne State University

ADVISORY EDITORS

Jeffrey Abt
Wayne State University

Fredric C. Bohm
Michigan State University

Michael J. Chiarappa
Western Michigan University

Sandra Sageser Clark
Michigan Historical Center

Brian Leigh Dunnigan
Clements Library

De Witt Dykes
Oakland University

Joe Grimm
Bloomfield Hills, Michigan

Richard H. Harms
Calvin College

Laurie Harris
Pleasant Ridge, Michigan

Thomas Klug
Marygrove College

Susan Higman Larsen
Detroit Institute of Arts

Philip P. Mason
Prescott, Arizona and
Eagle Harbor, Michigan

Dennis Moore
Consulate General of Canada

Erik C. Nordberg
Michigan Technological University

Deborah Smith Pollard
University of Michigan–Dearborn

David Roberts
Toronto, Ontario

Michael O. Smith
Wayne State University

Joseph M. Turrin
Wayne State University

Arthur M. Woodford
Harsens Island, Michigan

HOLLOWED GROUND

❖ ❖ ❖

*Copper Mining and Community Building
on Lake Superior, 1840s–1990s*

LARRY LANKTON

WAYNE STATE UNIVERSITY PRESS

Detroit

Library of Congress Cataloging-in-Publication Data

Lankton, Larry D.
Hollowed ground : copper mining and community building on Lake Superior, 1840s–
1990s / Larry Lankton.
        p. cm. — (Great Lakes books)
Includes bibliographical references and index.
ISBN 978-0-8143-3458-4 (cloth : alk. paper) — ISBN 978-0-8143-3490-4 (pbk. :
alk. paper)
1. Keweenaw Peninsula (Mich.)—History. 2. Keweenaw Peninsula (Mich.)—Social
conditions. 3. Keweenaw Peninsula (Mich.)—Economic conditions. 4. Copper mines
and mining—Michigan—Keweenaw Peninsula—History. 5. Copper mines and mining—Social aspects—Michigan—Keweenaw Peninsula—History. 6. Community life—
Michigan—Keweenaw Peninsula—History. 7. Mining camps—Michigan—Keweenaw
Peninsula—History. 8. Keweenaw Peninsula (Mich.)—History, Local. I. Title.

F572.K43L36 2010
977.4'99—dc22

2009047227

∞

Designed and typeset by Anna Oler
Composed in Trajan and Garamond

*To the memory of Bill Gregg, Michigan Tech professor of geological engineering.
While doing volunteer work for the non-profit Quincy Mine Hoist Association,
he became the last man to die in a fall at Quincy's No. 2 shaft, 6 December 2008.*

CONTENTS

❖ ❖ ❖

ACKNOWLEDGMENTS

❖ ❖ ❖

In the summer of 1978, while working for the National Park Service, I led an Historic American Engineering Record field team that documented the history of the Quincy Mine in Hancock, Michigan. Since then I have continued studying and writing about the copper-mining industry that once thrived on the south shore of Lake Superior. I've traveled to many libraries and archives and interacted with many students and scholars. If I thanked everyone who in the past thirty years has helped me, my acknowledgments would be nearly as long as the rest of the book. So I am obliged to keep things short. Still, it is necessary to thank people and institutions from decades ago, as well as people and institutions from today, because all of them contributed to the making of this book.

I thank the members of the 1978 HAER team who labored hard to collect data on the Quincy Mine, especially Charlie Hyde, Charles O'Connell, Richard Anderson, and Sarah McNear. Louis Koepel shared all the keys and combinations needed to access Quincy's record vaults, and he shared his love of Copper Country history. Meanwhile, Theresa Spence, then head of the Michigan Technological University Archives, made it easy to use the historical materials in her care.

Over time, the Smithsonian Institution, the Hagley Museum and Library, and the Dibner Institute supported me with research fellowships, and Michigan Technological University supported me with research grants. Michigan Tech undergraduate students marched off to the Archives, and as they learned how to be historical detectives, garnered useful information. William Gruhlke, David Kari, and Deanna Koryczan helped reconstruct nineteenth-century Hancock for me; Kathleen Dravillas, Eric Durkee, Keri Ellis, and Kim Wilmers did the same for the village of Calumet. Graduate students in our Industrial Archaeology program at Michigan Tech also contributed. Richard Fields, Nancy Fisher, Efstathios Pappas, John Griebel, Gary Kaunonen, and Scott See wrote helpful master's theses on Copper Country history. James Rudkin, Shannon Bennett, Stephanie Atwood, and Vanessa McLean, graduate students in my Industrial Communities class, contributed information on the Copper Range communities of Baltic, Trimountain, Painesdale, and White Pine. Similarly, grad students Seth DePasqual, Megan Glazewski, Sean Gohman, Jessica Montcalm, Chris Nelson, and Craig Wilson pointed me to historical sources and illustrations covering the Cliff and Phoenix mines and the waters, railroads, mills, and smelters of Portage Lake.

The Keweenaw National Historical Park contracted with me several years ago to write the Park's "Historic Resources Study," covering the Calumet and Hecla and the Quincy mines. In much revised form, parts of that study reside here. Lynn Bjorkman generously shared park materials she had put together concerning the mine locations around Calumet, and Jeremiah Mason provided historic photos from the park's archives. Back on the Michigan Tech campus, for years I have benefited from the knowledge and cooperation of Erik Nordberg and his staff at the University Archives and Copper Country Historical Collections. Most of the information in this book, and most of the illustrations, came from the Tech Archives. Erik, Julia Blair, and Christine Holland have always gone above and beyond the call of duty to help me resolve research questions and find historic photos and maps.

Graduate student Scott See saved his old professor much aggravation by skillfully manipulating digital images and helping assemble all the illustrations used in this book. Thanks also go out to Gary Kaunonen of the Finnish American Heritage Center for help with images and information. The Ontonagon County Historical Museum and the Lake States Railway Historical Association were also particularly helpful and cooperative.

I am also indebted to Jim Bekkala, Ron Whiton, and Larry Chabot, who reviewed the portions of this book covering the history of the White Pine mine and townsite. As longtime White Pine employees and residents, they all lived the history that I have written about, and I appreciated the extensive comments they provided on the White Pine story. If I still have any of that story wrong, the fault lies with me, not them. Charlie Hyde at Wayne State University read drafts of this manuscript and provided many helpful suggestions. I thank Charlie for his expertise and encouragement. I especially need to thank those at Wayne State University Press who moved this book project along: Kathryn Wildfong, Annie Martin, Kristin Harpster Lawrence, Maya Rhodes, Margaret Erdman, and copy editor John Flukas.

I would be remiss if I did not thank my departmental colleagues at Michigan Tech, historians, archaeologists, and anthropologists who have long heard me out and offered encouragement. Many have done their own work on copper or iron, on mining or other industrial communities. Any historian would be fortunate indeed to have offices down the hall occupied by faculty like Bruce Seely, Terry Reynolds, Pat and Susan Martin, Carol MacLennan, and Kim Hoagland. Thanks go out especially to Kim. While working on her own book about paternalism at the Lake Superior copper mines, *Mine Towns: Buildings for Workers in Michigan's Copper Country,* she graciously shared research findings and sources with me. I hope she believes that I returned the favor.

Finally, thanks to wife Rachel for her constant encouragement, and to the kids David and Laura, both born and raised on Michigan's Keweenaw Peninsula, a special place of beautiful shorelines and stamp-sand beaches, of extremely green summers and extremely white winters, of ice rinks and history that sustained us all very well.

INTRODUCTION

❖ ❖ ❖

Any visitor to Upper Michigan's Copper Country who is interested in the history of the Keweenaw Peninsula and its mines, communities, and people should go stand on the corner of Red Jack Road and Mine Street in Calumet. The visitor should slowly turn around to see the various parts of the landscape visible from this spot. One sees some of what's left of the great Calumet and Hecla (C&H) mine: its railroad roundhouse, machine shop, warehouse, gearhouse, and the smokestack over by the Superior engine's boilerhouse. This corner is near the old heart of the mine, once filled with hoists, compressors, man-engines, smoke, noise, and rail traffic in and out. Sitting adjacent this industrial tumult was the C&H library and administration building, plus the Miscowaubic Club, a gathering place for the area's well-to-do. Hard by the industrial core, and not tucked away in some quiet, protected spot, sat Calumet's high school and grade school. Look over to see these still-standing structures and also discover a nest of church steeples only a few blocks away, near the head of Calumet's Fifth Street, its main corridor of shops and stores. Housing, too, is found only a block from this corner, which seems to have been near the center of everything. The view reinforces a main historical point that this book makes about the Keweenaw (pronounced KEY-wah-naw) and its mines. Along the Keweenaw's mineral range, life and work and company and community were never far apart. They existed side by side. A major intent of this volume is to show how mining companies built both mines and communities, and to show how closely connected and overlapping their histories were, in times of economic growth and in times of decline and even closure.

The book has other aims, as well, and covers no little sweep of time. The narrative begins about a billion years ago, when the Keweenaw was being formed, and then pauses at about seven thousand years ago, when native peoples first began taking copper from the region. The bulk of the story covers a century and a half of copper mining, from the arrival of the first incorporated copper mines in the 1840s until the closing of the last copper mine in the mid-1990s. The book narrates the full life cycle not just of one, but of two, copper industries on Lake Superior. The first, the more famous one, mined *native copper*—metallic copper found underground, unalloyed with other elements. That industry chased copper deposits down mine shafts as much as 9,200 feet deep and produced approximately eleven billion pounds of the red metal from the 1840s until the late 1960s. About a decade and a half before the last native copper mines closed, in the 1950s about seventy miles south of the heart of the old mining district, a new copper mine opened at White Pine. This mine used vastly different technologies to exploit a very different type of copper deposit—a *copper sulfide* deposit, or *chalcocite,* found underground in a vast dome of shale. Over a life span of about four decades, White Pine contributed another 4.4 billion pounds of copper to Lake Superior's production. This book takes both the native copper and the sulfide copper–mining industries from inception to closure.

The performance of the overall industry across a century and a half is covered here, and many different mining companies are discussed or mentioned. This book makes no attempt, however, to deal with all the mines equally. The focus is on three mining companies in particular, each of which operated for about a century. The Quincy Mining Company, dating back to the 1840s, was the longest-lived of the first-generation producers and put together a remarkable string of dividend-paying years. C&H was by far the richest and largest Michigan copper mine in the nineteenth century and, after starting in the late 1860s, it mined until the late 1960s. Copper Range did not start up until about 1900 but quickly rivaled C&H as the district's major producer of native copper and then went on to start up the copper sulfide mine at White Pine in the 1950s that operated until the mid-1990s. These three companies, together, not only covered the full life span of the industry; they led the industry.

In addition to documenting the growth of these three companies' mines, mills, and smelters, *Hollowed Ground* examines their paternalistic involvement in community building. The Lake mines all controlled the ground immediately adjacent a line of shafts, where they built up a mixed industrial-residential landscape called a *mine location,* where many employees and their families lived. They all built, at one time or another, boardinghouses, single-family houses, neighborhoods, roads, hospitals, bathhouses, and libraries. They set the course of local schools, saw that churches got land to build on, and provided pasturage for workers' milk cows and space for vegetable gardens. Many companies also fostered the growth of a commercial village on the margin of a mine—a village that was not a company town, was not owned or controlled by a company, but one that offered workers a full range of goods and services unavailable at the mine locations proper.

Over a century and a half, while certain conditions persisted on the Keweenaw, much change occurred throughout the underground and across the social and physical landscape on the surface. The Lake copper mines offer a fine laboratory for studying change within an industry and within an industrial society because the mines started so early and lasted so long.

The mines started out new, shallow, and cool; many ended up being old, deep, and hot. They started with practically trained men at the helm and ended up with general managers who had never worked as miners, but who had university degrees in engineering. They started in an era when one company owned one mine and did one thing: produce copper. Over time, they dealt in an economic world of changing conditions, when companies became larger, when they merged or consolidated with others, when conglomerates sometimes took over, multiple mining properties were worked, and companies no longer just produced copper but diversified into manufacturing copper goods and other activities.

As soon as the Lake mines opened, and for forty years thereafter, they dominated U.S. copper production. Then in subsequent decades they faced competition from a growing number of copper districts, many of which enjoyed lower production costs because they worked higher-grade deposits or could use less costly technologies such as open-pit mining. Even though production at the Lake native-copper mines kept increasing until nearly 1920, their importance as a national copper producer kept eroding as new competition came from Montana, Arizona, Utah, Nevada, New Mexico, and Alaska. Starting in the last decades of the nineteenth century, the heart of American copper production moved way west of Calumet, Michigan, and stayed there.

To the Lake mines' benefit, the markets or uses of copper expanded a great deal over the decades. Copper is corrosion resistant and a good conductor of heat and electricity. It is easily soldered or brazed. It is easily alloyed with tin to make bronze and with zinc to make brass, and those alloys have an attractive appearance and patina. Copper and copper alloys, then, have had many technical and aesthetic uses. When the Michigan mines opened, their product was much used in the pot-and-pan industry. Rolled into sheets, it was used to roof buildings and to sheath the hulls of wooden ships. Copper alloys went into architectural hardware, such as doorknobs and hinges, into candlesticks sitting on mantels and dining-room tables, into buttons and statuary. It also went into machinery bearings, cannon, metal clockworks, and brewery kettles. Then the arms industry rolled and pressed brass into shell and bullet casings; the plumbing industry became a major new consumer of copper pipe and tubing; and the electrical industry demanded copper for transmission lines, household wiring, and motor and generator windings. On the heels of the electrical revolution, mechanical refrigeration and the auto industry placed new orders for copper radiators and heat exchangers. As technological change worked through the American economy, the uses of copper multiplied, and the copper industry expanded to meet the demand.

The mining companies, for decades, did pretty much what they wanted to do. They were unfettered and did not need to negotiate with or react to labor unions, unfriendly courts, government regulations, or environmental concerns. Then came state-mandated mine inspectors, lawsuits over injuries and deaths, the U.S. Bureau of Mines, worker compensation, child labor laws, an eight-hour workday, successful union drives, and the Environmental Protection Agency. The mining companies' hegemony and their power to control their own affairs declined as the society around them changed.

In a different sense, however, the mining companies' power grew. Human power initially did much of the work in the mines: drilling rock, pushing rock, and even lifting rock. Then came animal power, steam power, compressed air, electricity, and diesel engines. Virtually all work once done by hand came to be done by a machine for the purpose, generally, of upping production and worker productivity while lowering costs. Men once pushed rock in wheelbarrows to a shaft for hoisting. Then they pushed four-wheeled tramcars on tracks. Later, they loaded rock on conveyor belts and pulled it with electric locomotives or hauled it in four-wheel-drive trucks.

For decades the mines depended upon immigrants to do virtually all the work underground and much of the work on the surface in shops, in stamp mills, and in the woods cutting timber. Cornishmen, other Englishmen, Germans, Irishmen, French Canadians, and a smattering of Scots and Swedes first supplied the mines with labor. As traditional migratory streams dried up in the late nineteenth and early twentieth centuries, the mines employed a large number of Finns, Italians, and eastern Europeans. In the last years, many second- or third-generation Americans—many born and raised in the copper or iron regions of the Upper Great Lakes—replaced immigrants in the workplace.

Over the life span of the industry, mining companies always offered housing. That remained a constant. But the housing evolved from small log structures with wooden sills sitting on the dirt to houses planned by Sears-Roebuck to post–Second World War ranch houses and even a trailer court. The first houses had no utilities at all; the last houses had washer/dryers, refrigerators, automatic dishwashers, and central heat.

In the mid-nineteenth century the federal government helped create and bolster a new industry in a remote wilderness by sponsoring expeditions to the region, signing Indian treaties to acquire mineral and property rights, paying for land surveys, setting up a land office, building a fort and lighthouses, paying for the locks and canal built at Sault Sainte Marie with land grants, and establishing local post offices. In the twentieth century the federal government offered purchase contracts and price supports to the copper mines to keep them afloat and in production during times of war, and offered a $57 million loan that enabled the start-up of a major new mine while the copper industry, otherwise, was woefully in decline.

Since I have published articles and other books on the Lake Superior copper mines, some topics or themes are not addressed in detail in this book because they have been covered elsewhere. This book covers the introductions of numerous machines into the industry but does not always analyze what their effects were on work or workers (as opposed to their effects on the industry). The effects of mechanization on work are more thoroughly covered in my book *Cradle to Grave: Life, Work, and Death in the Lake Superior Copper Mines* (New York: Oxford University Press, 1991) and in the article "The Machine under the Garden: Rock Drills Arrive at the Lake Superior Copper Mines," published in the journal *Technology & Culture* 24 (1983): 1–37. *Cradle to Grave* also looks more analytically at the frequency and causes of fatal mining accidents than does *Hollowed Ground,* and that is also true of the article I coauthored with Jack Martin, "Technological Advance, Organizational Structure, and Underground Fatalities in the Upper Michigan Copper Mines: 1860–1929," published in *Technology & Culture* 28 (1987): 42–66. While *Hollowed Ground* deals very much with community building, and the development of mine locations and commercial villages—and deals with architecture, housing, and the immigrant workers who lived at and around the mines—it is not, at its core, a study in everyday life. That approach is taken in my book *Beyond the Boundaries: Life and Landscape at the Lake Superior Copper Mines* (New York: Oxford University Press, 1997), which more fully explores topics such as food, entertainment, religion, public health, and the daily lives of women and children.

Historians are often criticized for putting too much emphasis on growth and success, and for paying too little attention to failure. *Hollowed Ground* examines one mining district and three companies that indeed were great successes—yet the district and all its mines ultimately failed. They did not run out of copper; they ran out of copper they could mine at a profit. *Hollowed Ground* looks at success and failure, and at the end it looks a bit beyond the fall to see the environmental and social legacy the industry left behind. The mining companies all disappeared, but they left behind people, houses and streets, mine buildings, ruins, cellar holes, churches and schools, piles of poor rock, millions of tons of waste tailings, and memories good and bad. They left behind one artifact that is particularly worth noting, especially because so few people today know about it.

In 1850 the Cliff mine blasted out a 2,180-pound piece of mass copper and put it aboard the propeller ship *Independence* to begin its journey to the east. This mass never went into a furnace, was never rolled into sheets and put up on a roof. Instead, it was dressed into the shape of a regular block and polished, and craftsmen worked native silver, again taken from the Cliff mine, into a shield and letters applied to one side of the block. The shield was Michigan's symbol, and the letters spelled out "Michigan—An Emblem of Her Trust in the Union." The relatively young state of Michigan had selected this specimen of mass copper to serve as its memorial "stone" to be cemented in place facing the stairway rising up in the Washington Monument, still under construction in the nation's capital. After 1850 nearly fifteen billion more pounds of copper came out of Michigan, but none of it ended up in a place so indicative of how important Michigan copper once was to the state and the nation.

1

❖ ❖ ❖

KEWEENAW COPPER

Geology, Discovery, Dreams of Wealth

In the 1840s, a copper-mining district opened up on the Keweenaw Peninsula, located on the western end of Upper Michigan. No investors lived in this remote wilderness, which was distant from commercial markets. No overland roads—and certainly no railroads—reached up to this place. Only water routes connected it to the Lower Great Lakes, and winter closed them for nearly half of each year. No indigenous labor force lived here. Only a handful of voyageurs and trappers, several hundred Ojibwa, and a few Methodist and Catholic missionaries called this region home. But the Keweenaw did hold copper—people were sure of that. Though they didn't fully understand the geology, they migrated to the Keweenaw nonetheless, full of enthusiasm and hoping to profit from the red metal.

The struggling mining industry that barely survived the 1840s ended up producing copper for one and a half centuries. Over the first century of production, myriad companies mined lodes charged with nearly pure metallic or *native* copper, unalloyed with other elements. In the 1950s the last great mine to start up exploited not native copper, but a copper sulfide ore body. The last native copper mines closed in the late 1960s; the sole sulfide mine closed in the mid-1990s. As the industry died up and down a mineral range over a hundred miles long, it left behind the cities and villages and *mine locations* that had been built up to serve it. It left behind a population rich in ethnic heritage because the industry had drawn workers from dozens of countries. It left behind thousands of worker houses built by companies, as well as rockhouses, hoisthouses, machine shops, drill shops, and dry houses. It left behind remnants of stamp mills that had separated the copper from its host rock, and smelters, which had melted and refined the copper mineral. On the landscape, companies left behind piles of poor rock at their mines, stamp-sand beaches at their waterfront mills, and hillocks of slag beside their smelters.

Along the Keweenaw the fortunes of mines and settlements were tied to geology. The rock underlying the Keweenaw is some of the oldest in North America. About 1.1 billion years ago the earth's crust in this region thinned and tried to split. From deep in the earth's interior, molten rock rose up, erupted onto the surface, spread out, cooled, and solidified. Two to four hundred magma eruptions occurred here over a span of twenty-five million years. Each eruption deposited dark basalts (also called *traprock*) on the surface, and each flow overtopped the one before it.[1] Between magma eruptions, ancient streams and precipitation washed sand, rocks, and pebbles down on top of the basalts. This action formed a *conglomerate,* so called because it contained a conglomeration of materials. A later lava flow would overtop the conglomerate, compress it, and bind it up with the basalts being built up layer by layer. When the volcanic activity stopped, approximately twenty conglomerate layers were interbedded with several hundred lava flows.

After this rock was put down, geological change continued. The earth's surface underwent compression and was squeezed from two sides. The rock faulted, and along the faults rock slid under adjacent rock, lifting it up and bending it into a bowl shape. The lower part of the bowl became part of the Lake Superior basin. Along the bowl's raised rim, edges of rock strata once underground now outcropped, creating high ground on what became the Keweenaw Peninsula and Isle Royale. Many of the rim's exposed edges contained native copper.[2] This copper had not been present in the basalts or the conglomerates when they first formed. Instead, it had been carried in a hot solution that leached the copper from lower in the earth. Then, under pressure, the solution flowed upward, where it settled into fractures, fissures, and porous rock. Finally, the copper precipitated out of the solution in its metallic form.

The solution could flow into conglomerate rock because of the interstices that existed among the pebbles, stones, and sands that made it up. The solution entered the basalts because they contained many voids or vesicles. The lava, as it spread out, contained hot gases that migrated upward toward the atmosphere. As the lava cooled and then solidified, it trapped the gases. Gas bubbles became voids in the rock, especially near the top of the flow, which could have a frothy appearance. Also, after initial formation, many heavy lava flows slumped or dropped down, creating fractures or fissures that later served as conduits or receptacles for the copper-bearing solution.

Three types of underground cavities gave rise to three types of native copper. The copper in sedimentary rock strata came to be called *conglomerate* copper, named after its host rock. Geologists called the copper in the basalts *amygdaloid* copper. They took this term from the Greek word meaning "almond" because it described the shape of the cavities left behind by the gas bubbles. The larger specimens of copper found in fissures and fractures came to be called *mass* copper.

The copper did not disseminate evenly throughout the host rock. Impervious rock deflected the copper-bearing solution, keeping it out. Other rock was so porous that the solution passed right through it, leaving little copper behind. Consequently, the interbedded amygdaloid and conglomerate lodes running the length of the Keweenaw Peninsula differed considerably in their copper content. One lode might be rich, while adjacent rock strata were poor. This variation made mining here a "subterranean lottery." The mining firms could not tell if a piece of ground would *pay*—yield commercial quantities of copper—except by opening it up at considerable expense. And the fact that one property proved rich or poor was not a good predictor of a neighboring property's future.[3]

After the rock had been laid down, charged with copper, and bent into a bowl shape, glaciation took place, lasting from 1.8 million years ago until about ten thousand years ago. Glaciers as much as nine to ten thousand feet thick moved across the region, scouring the tops of outcropping copper lodes.[4] They sometimes snagged pieces of copper, moved them, and then laid them back down. This action left impressive pieces of *float* copper, as it came to be called, sitting on the surface of the ground, where humans eventually discovered it. Little is known about the first peoples who made their way to the Keweenaw, discovered the copper, and fashioned it into artifacts. They left little behind in terms of settlement remains. But before modern mining arrived to obliterate them, ancient pits or *Indian diggings* marked where native peoples had dug for copper.[5] They started taking Keweenaw copper seven thousand years ago and continued until the seventeenth century.[6] The aboriginal peoples picked up small pieces of float copper from the surface. If they found a large mass of copper aboveground, it typically had jagged, thin appendages that they bashed or twisted off, leaving them with pieces of workable size. If they encountered an upturned edge of a lode having visible copper, they hammered at the rock to free the copper, using hard, rounded hammerstones weighing ten to fifteen pounds. They wielded the hammerstones by hand, but some were hafted, probably with flexible thongs, so they could be swung with greater force against the rock. The existence of charcoal in some pits suggested that native peoples used fire to help break rock down, and they used wood, stone, or copper wedges to pry the copper loose. Most copper pits were shallow, but these first miners took at least one to a depth of twenty-six feet. When the Minesota [*sic*] Mining Company discovered this pit on its property in the mid-nineteenth century, a large mass of copper weighing six tons sat on the bottom.[7] Ancient miners had found it but had no means of cutting it up or lifting it out.

The first people on the Keweenaw and Isle Royale hammered the copper into different shapes by using stone tools to draw it out and form it. Cold-working the copper made it brittle. Although the early peoples had no furnaces to melt copper, they did use fire to anneal it—they heated the unfinished artifact, which recrystallized the copper and made it ductile again so it could be rolled or hammered some more. Besides hammering the copper flat, ancient artisans sometimes hammered it around or inside forms to give it shape. They used sandstone, sand, and wood ashes or other natural abrasives to grind

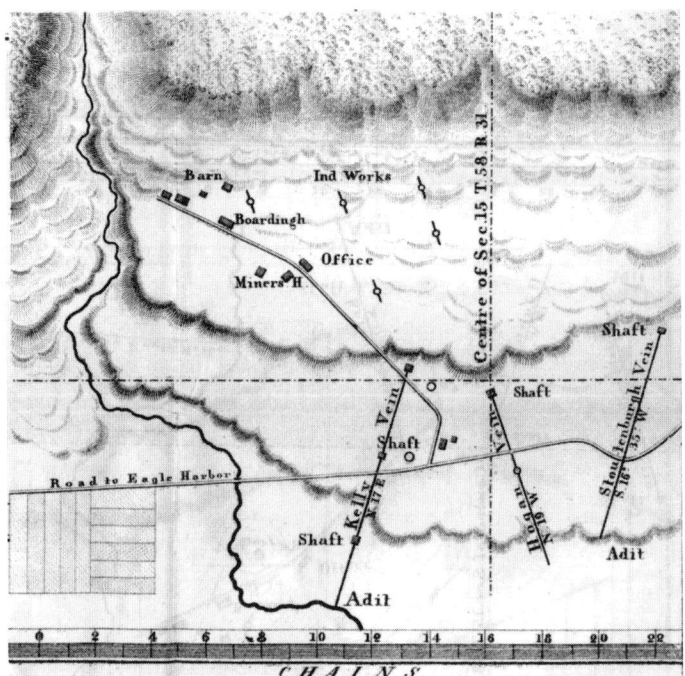

When Samuel Worth Hill surveyed the North West mine in 1847, he recorded the locations of five pits, or "Ind[ian] works," near three fissure veins being explored for mass copper. In an early stage of development, the mine had a boardinghouse and only a few other dwellings. Hill drew in the road that had been opened to the commercial village of Eagle Harbor and, alongside the road, garden plots had been cleared. *(Michigan Technological University Archives and Copper Country Historical Collections)*

the metal or polish it to a desired luster. They fashioned their copper into hooks, knives, chisels, awls, axes, scrapers, beads, and other decorative items and useful tools. They traded these artifacts with other groups, and Keweenaw copper diffused across much of eastern North America.[8]

In the seventeenth century, Europeans brought their metals to North America, and cultural contact and trade with the Europeans made Native Americans less interested in Keweenaw copper. At the same time, the British and French became interested in the metal when they saw specimens of it. Samuel de Champlain may have been one of the first Europeans to learn of the copper when he encountered two Indians on the St. Lawrence River in 1610:

> After conversing with them a short time about a number of things touching their wars, the Algonquin Indian, who was one of their chiefs, drew out of a sack a piece of copper a foot long, which he presented to me. It was very fine and pure. He gave me to understand that the metal was abundant where he had obtained it, which was on the bank of a river near a large lake.[9]

A half century after Champlain first heard of the copper, French explorers searched for it on Lake Superior. These early expeditions netted no significant finds. The French returned in the eighteenth century and did some experimental mining along the Ontonagon River in 1737–38. After gaining control over French Canada in 1763, the British searched for

Lake Superior copper, too. Alexander Henry led an expedition to the Ontonagon Boulder in 1766. This "boulder" was a large piece of float copper that a glacier had long ago dropped off alongside the west branch of the Ontonagon River. This single specimen—the most famous piece of mass copper in the region—confirmed all the tales that Europeans had heard about metallic copper sitting atop the ground near Lake Superior. Early explorers for copper, usually with the aide of Ojibwa guides, made difficult, yet almost obligatory, pilgrimages to see the Ontonagon Boulder. After first visiting the boulder in 1766, Henry led a small party of miners into the Ontonagon River valley in 1771–72.[10]

The French and British never succeeded at mining Keweenaw copper. They arrived too early, with too few men, inadequate knowledge of geology, and inadequate technologies and transportation. After Alexander Henry's expedition in the early 1770s, the American Revolution delayed further attempts to wrest copper from the Keweenaw, as did subsequent disputes between the United States and Britain over the Great Lakes and the War of 1812. By 1820, however, a chain of events began that led to the opening of a mining district in the early 1840s. Several expeditions made their way to the Keweenaw. Each one resulted in more news regarding its copper, which in turn generated more enthusiasm for another expedition and, eventually, a rush of miners.

In 1820, Lewis Cass, governor of Michigan Territory, and John C. Calhoun, the U.S. Secretary of War, sent an expedition to Lake Superior to discover what resources it held that could encourage settlement in the region.[11] Cass himself led the forty-man party that left Detroit in May in three large canoes. Chief among his companions was Henry Rowe Schoolcraft, recruited for the journey because of his expertise in mineralogy and his interest in mining. A month after leaving Detroit, the party arrived at the mouth of the Ontonagon River. A portion of the crew made the arduous twenty-mile trip up the Ontonagon's valley to see the copper boulder, only to be disappointed. Expecting to see a wonder of the natural world, they discovered a specimen only four feet wide, weighing three to four thousand pounds. David Bates Douglass, a West Point engineer, called the Ontonagon Boulder "a mere stone, a large pebble."[12] Yet the party's disappointment did not dampen public interest in the future of Lake Superior copper. In a letter to Calhoun (which later served as his official report of the expedition), Schoolcraft admitted the boulder was smaller than legend had it. Still, he called it "one of the largest and most remarkable bodies of native copper upon the globe."[13]

Schoolcraft, while working out of Sault Sainte Marie, Michigan, as the Indian agent for the Upper Great Lakes, commanded additional westward expeditions in 1831 and 1832.[14] Importantly, he chose twenty-one-year-old Douglass Houghton to accompany him. Like Schoolcraft, Houghton was a New Yorker of many talents who had moved west to seek fame and fortune. Professional science was in its infancy, and few careers exhibited narrow specialization. Houghton, a well-educated generalist, later served as Michigan's foremost geologist. He also served as a medical doctor and as mayor of Detroit. But in 1831–32 he was a young man just getting started, who had a chance to travel to a place rumored rich in copper. While mineralogy was becoming less important to Schoolcraft's career, Keweenaw copper would become more important to Houghton's.

The 1831 expedition coasted down the northwestern shoreline of the Keweenaw, occasionally landing to collect green and black copper ores. Schoolcraft chose not to revisit the Ontonagon Boulder this time, but Houghton made the pilgrimage. When Houghton returned to Detroit in the fall, local papers pressed him for copper news and ran articles on his discoveries. Later, in November, Houghton wrote a letter to Lewis Cass, titled "A Report on the Existence of Copper in the Geological Basin of Lake Superior." The letter (later published as a widely read congressional document), made Houghton the leading expert on the Keweenaw's native copper and its deposits of copper oxides and carbonates.[15]

The 1832 Schoolcraft expedition traveled farther afield, all the way to the Mississippi's headwaters in present-day Minnesota. Houghton again went along, this time as doctor. He vaccinated many Indians against smallpox, performed surgeries, and explored Upper Great Lakes botany. On the way back east, Houghton returned to the Keweenaw. He removed some jagged appendages from the Ontonagon Boulder and collected more copper specimens. This expedition gained considerable notoriety. The public hailed Schoolcraft as the discoverer of the source of the Mississippi River, and Houghton's letters and geological reports ended up in congressional documents and in newspapers. A Detroit newspaper trumpeted the prospects of the northern region where copper was abundant. Eventually it would "reward future exertions [explorations and mining]. It may have a mark'd influence on the coming prosperity of Lake Superior."[16]

After 1832, Douglass Houghton enjoyed success as a doctor and real estate investor. He became a professor of geology at the University of Michigan; the first president of the Detroit Board of Education; and, later, mayor of the city. He also became Michigan's first State Geologist. In 1837, Michigan became a state, but only after a boundary clash with Ohio had been resolved. Michigan and Ohio both claimed Toledo and a strip of land running west from that port city on Lake Erie. The federal government awarded Toledo to Ohio and compensated Michigan for the loss by granting it the western three-fourths of the Upper Peninsula—which, unlike the eastern end, near Sault Sainte Marie, had not been part of Michigan territory.[17] Michigan's new governor, Stevens T. Mason, then named Douglass Houghton his State Geologist.

In this new job, Houghton did not pursue geological knowledge just for the sake of learning it. He used geology to discover what resources could be exploited to produce new industries and wealth. In 1840, Houghton extensively surveyed the Keweenaw copper deposits, and the following year he reported on them to the state legislature. Houghton still lacked a full understanding of the geology. He was uncertain about what sorts of copper deposits might best support a mining industry. Knowing that no successful mining industry anywhere had exploited native copper alone, he wrote a report that was at best guardedly optimistic about mining prospects:

While I am fully satisfied that the mineral district of our state will prove a source of eventual and steadily increasing wealth to our people, I cannot fail to have before me the fear that it may prove the ruin of hundreds of adventurers, who will visit it with expectations never to be realized. . . . I would by no means desire to throw obstacles in the way of those who might wish to engage in the business of mining this ore . . . , but I would simply caution those persons who would engage in the business in the hope of accumulating wealth suddenly . . . , to look closely before the step is taken, which will most certainly end in disappointment and ruin.[18]

Despite its cautionary words, Houghton's 1841 copper report stirred interest in a mine rush up to the Keweenaw. Three other events, all in 1843, also served as catalysts: the Treaty of LaPointe went into effect; the federal government opened a mineral land office at Keweenaw Point; and the Ontonagon Boulder, finally snatched from the Keweenaw, made a well-publicized trip to the east coast.

The federal government took several steps to encourage copper mining on Lake Superior. It funded expeditions to the region in 1820, 1830, and 1831. It signed two treaties with the Chippewa: the Treaty of Fond du Lac in 1826 and the Treaty of LaPointe, negotiated in 1842 and made effective in 1843. Thanks to these two treaties, the U.S. government possessed both mineral and property rights to the copper lands, and in 1843 it opened a mineral land office in Copper Harbor.[19] Investors or prospectors in Copper Harbor and in Washington, D.C., began leasing Keweenaw mineral lands.[20] That same year, the Ontonagon Boulder spurred additional enthusiasm for Lake Superior copper while slowly making its way from the Keweenaw to the east. Many early visitors had wanted to claim the Ontonagon Boulder as a prize, but moving it proved a daunting proposition. It sat twenty miles up the Ontonagon River, above three rapids totaling over seventy feet of fall, and in a steep valley surrounded by rugged, wooded terrain. In 1826, Lewis Cass sent a twenty-man crew well supplied with tools (or so they thought) to snatch the boulder and remove it to the nation's capital. The work party failed and abandoned the effort.[21]

In 1843, a Detroit hardware-store merchant, Julius Eldred, finally removed the boulder. His men used a capstan to lift it fifty feet up to the top of the adjacent bluff. They loaded it on a small railcar, and for four miles they inched their prize along. They cut a swath through the woods, laid a short stretch of rail, pushed the car to the end of the short line, picked up the rail behind the car, and laid it out front again. Once below the rapids, the boulder went on a raft; at the mouth of the Ontonagon, it went aboard a schooner. Even on Lake Superior, Eldred's expedition found no smooth sailing. Eldred thought he had legally purchased the boulder from the Ojibwa, but when his schooner landed at Copper Harbor, he was informed that the U.S. Secretary of the Treasury had instructed the Secretary of War to claim federal ownership of the copper boulder, seize it from Eldred, and ship it to Washington, D.C. Eventually, Eldred surrendered the specimen, and the federal government paid him $5,665 for his efforts.[22] On the way to Washington, via Detroit, Lake Erie, the Erie Canal, and New York City, the Ontonagon Boulder received much public attention. The publicity generated by the boulder's journey sparked further interest in participating in a mine rush to the Keweenaw in the mid-1840s. The explorers, geologists, miners, and investors who participated in that rush soon discovered just how hard it would be to transform a remote wilderness into a mining frontier. Douglass Houghton, who warned that mining Keweenaw copper would be risky, paid the ultimate price for this quest. In 1845 a violent storm overturned his small boat, and he drowned in Lake Superior near Eagle River.[23] Just as mining started to take root, the man thought to best understand Keweenaw copper was dead.

Copper Harbor, located on the northern tip of the peninsula, called Keweenaw Point, served as the mine rush's jumping-off place. Here, the land office first leased and later sold mineral lands to adventurers, while the Army's Fort Wilkins stood as a symbol of law and order in the region, both for rowdy explorers and Native Americans.[24] The copper seekers reached Copper Harbor by boat, often a small one. Prior to the Soo Locks opening in 1855, large lake boats could not sail from Detroit, up Lake Huron, along the St. Marys River and on to Lake Superior because they could not navigate the rapids at Sault Sainte Marie. Only a few large boats had been built on Superior or portaged around the rapids to be put on the big lake. So, many travelers arrived at Copper Harbor in small open craft that had coasted along 250 miles of Superior's southern shore. The travelers themselves were poorly equipped to navigate these sometimes treacherous waters or, once on land, to lead themselves through the Keweenaw's swamps or forests. They turned to colorful voyageurs, who harkened back to fur-trading days, as guides and transporters. The voyageurs—often French Canadians, Native Americans, or some mix of the two— were stereotyped as bawdy, hard-drinking men, barely civilized. But on the water or in the woods, nobody proved better or stronger when it came to delivering people and cargo safely.[25]

In the mid-1840s, Copper Harbor wore the trappings of colorful tent city. Adventurers camped near the water and marched into the hinterland in search of copper. They trekked through swamps, thickets, and forests, while suffering the siege attacks of blackflies in early summer, followed by swarming mosquitoes. Few explorers relished the idea of wintering on the Keweenaw, cut off from supplies, living in isolation, and squatting in ice-covered tents or crude huts. Only as productive mines opened did a year-round population build up, reaching a thousand permanent settlers by 1850.[26]

In the 1840s, when the hunt for copper began, the market for the metal was important, but not vast. Because of its heat conductivity, manufacturers (many located in Connecticut, which had become a center of copper and brass production) formed copper into pots and pans. Because of its corrosion resistance, sheet copper sheathed wooden ships' hulls and served as roofing and gutters for buildings. Because of the attractiveness of copper and its alloys, the metal found its way into hardware, such as doorknobs and hinge plates, and into lamps, buttons, statues, and other decorative items. Many shafts and spindles in early American machines turned in brass bearings.[27] Michigan copper speculators knew that in trying to break into these markets, they would not have to pit their nascent industry against many well-established firms. Americans had never mined copper on a large scale. In the colonial period, an early copper mine operated at Simsbury, Connecticut, and the largest producer had been the Schuyler Mine, near Newark, New Jersey. Production lagged in the late colonial and early federal periods, and American markets relied heavily on imported copper. About the time that School-craft and Houghton made their travels to the Keweenaw, new copper mines opened in Vermont, Connecticut, Maryland, New Jersey, and Tennessee. All, however, were small operations.[28]

Michigan's copper deposits seemed to promise a future well beyond that of estab-lished producers, but it soon became apparent that a substantial investment would be needed to turn any property into a paying mine. Copper was not a precious metal; it did not bring as dear a price as gold or silver. Copper profits would come from large-scale production, which required a sizable labor force and investments in technology, housing, and other site developments. Turning a wilderness site into a productive mine usually entailed years of risky development and expenditures of at least $50,000 to $200,000.[29] Because of the risks and costs involved in opening these copper mines, corporations became key players in the region. Individuals and small groups participated in the mine rush, but corporations led the way. Their stock subscriptions raised the capital needed to sustain several years' work. Also, the corporate form attracted investors who wanted to limit their financial risk. They spread their investment over several companies, hoping one great success might repay them for other failures.

By the last quarter of the nineteenth century, Boston became the hub of investment in Michigan copper. But at the start, many investors and companies came from Lower Michigan and the Midwest, from the Pittsburgh area, and from the east coast in gener-al.[30] Often, *families* of investors formed, typically men who already shared business inter-ests and who combined again to invest in Michigan copper. Often these investors were friends or even related to one another through blood or marriage. Thanks to investment families, many ventures on the Keweenaw shared the same capitalists, corporate directors and, in some cases, mine managers, who looked after two or more properties at once.[31]

The earliest mining companies searched for deposits of mass copper found in fissure veins, which tended to run across, rather than along, the peninsula. The lodes charged with amygdaloid and conglomerate copper ran the full length of the Keweenaw's central spine on a southwest to northeast line or strike. These lodes held the vast majority of the peninsula's copper, but mass copper had greater allure. It seemed so simple. Remove big pieces of copper from the ground, transport them to a smelter, melt them down, and cast the copper into ingots for sale.

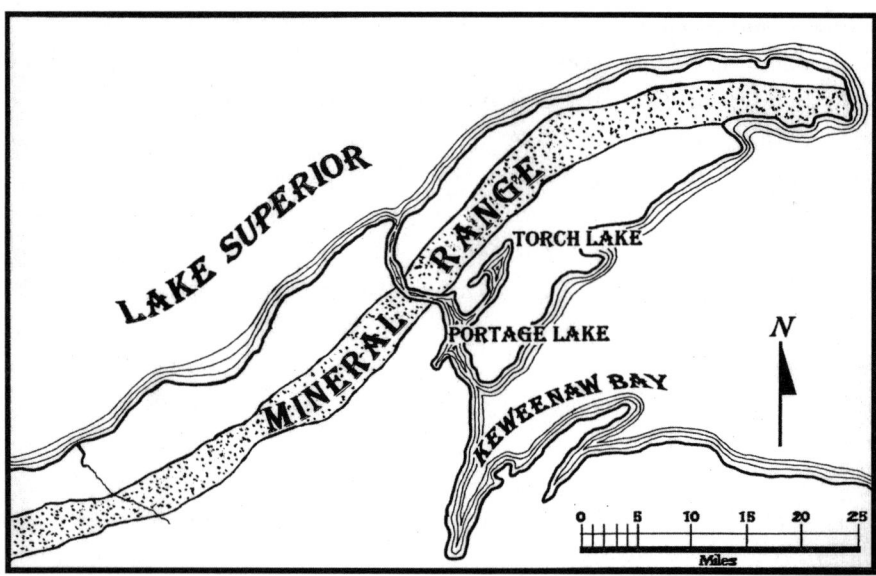

Copper lodes on the Keweenaw Peninsula ran on a southwest to northeast strike along the mineral range. They paralleled each other and tended to run long distances. Usually, however, only a short run of any lode was sufficiently charged with copper to support profitable mining. *(Society for Industrial Archeology)*

The first five or so years of mining on the Keweenaw proved frustrating. The mines consumed wealth without producing new wealth. Then, in the late 1840s, the region finally boasted a success. In 1844–45 the Pittsburgh and Boston Mining Company explored around Copper Harbor. Then the company relocated southward to a more promising site dominated by a steep cliff. In 1849, sales of mass copper taken at the Cliff mine enabled the Pittsburgh and Boston Mining Company to reward its stockholders with the region's first dividends.[32] These dividends proved that if a company found the right property, money could be made. After the Cliff mine delivered a crucial success near the northern tip of the peninsula, in short order the southern base of the mineral range provided a major success. In 1847 the Minesota [*sic*] Mining Company started working near ancient Indian diggings in the vicinity of Ontonagon. The Minesota Mining Company's production reached four million pounds per year by the late 1850s. The Cliff and the Minesota were the two best mines in the era before the Civil War. Investors who paid in $110,000 of stock assessments at the Cliff received dividends totaling $2.5 million; an investment of $366,000 at the Minesota mine returned $1.8 million.[33]

On the mining frontier, change was constant. A few mine locations became thriving communities of fifteen hundred or more settlers, while many others went broke and sat abandoned. The center of mining activity moved around. Mining started on the Keweenaw's northern end (Keweenaw County). By the early 1850s many mines clustered around the Minesota mine, on the southern end of the range (Ontonagon County). The

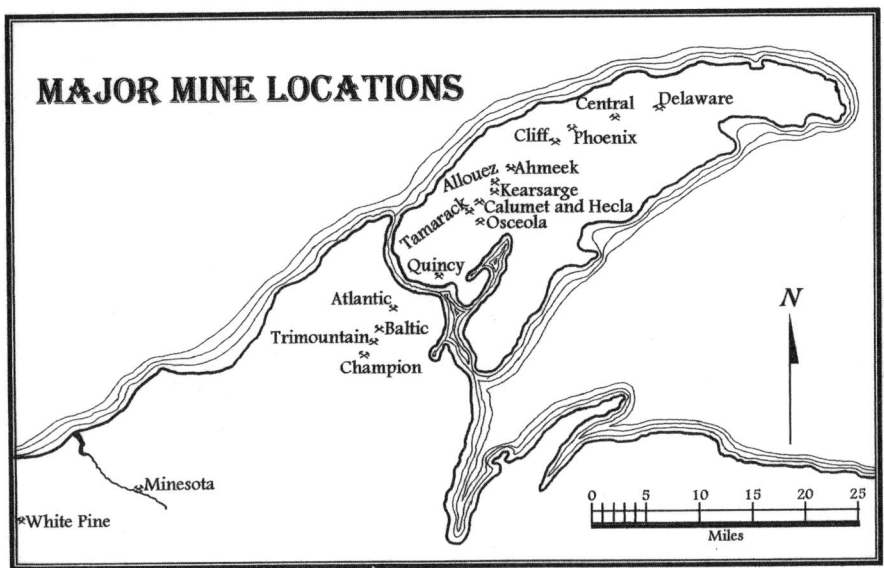

The Cliff mine, in present-day Keweenaw County, was the first Lake mine to turn a profit in the 1840s. To the south, the Minesota [sic] mine in Ontonagon County was the second great success story in the district. But these two mass-copper mines played out by 1870. From the 1860s onward, amygdaloid and conglomerate mines in Houghton County, especially Calumet and Hecla, Quincy, and Champion, dominated copper production. *(Society for Industrial Archeology)*

middle stretch of the peninsula, near Portage Lake, developed a bit later. But while the Portage Lake district opened last, it proved the greatest long-term success. The mines from ten miles south to fifteen miles north of Portage Lake (in Houghton County) produced some 95 percent of the native copper taken from the Keweenaw. Most of this product was not mass copper, but amygdaloid and conglomerate copper.[34]

Despite the great success of the Cliff and Minesota mines, mass copper did not prove a long-term bonanza. Rich fissure veins were few and far between. Mass copper, once found, was expensive to cut up and remove from the mine. Also, fissure veins tended to peter out rapidly. Because of these problems, companies—such as those along both sides of Portage Lake—began searching more intensively for amygdaloid and conglomerate lodes.[35] South of the Portage, above the tiny village of Houghton, the Shelden Columbian, Isle Royale, Albion, Huron, and Grand Portage mines strove to become productive. On the north side, above Hancock, stood the Quincy, Pewabic, and Franklin mines. None of these mines succeeded as quickly as the Cliff or Minesota, and many never made a profit. But in the late 1850s, after a decade of trial-and-error development, the fortunes of this district improved. The Pewabic mine discovered a rich amygdaloid lode that ran across its land and onto Quincy's property. Quincy sank shafts into the Pewabic lode in 1856 and paid a dividend for the first time in 1862. The Pewabic lode was the kind that investors dearly sought. Quincy mined it continuously from 1856 until 1931 and then from 1937 to 1945. The first run of years produced a long string of annual dividends that earned Quincy the moniker of "Old Reliable."[36]

By the Civil War, mining companies had explored more than a hundred miles of the mineral range. The northern end had the great Cliff mine, the productive Central, the ever-promising Phoenix, and a host of others. On the southern end, the Minesota dominated, and young Ontonagon thrived as a commercial village. In the middle, the Portage Lake district stood on the verge of a great takeoff, as did the villages of Hancock and Houghton. Already—and for some time—the Lake Superior mines dominated American production. In 1847, two years before the Cliff paid its first dividend, Michigan production reached 672,000 pounds, or 71 percent of total U.S. production. The mines produced 1.3 million pounds in 1850, 5.8 million pounds in 1855, and 15 million pounds in 1861. In these years the Keweenaw accounted for 86 to 90 percent of America's new copper.[37]

The Civil War promised to be an economic bonanza for the copper industry because demand and prices ran high. Copper jumped from nineteen cents per pound in 1861 to a wartime high of forty-six cents in 1864. But 1864 did not prove to be a banner year for the Lake mines, because they produced only 12.5 million pounds of copper, or about 65 percent of the nation's total.[38] A local labor shortage contributed to the decline, as men left to serve in the Union Army. Houghton County sent 481 men off to war; Keweenaw County, about 120; and Ontonagon County, about 250.[39] Another, perhaps more important cause of the labor shortage and production decline was that high copper prices encouraged new mines to start up and closed mines to reopen. In 1860, of twenty-three companies actively mining, only one produced as much as two to three million pounds of ingot copper, four produced one to two million pounds, two accounted for one-half to one million pounds, and sixteen mines managed less than a half-million pounds each.[40] By 1865 the number of small producers (less than a half-million pounds) had jumped to twenty-nine. All these marginal mines robbed good mines of the workers they needed to step up—or even maintain—their prewar production levels.

Producing copper and making profits were two very different things. In 1841 Douglass Houghton had been correct in his fear that the pursuit of Lake copper would be the ruin of many. Over nearly a quarter century, some three hundred mining ventures had been launched, including ninety-four incorporated firms. Of the incorporated companies, only eight paid dividends by the end of the Civil War.[41] Fewer than one in ten mining corporations had improved from running in the red to running in the black, and more money had been lost than made.

2

❖ ❖ ❖

GETTING THE COPPER OUT

Exploration, Development, and the Tools of Production

The successful mines on Lake Superior passed through three stages of growth and change. In the *prospecting stage,* a small workforce explored company lands, seeking copper. In the *development stage,* an enlarged crew opened up the most promising finds. If no finds bore commercial quantities of copper, a company often went out of business at this point. But if a fissure vein or lode showed real promise, an optimistic company moved into the *production stage.* It brought in more people and technologies and spent more money underground and on the surface.

A smart mining company conserved capital. It did not send an army of prospectors into the woods to search for copper. Instead, it typically employed fewer than a dozen men at the start—men who lived in rude shelters and used simple tools. Using axes, saws, and fire, they cleared timber and undergrowth. With shovels and picks they crisscrossed the property with trenches in hopes of exposing the upturned edges of copper-bearing lodes. They sank test pits and used sledgehammers, hand steels, and black powder to blast free copper specimens and expose the rock for closer examination.

The earliest companies leased from the federal government up to nine square miles of mineral land to explore. Leases soon gave way to federal sales of mineral lands, and companies usually purchased a one-square-mile section.[1] There was no easy way to find copper on a square mile occupied by a dense forest, swamp, cliff, or ravine. Early prospectors did not really know the geology of the district, which was still little understood and would be revealed only as companies succeeded here and failed there in the search for copper.[2] Many companies looked to ancient Indian diggings as signposts for where to start looking. At least sixty mines up and down the Keweenaw, and another five on Isle Royale, explored ancient, long-abandoned pits on their properties. On the northern end of the Keweenaw, the Copper Falls mining company "made a great start by the discovery, last fall, of a new vein with extensive ancient works upon it." In the central Keweenaw, the Quincy mine discovered "a line of pits," while the Pewabic mine reported "numerous pits." Nearby, the pits at the Pontiac mine were "extensive" and showed "a regular system." To the south, at the base of the peninsula, on the Adventure mine's property "the pits and diggings of the ancient miners are found in great numbers," and from one they were "taking a mass of almost pure copper which . . . will weigh five tons." Besides finding depressions in the ground, new companies sometimes found "barrels full" or "cart loads" of well-worn hammerstones piled nearby, sometimes accompanied by cedar shovels, copper and wooden wedges, and even primitive ladders or scaffolding.[3]

Ancient pits and tools, however, were not foolproof markers. Looking at the copper found in shallow pits or trenches did little to forecast the ground's true potential. A company that turned up copper near the surface had to open up the fissure vein or lode, and this development work cost more than prospecting. The mine expanded its labor force, increasing it to fifty or as many as two hundred men.[4] Miners sank several shafts to test the copper content of a lode as it dipped deeper into the ground. To test the copper rock from one end of the strike to the other, they drove *drifts* (horizontal tunnels) from one shaft over toward the next. Sometimes they used *adits* or crosscuts to test adjacent ground. Running into the base of a hillside, a tunnel-like adit might intersect and uncover a new lode as it ran underground to connect with established shafts and drifts. Similarly, crosscuts driven perpendicularly from drifts helped determine whether the mine was working the main lode or whether better ground was beside it in either direction.

If development work showed a lode to be too narrow or little charged with copper, a smart company shut down operations and abandoned or sold the property. But if the ground proved to be well charged with mass, amygdaloid, or conglomerate copper, the opposite followed: the company accelerated work and moved into full production. This decision—to stop work or push it—was often hard to make because a company could not see exactly what the underground held. Few copper strikes were so rich right from the start that a company could decide with total confidence to go into production. Many properties proved just rich enough in copper to keep companies limping along, but too poor in copper to ever repay the investment sunk into the mine.[5]

The decision to go into production triggered much change. Company directors called for investors to pay in more assessments on their stock, which made more capital available for construction and wages.[6] Employment might rise to three to five hundred workers while a rudimentary mine camp evolved into a community, called a *mine location,* with women and children and churches and schools and single-family houses. Simultaneously, a company built up its industrial infrastructure. It acquired boilers and hoisting engines and erected blacksmith and machine shops.

The Lake mines opened while the American economy was in the early stages of industrialization, which changed not only the workplace, but everyday life. With industrialization came factories, mass production by machine, new mechanical sources of power, a greater division of labor, and the rise of a working class that toiled for wages and then used money to buy things they had previously grown or made for themselves. Several manufacturing industries were in the vanguard of industrialization because of their new technologies or products. These included the textile and firearms industries and the manufacturers of machine tools, sewing machines, farm implements, and other goods. In 1851, at the Crystal Palace Exhibit in London, American machines and products caught the attention of the British and the Europeans, who recognized that America, once disparaged as a borrower of technologies, was now an innovator to be studied.[7]

The mining industry was not at the forefront of the Industrial Revolution in America. Mines produced the coal, iron, and copper that manufacturers needed, but they were not so much leading the industrialization parade as following it. Mines benefited from technological changes initiated by other industries. For instance, the nineteenth century was the "Age of Steam," and steam power transformed the mining industry.[8] But mines were the buyers of engines, not the designers, not the makers. The first Lake copper mines worked at the boundary between a preindustrial, craft tradition of mining and the more advanced industry to come. While engines and machines soon labored on the surface, for many years no engines or machines toiled underground. Underground, human labor built the industry.

Hard-rock mining proceeded slowly in the premechanized era, and miners made small advances daily in sinking, drifting, or stoping. Since production was limited at each working face, to extract much product an early mine attacked its lode at many points. A company sank several shafts, spaced only two or three hundred feet apart, along the lode. This provided more entryways into the lode, more places for men to work.

The inclined lodes on the Keweenaw dipped into the ground on angles from 30 to 80 degrees. A mine usually sank its shafts right through the lode rather than in the hanging wall (above the lode) or the footwall (beneath the lode). Shafts opened the mine from the top down and served as thoroughfares for transporting men, tools, and copper. At set distances apart, tunnel-like drifts headed off from one shaft and ran horizontally through the lode toward adjacent shafts. Early drifts were typically sixty feet (ten fathoms) apart.[9] Shafts and drifts set off rectangular blocks of ground for miners to stope out. Stoping miners started drilling and blasting right next to one shaft, and as they moved out along the drift, they also worked *overhand* and moved from one drift up toward the next. Overhand stoping used gravity to advantage. After every blast, the broken rock tumbled down the stope to the drift below and was ready for tramming.

Most lodes on the Keweenaw were narrow, meaning that, on a given level, only so much copper could be taken. To sustain production a mine had to sink its works another level or two each year. An underground landscape had three regions. Abandoned works

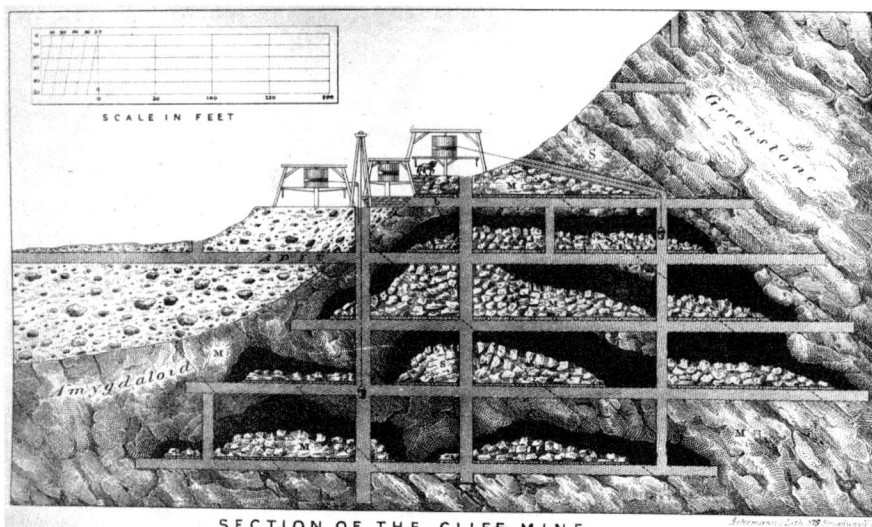

SECTION OF THE CLIFF MINE

Geologists J. W. Foster and J. D. Whitney visited the Pittsburgh and Boston Mining Company's Cliff mine in 1849—the year in which the company paid out the district's first-ever dividends to shareholders. Published a year later, this sectional view by Foster and Whitney shows three shafts, five horizontal levels or drifts, and three horse-whims that raised copper in wrought-iron kibbles or buckets. *(Michigan Technological University Archives and Copper Country Historical Collections)*

occupied the top of the mine, where the drifts and stopes stood empty of men, tools, and activity. Below, one found the producing levels of the mine, where most of the workforce labored, stoping out and tramming copper rock. At the bottom of the mine, men sank shafts, drove drifts, and put in *winzes* (openings that ran from one level to the next and facilitated airflow in the mine). This development work readied blocks of ground to be stoped out in a year or two. In the three regions—stoped-out ground, producing levels, and development levels—one saw a mine's past, present, and future.

Miners used drills and powder to open up all parts of a mine. Because hand drilling was slow, skilled miners first read the mine rock and placed the shot holes where the fewest blasts would bring down the most ground. Having determined the hole locations, one miner held the drill's sharpened chisel bit against the rock face. His partner, or partners, then hammered the drill home with eight-pound sledges. After each strike, the holder lifted the drill just a bit and turned it slightly so that the next blow cut a better chip.[10] Since the angle and location of the holes differed from one to the next, miners needed the physical dexterity to swing their hammers through a variety of arcs, always striking the drill head squarely while missing the holder's hands and wrists. The number of men on a drill team varied, depending on the size of the opening. In a large stope, often one man held the drill steel while two men wielded sledges. In a small drift, as little as six feet high and five feet wide, there was no room for a third man, so one man "jacked" (hammered) the drill while his partner held it.[11]

Using a succession of hand steels of increasing length, miners drove shot holes to a depth of perhaps four feet. As the end of the shift neared, they cleared the holes of chips and charged them with explosives and fuses. For the first two decades the mines used black powder made up of saltpeter, sulfur, and charcoal.[12] Mines stored explosives in magazines on the surface, and some mines transported powder underground on Sundays, when fewer men would be around if a big, accidental explosion occurred. The mining teams had lockboxes underground to store explosives and drew from their own powder kegs to charge holes. After lighting fuses, they retreated a safe distance and waited for the explosions. When a burning fuse ignited a charge, the explosive detonated at a rate of about fifteen hundred feet per second. The expanding gases heaved the rock, breaking it free. From their place of retreat, the miners counted the blasts. If they set eight charges and heard eight explosions, they were done for the day. If not, they had a missed hole—a charge had not fired. In that case, they waited around long enough for it to be safe to go back in and try to fire that hole again. Unexploded charges were accidents waiting to happen and needed to be dealt with.[13]

When miners discovered a large mass of copper, they drilled and blasted all around it. To free the mass from the last ground holding it, miners tucked full, twenty-five-pound powder kegs around the point of contact and used sandbags to help direct the force of the blast. Copper is a gummy, malleable metal. Explosions did not fracture it, and no machine existed to cut it up as it sat on the mine floor, so men cut it up by hand. Knowing how large a piece of copper that men could transport out of the mine, a mining captain drew lines across the mass, showing copper cutters where to part it. One man held a chisel to the copper; others struck its head with sledges. The chisel holder walked the chisel along the cut line, and the tool produced a long chip as it moved from one side of the mass to the other. The men repeated this operation time after time. As the channel deepened, the holder reached for a longer chisel. Finally, the men cut though the bottom of the mass.[14] They parted it again, if necessary, and then others transported the pieces of mass copper to the shaft for hoisting to the surface.

Some pieces of copper indeed earned the name *mass:* they were indeed massive. The Cliff mine encountered a mass-copper specimen weighing fifty tons, and the Minesota mine found even larger ones. In the early 1850s large pieces of mass copper proved so numerous that they hindered the sinking of one of Minesota's mine shafts, and development work soon uncovered over one thousand tons of mass copper. The single biggest specimen, forty-five feet long and up to eight or nine feet thick, weighed an estimated five hundred tons.[15] It took more than a year to cut up the largest pieces of mass copper. Because parting large finds required so much human labor, mass copper was not the economic bonanza companies had hoped it would be. They discovered they could mine copper rock from an amygdaloid deposit—one that was only 2 percent copper—and make that pay as well as mass copper because copper rock was easier and faster to drill, blast, and transport.

Only the men who drilled and blasted rock were called *miners.* That class of workers

Mines that primarily worked amygdaloid or conglomerate copper deposits occasionally encountered large masses of copper. Early in the twentieth century, these Quincy
miners posed with a piece of mass copper they cut up using sledge hammers and chisels, the same tools used at the Cliff mine more than a half century before. *(Michigan
Technological University Archives and Copper Country Historical Collections)*

reserved the term for themselves because it signaled their higher status among underground laborers. Men performing other tasks had other titles. *Trammers* mucked up
broken rock and pushed it to the shaft to be hoisted. They picked up the smaller stuff
by using short D-handled shovels. They picked up larger rock—pieces weighing up to a
hundred pounds or so—with their hands. If they couldn't lift the rock, they skidded or
rolled it up a stout plank and into a tramcar. If the rock was too big for that, trammers
called in a *block holer,* a miner who drilled a hole in the rock and charged and fired it,
reducing it to rubble.[16]

In the first years of copper production, trammers moved rock in wheelbarrows, but
before the Civil War, most mines replaced wheelbarrows with four-wheeled tramcars
running on light-rails laid along broader drifts.[17] Tramcars carried bigger loads, two or
more tons, and the rails reduced friction, making the cars easier to push. Once they
pushed the rock to the shaft, trammers often dumped it directly into a bucket or box to
be hoisted to the surface, but sometimes they dumped it onto a flat floor, called a *trip
plat,* beside the shaft. In early mines, hoisting, like tramming, was an intermittent thing.
Companies stockpiled rock at the stopes before tramming it; they also stockpiled rock
next to shafts until there was enough to merit hoisting. Initially, rock traveled to the

surface in wrought-iron buckets called *kibbles,* which were drawn up by heavy Manila rope or hoisting chain. In a vertical shaft the kibble hung free and rose without the aid of guides. In the more common inclined shafts, side-by-side longitudinally strung timbers formed a skid road that cradled the kibble and reduced friction.[18] By the early 1860s the region's most productive mines replaced kibbles with *skips,* which were four-wheeled, wrought-iron boxes, open on the upper end. The skips, rolling over iron rails laid in the shaft, were raised by wire rope, instead of chain, and carried heavier loads at faster speeds.

Early mines often used three generations of hoisting technologies as they moved from exploration to development to production. Initially, men on the surface turned windlasses to raise the copper. Companies replaced human power with *horse-whims,* which had a winding drum on a vertical axis connected to booms or sweeps. Horses, harnessed to the sweeps, walked in circles, which rotated the winding drum and drew up the copper rock. Next came steam power.[19] A reciprocating steam engine produced rotary motion at its crank, which was transmitted to a cylindrical winding drum on a horizontal axis. The drum usually carried wire rope by this stage, which drew up a rock skip instead of a kibble. In a nine- or ten-hour workday, two windlassmen could raise about four tons of rock from a depth of one hundred feet. A horse harnessed to a whim could raise nine tons of rock from 150 feet per shift. Steam power offered great advantages of speed and capacity. Even the earliest steam engines raised two or three tons of rock per lift, moving the skip at five hundred feet per minute.[20]

The first steam engine arrived on the Keweenaw in 1845 to work at a mine near Eagle River. By 1857 twenty different companies had acquired a total of forty-eight engines, having 1,300 aggregate horsepower. The engines mainly hoisted and stamped rock and pumped water. Some engines had vertical cylinders and walking beams—a *British-style* engine. Others were horizontal, high-pressure engines fitted up in the *American style.* A minority were small portables, used to buzz wood or grind grain.[21]

Into the 1870s, as long as miners still used hand drills and black powder, each mine shaft produced too little rock to merit constant hoisting. So, many companies installed their steam hoists between shafts, where they could lift rock from two or three different ones. Making each hoist do double or triple duty saved the company capital because it didn't have to invest in additional engines, winding drums, boilers, and buildings. The winding rope from an engine ran over pulley stands to whichever shaft had enough rock stockpiled underground to need lifting.[22] Once the engine raised that shaft's rock, surface workers rerouted its hoisting rope to a neighboring shaft so rock could be lifted there.

Until the mid-1860s, men transported themselves in and out of a mine. In a vertical shaft, they sometimes rode out of the mine on a rope sling or in a kibble. In an inclined shaft, they hitched rides on rock skips. But wooden ladders remained their principal means of getting to and from work. Rough planks often separated the ladderway from the shaft's adjacent hoisting compartment. Men risked falls from ladders, whose rungs could be damp and slippery or in poor repair. In winter, close to the frigid surface, a ladder might even be covered in ice. Ladders also taxed the hearts and muscles of men. In the 1860s a writer for the *Portage Lake Mining Gazette* made clear the travail of ladder climbing:

> No person, who has ever been obliged to make the ascent from the bottom of even a tolerably deep mine up to grass and sunshine but has heartily wished there were some less laborious manner of accomplishing the journey. Especially it is the case with such persons as ourself, who Nature has endowed with a light physique instead of the thews of an ox, and we have willingly imperiled our life in an old-fashioned kibble, which took us rolling along to the surface in five minutes, rather than perhaps make a thousand steps upon perhaps safer ladders in half an hour. How much more laborious must it be then to the miner who, after a hard day's work under-ground, is compelled to climb up carrying from twenty to forty pounds of [drill] steel on his back?[23]

In 1864 the Cliff mine, already a thousand feet deep, became the first Lake mine to transport men using a man-engine. German and Cornish mines had employed man-engines since the 1840s, and Cornishman John Rawlings designed and installed the Cliff's. Others soon followed at other mines.[24] The man-engine was a mechanical ladder, moved by a steam engine on the surface and assisted by counterweights. They all worked in the same manner as the one installed at the Pewabic mine in 1865:

> [The man-engine] consists of two lines of heavy timbers bolted together continuously . . . , resting on a steep incline on rollers, with steps at certain distances the entire length. When the propelling power at the top starts, one of these timbers moves up the incline and the other down, [a] measured and equal distance, until the steps come opposite each other, when both halt for a second, and then move in reverse directions—up and down, up and down, continuously. The mining captain takes his position on a step, which comes up to a level with the floor. . . . As the timbers move on their measured tread, and the step on which the leader stands moves down to a vacant step on the opposite timber, he changes from one to the other, and gives the word of command, "Step over," and all the party do the same, and are carried down gradually in this way, until the bottom of the moving timbers are reached.[25]

At the end of a shift, underground workers reversed the procedure to get up to grass. Instead of stepping over the rod poised to go down, after each pause of the man-engine rods, they stepped over the rod ready to move up.

A mine's underground openings increased in volume every year, and the *hanging*

After the Cliff mine installed the region's first mechanical ladder or man-engine in 1864, many other companies followed. Quincy's man-engine, shown here, operated from 1866 until 1895. Despite the seeming hazard of workers having to step from one platform over to another in order to go up or down, very few men died from falls while using these contrivances. *(Historic American Engineering Record Collection, Library of Congress)*

wall—the rock roof perched over the stopes, drifts, and shafts—increased in area. In the first few decades of mining, most companies and their men discovered that the rock here was surprisingly strong and secure.[26] Consequently the mines left no regularly spaced rock pillars alongside shafts or up in the stopes to support the hanging. Instead, they stoped out rock rich in copper, regardless of where they found it. A large, open stope might extend up and down several levels or from one shaft across to another. The companies left some ground unmined where it was deemed too little charged with copper to merit taking. These random, poor-rock pillars helped support the hanging. Besides eschewing a regular pattern of protective pillars, most companies found no need to timber their openings from top to bottom or end to end. Instead, workers called *timbermen* propped up the hanging here and there, where it was loose and hazardous, by using *stulls,* which were sections of tree trunks.

Companies almost wholly depended on natural ventilation to move fresh air into and through their mines. Shafts filled with air columns of different heights, coupled with temperature gradients between the surface and the underground, kept the air moving. Certain shafts were downcast and carried a strong draught of fresh air underground.

Other shafts served as chimneys. Upcast shafts exhausted stale air, blasting gases, and dust into the atmosphere at grass. For decades the Lake mines resorted very little to mechanical fans to augment their system of natural ventilation, and all the while, experts lauded the mines for their clean and cool air.[27]

To light their way underground, each man carried a shift's worth of tallow candles with him, using their extra long wicks to tie them to his belt. He walked along a drift by the light of a candle held to his helmet with a ball of clay. A mining team drilled rock by the light of a few candles spiked to the wall or held to the rock with the clay balls taken from helmets. The men worked in a dim pool of light, and a vast darkness spread from one team of men over to the next. Looking over, a man saw a pinpoint candle flame in the distance, but little in the way of human features.

From the time they left the surface until their return, underground workers toiled nearly ten hours a day. Once underground, most men stayed there for the entire shift. They ate a meal underground; one item of choice was the *Cornish pasty,* a meat and vegetable pie wrapped in a folded-over pastry crust. Men carried their victuals below in tin pails and reheated them over candles. Because their shifts were so long, men inevitably had to answer nature's call while underground. Up on the surface, companies erected privies near major surface facilities. Such niceties did not exist underground. Men wandered a short distance from their workstations to urinate at any convenient spot. For more serious business, they squatted over an empty powder keg. For decades, underground sanitation remained primitive. About the best the mines did was to bury offensive wastes under poor rock, perhaps after throwing some lime onto them.[28]

The underground was a world unto itself: dark, rough, hazardous, enclosing, steeply pitched, filled with the sounds of hammer blows and the smell of spent blasting powder. It was a place of work with a single purpose: to liberate copper and get it to the surface. Atop the mine, a very different world existed.

3

❖ ❖ ❖

ISLANDS OF INDUSTRY IN
A SEA OF TREES

W hen a company progressed into full production, on the surface it tended to industrial and domestic needs. It built houses for families and hoisthouses for steam engines as its mine camp evolved into a mine settlement. The mining companies were not alone in building the infrastructure of an industrial society on Lake Superior. While they built mines, others established commercial villages and erected hotels and stores. Mining companies, small businesses, and professionals needed one another on the Keweenaw, and together they changed the landscape.

Henry Rowe Schoolcraft described the Keweenaw as being "beyond the boundaries appointed for the residence of man."[1] In the summer, choking vegetation slowed travel and work while blackflies and mosquitoes exacted a toll on settlers, beleaguering them with nasty bites. Then winter arrived by late November and lingered until late April. The green landscape of summer became a white landscape of deep snow, thanks to the *lake effect* and its influence on the climate. Air passing over Lake Superior picked up moisture that fell as snow over the cold Keweenaw land mass. Cold air and winter ice interrupted transport to and from the Copper Country (the name given to this area in the Upper Peninsula of Michigan); it closed shipping lanes and locked settlers out from the world below. Few people got in or out of the Keweenaw during the winter unless they were willing to walk south to Wisconsin. Winter cut off the flow of food and other supplies, so settlers had to make do with whatever provisions they started the season with. The sense of isolation grew as the lake boats stopped coming, and only an occasional mail delivery arrived by dogsled from Green Bay, Wisconsin.[2]

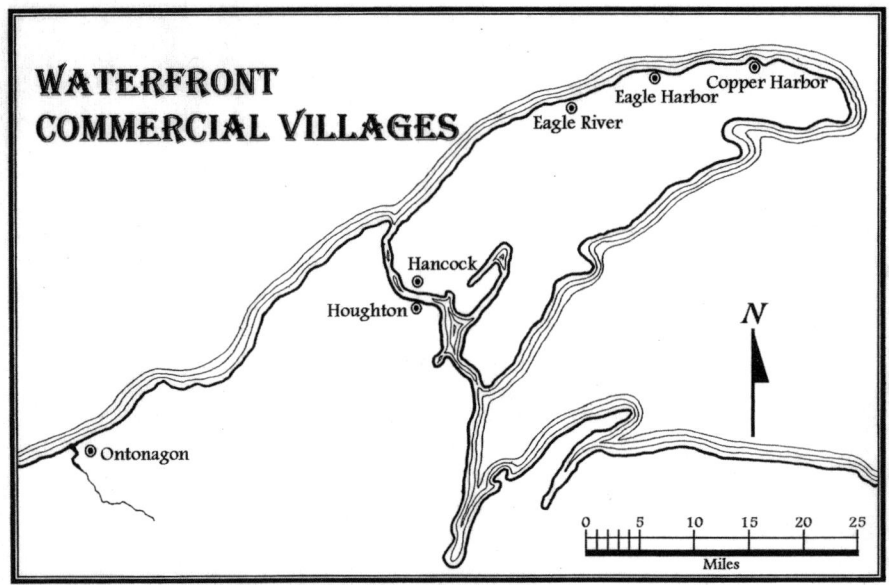

Commercial villages on the water played a key role in developing the Keweenaw.
Lake boats arrived at their docks to deliver new settlers and all the goods, supplies,
tools, and machines that the mine settlements needed. Lake boats leaving these villages carried the copper that hinterland mines had produced. *(Society for Industrial
Archeology)*

Lake Superior imposed hardships on settlers but also promised opportunity, not only
for mining firms, but for all kinds of workers, small businessmen, and families. Many
came to work underground in the mines. Cornishmen and their families left their traditional copper- and tin-mining district in the southwest corner of England because that
industry at midcentury was in decline. To seek opportunity and a better future, Cornishmen booked passage on ships that carried them to many new mining districts around
the globe, including Lake Superior.[3] Others came to the Keweenaw not to work *in* the
mines, but *at* the mines as managers, clerks, engineers, geologists, doctors, blacksmiths,
or machinists. Still others came to help establish commercial villages and do business
with the mines. This was especially true of Americans, who did not migrate to Lake
Superior to work underground. In the 1830s the New England to New York to Michigan migratory path first carried many settlers to the Lower Peninsula; in the 1840s and
1850s it carried a smaller number of Americans to the Keweenaw, including C. C. Douglass, Samuel W. Hill, James North Wright, and Daniel Brockway.[4] These transplanted
Americans—often connected to one another by kinship ties—migrated to Lake Superior
and made their mark and found success by establishing mines, commercial villages, and
businesses.[5]

While mines located along the central spine of the Keweenaw, commercial villages

In 1859, looking eastward at "downtown" Houghton, one saw little more than a dirt lane paralleling the shoreline of Portage Lake, where businesses were interspersed with residences. While merchants set up stores, several mines south and east of Houghton struggled to make a go of it. *(Michigan Technological University Archives and Copper Country Historical Collections)*

sprang up on the shorelines of Lake Superior and astride Portage Lake, where people and goods arrived and copper left on boats.[6] The earliest important villages were Copper Harbor, Eagle Harbor, and Eagle River, along the northwestern shore of Keweenaw Point. On the southern end of the mineral range, Ontonagon became the most important center of commerce and trade. In the 1850s and early 1860s, Houghton and Hancock developed on opposite sides of Portage Lake, just when a cluster of nearby mines went into production.

Copper Harbor was the region's first settlement of note. In 1843–44 it started as a picturesque encampment. Transient explorers lived in tents alongside the harbor and prepared for forays into the field or waited for boats to return them to the world below. Within a short time, log houses, small shops, and the Brockway Hotel brought a greater appearance of permanence to this outpost. The federal government, especially, made Copper Harbor the premier location on Lake Superior by putting its land office there to handle mining claims, by erecting and garrisoning Fort Wilkins, and by a having the U.S. Lighthouse Establishment build the region's first lighthouse there in 1849.[7]

Despite its early prominence, Copper Harbor's growth soon stalled. The U.S. Army delivered a major blow when it abandoned Fort Wilkins in 1846, just two years after building it. The biggest blow to Copper Harbor's fortunes, however, was that no major mines developed nearby. The best of the early mines sat closer to other villages further south on Keweenaw Point, such as Eagle Harbor and Eagle River. The Cliff, Phoenix, and other neighboring mines depended on Eagle Harbor for shipping and provisioning. Edward Taylor built the first pier at Eagle Harbor in 1844. William Raley later enlarged the village's dock and warehouse facilities, and a lighthouse was built in 1851.[8] By the 1850s, Eagle Harbor boasted more hotels, stores, churches, and drinking establishments than any hinterland mine location. Eagle River, too, became a coastal trade center, even though it did not have a fine natural harbor. Eagle River occupied land obtained from the federal government in 1843 by the Lake Superior Copper Company, which tried to start a paying mine three miles away. Besides becoming the seat of Keweenaw County, Eagle River boasted one of the early manufactories on the Keweenaw—a fuse factory. Settlers at mines traipsed to Eagle River seeking goods and services or entertainment. Henry Hobart, who taught school at the Cliff mine during the Civil War, trekked to Eagle River to shop, and his students took sleighs to Eagle River to compete in spelling bees against that settlement's young scholars.[9]

On the southern end of the mineral range, Ontonagon became the key commercial village. Prior to the mine rush, Indians camped and fished at this site located on Lake Superior at the mouth of the Ontonagon River. After the rush, some Ojibwa stayed. They caught and sold fish to settlers, loaded and unloaded lake boats, and transported people and goods up the Ontonagon River, toward the mines. James Paul (a rare Virginian who migrated to Lake Superior) helped found Ontonagon, starting in 1843. The village became the jumping-off point that serviced as many as thirty mines in the hinterland. As the only major port of call on its end of the mineral range, Ontonagon thrived, particularly in the 1850s, and for a time was the largest settlement on the Keweenaw.[10] Ontonagon improved its channel and docks, received a lighthouse, built hotels, welcomed druggists, erected schools and Protestant and Catholic churches, and served as the county seat. What limited the continued growth of Ontonagon was that mining declined on this end of the mineral range by 1870. At about the same time, the fortunes of the Cliff and other Keweenaw County mines declined, too, which capped the growth of Eagle Harbor and Eagle River.[11] Meanwhile, the center of mining and population increase shifted to the middle of the Keweenaw, which gave rise to Hancock on the northern side of Portage Lake and to Houghton on the southern shore.

Hancock's growth after the late 1850s was particularly tied to the discovery of the nearby Pewabic lode and to the increase in operations and employment at the Quincy, Pewabic, and Franklin mines, which sat on the hilltop above Portage Lake. Houghton had several mines on its side of the lake, including the Isle Royale, Huron, and Shelden and Columbian. Ransom Shelden, merchant, mine developer, and entrepreneur, originally acquired the land that comprised Houghton in 1852. He platted part of it for a commercial village and reserved other portions as mining lands.[12] In both Hancock and Houghton, the main commercial streets filled with shops, stores, hotels, saloons, and boardinghouses. Meanwhile, their shorelines filled with docks, warehouses, shipping offices, stamp mills, copper smelters, and iron foundries.

To promote growth and trade, the region desperately needed to improve water and overland transportation. Waterfront villages needed better channels and docks and better roads to connect to mines. A considerable mix of peoples and institutions brought about these transportation improvements, including one located on the far-eastern end of the Upper Peninsula. The Keweenaw benefited from the opening, in 1855, of a federally sponsored canal and locks at Sault Sainte Marie, Michigan. Here, the St. Marys River connected Lake Superior with the Lower Great Lakes, but rocky rapids prevented boats from freely passing by; cargo had to be unloaded from one vessel, portaged around the rapids, and reloaded onto another. Building a canal and locks at the Soo—Sault Sainte Marie—eliminated the portage and made the transport of people, goods, and copper both faster and less expensive. The locks at the Soo originally proved a hard sell in Washington, D.C. Senator Henry Clay scoffed at the project, saying it would be "a work quite beyond the remotest settlement of the United States, if not the moon." After Michigan became a state the federal government took a decade and half to commit to the project. In the end, the United States gave a land grant to Michigan, which the state transferred to the canal's builder as payment for the work. The St. Mary's Falls Ship Canal Company received 750,000 acres to sell or develop as it saw fit, and in return over a two-year span it built the locks and canal.[13]

With the opening of the Soo Locks, the largest lake boats sailed onto Lake Superior and then another 250 miles over to the Keweenaw. They could not, however, dock at Houghton and Hancock, deliver cargo, pick up cargo, and leave. Lake boats could not navigate the meandering Portage River, in places only three or four feet deep, that connected Keweenaw Bay and Lake Superior to Houghton and Hancock on Portage Lake. All cargo entering or leaving Portage Lake—including the first copper shipped out of the Portage in 1854 by the Isle Royale mine—required loading onto two vessels: a small boat (*lighter*) or barge that covered the fourteen miles between Keweenaw Bay and Houghton-Hancock, and a large lake boat that traversed Lake Superior and beyond. It cost mining companies and merchants a high shipping fee of four dollars to move a ton of cargo just fourteen miles. To cut this cost, in 1859 they took action. Instead of waiting for government to improve navigation, they raised $30,000 to remove the sandbar at Keweenaw Bay and dredge and straighten the channel to Portage Lake. The improved shipping lane opened to the biggest lake boats in 1861, and the mining companies and merchants created the Portage Lake and River Improvement Company to charge boat tolls to pay for the work already done and to generate revenue to pay for future dredging.[14] Similarly, on the northern end of the mineral range, mining companies organized a corporation that cut a shipping channel from Keweenaw Bay into Lac La Belle, making that lake a better, more useful harbor.[15]

In the 1870s, two more important improvements were made to water transport. One followed the government model, after the fashion of the Soo Locks; whereas the other was the "do-it-yourself" model of the mining companies. After the Portage River

Improvement Company dredged and straightened what would later be called the *lower* or *southern* entry from Keweenaw Bay to Portage Lake, on the opposite side of the Keweenaw it became desirable to create an *upper* or *northern* entry to Portage Lake from Lake Superior. There was no river to enlarge on this side of the Keweenaw; steam dredges would have to eat away three miles of ground to create a canal. In 1864 the Portage Lake and Lake Superior Ship Canal Company organized under the laws of Michigan to do this job.[16] The canal would be at least one hundred feet wide and thirteen feet deep; at the new upper entry, a breakwater and safe harbor would be constructed. Cut through between 1868 and 1873, the new ship canal cost about $2.5 million; the canal company did the work in return for government land grants totaling 450,000 acres in the Upper Peninsula.[17] At about the time this project was being completed, Calumet and Hecla and neighboring mines, following the do-it-yourself model, subscribed to a canal company that dredged a new two-mile-long channel connecting Torch and Portage lakes. This channel made Torch Lake accessible to the biggest lake boats and opened its shoreline to large industrial and commercial development.[18]

The region needed roads to transport steam engines or barrels of salt pork from dock to mine and to carry copper from mine to dock. But the construction of good roads lagged in this environment. In the 1840s and 1850s, new settlements were served by "mere trails, over which everything was carried to the mine by half-breed packers."[19] Footpaths linking the settlements gave way to horse paths and then wagon roads. The mining companies took charge of building early roads because of a lack of public funding. Not until the 1860s did federal or state money boost road building on the Keweenaw. During the Civil War era the Michigan legislature finally awarded land grants to support local road construction, and it also returned to local counties revenues accruing from taxes on copper output. The counties then applied this revenue to internal improvements. Meanwhile, the federal government, through an act of Congress, granted land to Michigan and Wisconsin to fund a military wagon road running down the mineral range from Copper Harbor into Wisconsin.[20]

In the early 1850s, Ontonagon County mines combined to build a road surfaced with boards laid side by side and spiked to longitudinal stringers. This plank road expedited the cartage of heavy loads and reduced costs. The Ontonagon Plank Road led from the commercial village of Ontonagon out to a dozen mines, including the mighty Minesota. For the Toltec mine (which invested $12,000 in the road) the cost of shipping a barrel of mass copper to Ontonagon fell from $1.50 to just fifty cents between 1851 and 1855. The plank road benefited settlers, as well, because stagecoaches provided regular passenger service over it.[21]

Regular dirt roads were not nearly as well engineered as the Ontonagon Plank Road. To save on the cost of felling trees, woodsmen cut narrow roadbeds through forests, leaving stumps and rocks in the roadway, which rarely was well graded or ditched. In the spring, snowmelt and rain turned roads into impassable quagmires. In the drier months of

NONESUCH MINE, L. S.

Folk artist Agnes Hathaway's 1884 drawing of the Nonesuch mine in Ontonagon County depicts how this mine and its settlement once existed in a sea of trees. Here, a pond along a dammed stream sends water coursing past a shafthouse and on to a stamp mill. Houses sit amid tree stumps, and large stockpiles of cordwood indicate that soon many other trees around the site will be harvested for human and industrial use. *(Michigan Department of Natural Resources)*

summer, the same roads became hard, rutted, and dusty. Often, winter actually improved overland travel. Snow filled in the ruts, covered the rocks and stumps, and provided a smoother surface to glide over. Winter also froze lakes and rivers, creating natural bridges and allowing for shortcuts unavailable during summer. Some mines stockpiled their copper during the summer and waited until winter to haul it in heavy sleighs to a shipping dock.[22]

Primitive trails and roads ran to primitive mine locations carved out of the wilderness. A typical mine location did not sit next to a commercial village having a host of stores, shops, churches, or bars. Instead, it sat out in the woods, an island of industry in a sea of trees, which in time, as the mine expanded, became a sea of tree stumps.

At a mine camp, a small crew looked for copper by using picks, axes, shovels, hand steels, sledgehammers, black powder, and wheelbarrows. Holes in the ground and piles of rock outnumbered buildings, which might have included storage sheds for tools and copper specimens, a powder magazine, and a blacksmith shop, where a smithy forged and sharpened tools such as drill steels. The first "dwelling" at a camp was often a squatter's tent, or a tepeelike hut erected out of poles and faced with cedar bark. The hut had a dirt floor, a circle of rocks for a fire, a hole in the roof to let smoke out, and a bed of moss and tree boughs.[23]

Next came boardinghouses, which sheltered a half-dozen to twenty young, single men, in one- or one-and-a-half-story structures up to twenty-eight feet wide and thirty to forty feet long. The men slept in hard, tiered bunks that ran alongside the walls and sometimes filled a loft under the roof. The earliest boardinghouses were built of logs cut and dressed at the site, but by the late 1850s, frame construction and clapboard siding became fairly common. Boardinghouses usually had wooden doors, windows, flooring, and a cast-iron stove for heat. At some camps, each boardinghouse had its own cooking and eating area; at others, the men bunked in one building and ate in another.[24]

When a company entered into serious development and opened the underground with new shafts and drifts, it still held off on building an extensive surface plant. Rudimentary structures sufficed, most constructed of wood, perhaps a few of poor rock. Machinery remained little in evidence. Simple wooden head frames stood over the shafts. Men or horses, not steam engines, hoisted the rock. Piles of poor rock grew larger. More trees disappeared as they were taken for fuel, structures, or mine supports. A single wagon road meandered to and through the mine camp, following the easiest path and avoiding the biggest obstacles. To house upward of a hundred young, single men, a company multiplied its number of boardinghouses. To live communally in this rugged mine camp, a man paid back to his company one-quarter to one-third of his monthly earnings.[25]

If pioneering shafts, drifts, and stopes exposed enough copper to encourage full production, a company transformed its camp into a mine location having a more extensive range of industrial and domestic structures. It usually sank additional shafts to expedite the taking of copper. At each shaft collar, it built a head frame of heavy-timber construction covered by a sheathing of rough vertical planking, unpainted. No machinery operated inside a shafthouse. A pulley at the top carried the hoisting rope or chain, which ran down into the mine and then lifted the kibble or skip. Now the company used steam power, rather than men or horses, to raise copper rock. Along the line of shafts stood two or three hoisthouses with poor-rock foundations and wooden superstructures. A tall, riveted iron smokestack marked each hoisthouse having a steam engine and boiler. Near the hoisthouse sat impressive stacks of cordwood to be burned as fuel. Because the boiler and stack posed a fire hazard to a wooden hoisthouse, sometimes a catwalk led up to and along the ridge line of the building's shingled roof. On the catwalk stood water barrels that men could dump over the roof to help fight a blaze.

When the hoist delivered the skip to the surface, it tipped and dumped its contents, which men put into wheelbarrows or small cars running on light-rails, which they pushed to a nearby sortinghouse. Men dumped the material onto the floor and hand-sorted it by quality and by size. They tossed smaller pieces of mass copper into barrels

Companies built their first-generation mine buildings almost entirely and universally out of wood. At the Wolverine mine, an endless-rope tramroad moved rockcars back and forth along the line of shafts, passing by a boiler- and hoisthouse. The view shows how a hoisting rope passed over pulley stands along the top of the lode and then wrapped part way around a head sheave, or pulley, to head down a shaft. *(Michigan Technological University Archives and Copper Country Historical Collections)*

ready for shipment to a smelter. (Hence the term for these pieces: *barrel copper.*) Larger pieces of mass were warehoused at the mine before being hauled away in heavy wagons or sleighs.[26] They relegated poor rock, carrying little or no copper, to waste piles ringing the mine. Good copper rock from amygdaloid or conglomerate lodes—with a copper content of about 2 percent or more—had to be stamped and concentrated before smelting. These processes mechanically liberated the copper from the rock matrix and then separated the two materials to produce a mineral concentrate that was 60 to 80 percent copper. The stamps at a concentrating mill could accept rock no bigger than three or four inches in diameter. So, at a sortinghouse, men separated good copper rock by size. They sent pieces under four inches straight to a stamp mill. They put oversized rock into cars or wheelbarrows that men pushed to nearby kilnhouses. At a kilnhouse, up to twenty-five laborers burned and dressed copper within a stone-lined pit about seventy-five feet long that was floored with cast-iron plates and covered with a roof. The men built a bed of timber about twenty feet square and four feet high, which they covered with four to six feet of copper rock before lighting. The heat calcined and cracked the rock; sometimes men aided this process by pouring water on the rock, causing it to cool rapidly and split.

A few years after the Civil War, the South Pewabic Mining Company (later taken over by the Atlantic Mining Company) opened this major stamp mill on the south shore of Portage Lake, west of Houghton. Originally equipped with four Ball steam stamps, this mill sent tailings out into the Portage until the mid-1890s. When the Atlantic mine had to cease dumping stamp sands into the small lake, it opened a new mill on Lake Superior at Redridge. *(Michigan Technological University Archives and Copper Country Historical Collections)*

When the rock fully cooled, laborers used hammers and picks to break up any pieces still too large to send to the stamps.[27]

Stamp mills required large quantities of water. Flowing water transported rock and copper from operation to operation within the mill. With the help of various washing machines, water flow also separated the lighter rock (which tended to be carried along by the water) from the heavier copper (which tended to sink or settle out). Many early stamp mills were located close to the shafts and were part of the mine location. Companies supplied water to these mills by tapping water pumped from the mine itself and by damming a nearby stream to create a millpond. But as mines became larger, local water supplies proved insufficient. Companies moved mills to larger lakes: to Portage Lake, Torch Lake, and Lake Superior. While providing the water that mills needed, the lakes also served as dumps for the mill wastes, known as *stamp sands.* To connect mine and mill, companies built a rail line or tramroad between the two.[28]

From 1845 to 1855, mills used batteries of *drop* or *gravity* stamps like those found in Cornish tin and copper mines. Each individual stamp was a block of cast iron affixed to the bottom end of a rise-and-fall vertical stem—and each battery carried four or five stamps working in a common mortar box. Copper rock up to four inches in diameter was fed into the mortar box, as was a strong stream of water. A steam engine, through belting and gearing, drove the main shaft on each battery. This shaft carried cams, one for each stamp. As the shaft rotated, its cams sequentially lifted each of the machine's stems. As the shaft rotated further, each stem fell off its cam, dropped by gravity, and its stamp shoe fell onto the rock, breaking it up. A perforated plate formed one or two sides of the mortar box and served as a screen controlling outflow from the stamp. When crushed finely enough, down to about three-sixteenths of an inch, the rock and liberated copper particles washed out through the screen and flowed to devices that separated the two materials. By the late 1850s some companies adopted the newer technology of steam stamps. These machines reciprocated a single, heavier stamp in a very different way. The stamp's stem ran up to an overhead steam cylinder, passed through a steam-tight packing gland, and connected to a reciprocating piston. The pressurized steam that moved the piston up also helped propel it downward, enabling the machine to strike a more power-ful blow against the rock.[29]

At the earliest mills, downstream from the stamps men and boys separated rock and copper by hand using jiggers, as well as pools, troughs, and gates laid out along the mill floor.[30] The *jigger* was a low tub with a perforated brass bottom. A boy shoveled up water, rock, and copper from a trough or pool and swished it around before setting the jigger on the floor. The heavier copper settled at the bottom of the jig, with the rock and lighter, smaller pieces of copper above it. Using a scraper, the boy separated the two layers, collect-ing the copper and sending the top layer of material off for further washing and separating in troughs and *buddles,* which were like miniature rapids or waterfalls, with gates going across the water flow. If the flow of water was just right, the copper settled out along the bottom of a trough, or behind a buddle's gate, while the water carried the rock along.

Cornishmen brought over the knowledge needed to set up the first stamp mills, whose technologies were not static. They rapidly changed over the first few decades, especially as more companies started working amygdaloid and conglomerate deposits. Companies that tied their futures to the recovery of fine pieces of copper needed to mill great tonnages of rock. They fine-tuned their milling practices to capture more and more copper, and a home-grown style of milling arose on the Keweenaw. New machines replaced steps previously done by hand or with simple troughs. Companies no longer hired boys to jig copper by hand. They turned to mechanical jigs powered by a steam engine. In these jigs, plungers agitated a watery suspension of copper and rock particles in a chamber with a sieve-plate bottom. When the agitation stopped, the copper settled to the bottom. At mechanical buddles, a slimy mixture of rock and copper flowed out slowly from the center of a large, rotating disk. The copper stayed on the disk while water carried the rock particles across the disk's surface and tailed them over its edge. With the addition of steam stamps, jigs, and rotating buddles, mechanization proceeded faster at the mills than at the mines.[31]

All successful mining companies operated their own mills, but few erected a smelter. The output of a single company, unless it was very large, did not justify such an investment. In the pioneering days of mining on the Keweenaw and Isle Royale, a few companies tried to smelt their own product but gave it up. The region could not boast of having a successful smelter until 1860. Before then, Lake Superior copper first traveled to faraway Boston and Baltimore to be melted and cast into ingots. A bit later, smelters arose closer to home, in Pittsburgh, Cleveland, and Detroit.[32] The Waterbury and Detroit Copper Company (in Detroit) smelted the product of many Lake mines (and put each mine's individual trademark on each ingot). In 1860 the Portage Lake Copper Company finally opened a smelter near Hancock.[33] In 1867 the smelting companies in Detroit and on Portage Lake merged to form the Detroit and Lake Superior Copper Company, which continued to handle the products of many mines at smelters at Detroit and on Portage Lake.

Because of native copper's purity, smelters had to do minimal refining to rid it of undesirable elements. They basically just melted it in a reverberatory furnace. The design of the furnace protected the metal's purity; it had a low bridge wall that kept the fuel (coal) in the firebox end of the furnace, segregated from the mineral. Heat and flame swept over the bridge wall and reverberated along the hearth before going up and out the stack. Small doors in the masonry furnace gave men access to the melt. After the copper reached its melting point, workers skimmed or ladled off the slag of molten rock, loading it into buggies that others wheeled away and dumped. Then furnacemen splashed the copper around with a paddle called a *rabble*. *Rabbling* mixed air into the melt to oxidize low levels of impurities. Next they eliminated excess oxygen by plunging hardwood poles into the melt. The poles combusted immediately, agitated the melt, and produced carbon that combined with oxygen to produce a gas that left the melt and passed up the stack. Furnacemen skimmed off remaining slag and then ladled copper into molds to produce copper ingots and cakes of different sizes and shapes for different markets. When done, they recharged the furnace with mineral and mass, through the top, to start another run.[34]

The mine camps along the mineral range evolved into real industrial sites as they cut back the forest and added steam engines and boilers, shafthouses, sortinghouses, and kilnhouses, stamp mills, and blacksmith and machine shops. By 1864 the region boasted its first railroad, a short line built in 1864 to move copper rock back and forth across the Pewabic and Franklin mines.[35] While companies invested in technologies, they tended to social needs, as well. The mine operators weren't utopians or great community planners, who went in with a well-thought-out, unwavering vision of what life should be like at a mine on Lake Superior. But they knew that squatter's huts and boardinghouses filled with single men out in the woods did not suffice. They needed a more complete social infrastructure, so they tackled the building of houses, neighborhoods, churches, and schools.

4

❖ ❖ ❖

OUT AT THE LOCATIONS

From Camps to Communities

While a company prospected for copper, its small force of workers eked out a marginal existence. They lived without women, bunked in a boardinghouse, and ate a repetitive diet of bread, potatoes, and salt fish or salt pork. (One company's men got a taste of "fresh" red meat only when a work ox died.)[1] Their mine camp was blighted by tree stumps and pockmarked by holes; it sat alongside a rude trail leading off toward some waterfront village. Nothing had the look of order, the feel of comfort, or the air of permanence.

At successful mines, rude camps evolved into settlements. A company owned at least a square mile of land. It reserved a fraction of that, near its lode, for industrial needs. It set aside other land, when cleared of trees and brush, as pasturage or agricultural fields.[2] That left considerable property to be developed as a place to live. The mining companies built their communities up pragmatically. Social needs were felt and were met, but not in accordance with any grand plan. Clearly, though, by the early 1850s the mining companies had worked out a mental template of what a mine location was supposed to be, what it would include and exclude. Once formed, that mental template remained consistent over decades from one company to the next.[3]

According to this template, a mine location was primarily a place to reside and work in, and only marginally a place of commerce. The industry did not call its settlements *company towns* or *villages*. They were unincorporated *locations*. Companies did not build complete towns, nor did they want to. The typical mine location perhaps had a store or two, but no main street, no downtown. No run of shops, no bars or billiard halls, no bank, no newspaper office, no druggist or millinery. Mining companies understood that first and foremost they were in the mining business. They recognized the need for trade, for support services—but they did not want to tend to every human need. They left much for others to do, and many companies—while they eschewed building a town right at their mine—encouraged the development of a town on the margin of the mine. That town would attract more business and settlers, in fact, if it were not tied to one mine but sat within reach of a cluster of mines whose employees needed a choice of places to live, entertainment, and myriad shops and stores.

The Minesota mine set off eighty acres into a village plat and sold off lots to start a commercial settlement. Not far away, the National mine followed suit and sold lots to create a place called Webster; within a year the lots were occupied by four stores and forty houses. The Mendota Mining Company surveyed some of its land to create the village of Mendota and sold off lots valued at $13,600. The Northwest Copper Company created Wyoming, where it sold at least seventy-seven lots.[4] The Phoenix Copper Company, while erecting a mine location over its lode, platted and sold off lots three miles away, at Eagle Harbor. Phoenix was very willing to encourage the growth of that commercial village even as it recognized that "our object is to make money by mining and not to speculate in house lots."[5] The Quincy Mining Company made no attempt to build a town at its mine on the hilltop overlooking Portage Lake but owned property running down the hillside to the lakeshore. In 1859, Quincy platted and sold some of its ground near the lake. It laid out an orderly street grid, originally three blocks by four blocks, with single and double lots marked off. Later, it added more land to what became the important city of Hancock.[6]

At all mine locations, companies built worker houses because they had to and not because they sought substantial profits by being corporate landlords. In this harsh environment, where many early mines sat in isolated spots, it was important to put housing near the line of shafts, convenient to work. The companies owned the land; they were the ones bringing in hundreds of employees, including men with families; and they had the financial resources to build houses, more so than arriving immigrants. Erecting company housing became part of the cost of doing business on this mining frontier.

The mines sometimes placed their first houses haphazardly on whatever ground was already cleared of trees. These houses might have piles of poor rock or shaft and engine houses for neighbors. On old mine maps, houses exist as little rectangles found here and there alongside the trail or road running through the location.[7] But that situation changed. When a company entered into production, a growing sense of orderly development took over. Near the heart of the mine, the company cut unpaved lanes or roads (calling them "streets" would be a misleading overstatement) to create lots for houses and neighborhoods. Along these lanes, new houses represented the arrival of civilized order. The houses along a given lane, built at the same time, usually were identical in size, form, and materials. They stood on lots of uniform size (often 50 by 100 feet), which were often fenced.[8] The houses were oriented in the same manner toward the lane (such as all

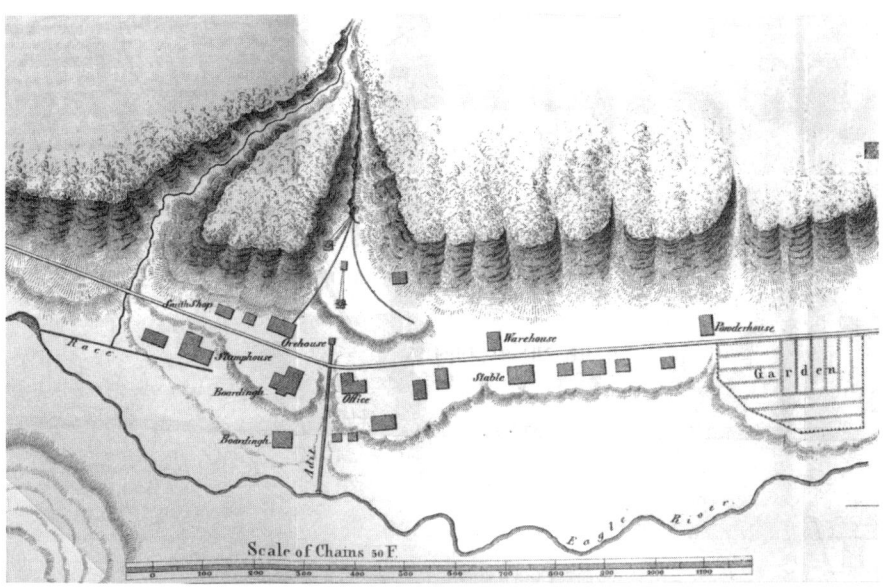

When surveyed by Samuel Worth Hill in 1847, the Cliff was a mine camp on the verge of becoming a mine location or community. The Cliff had garden plots but lacked a school, churches, and single-family dwellings set off into a neighborhood. Most men resided in boardinghouses while the company built up its technological infrastructure. *(Michigan Technological University Archives and Copper Country Historical Collections)*

ridge lines parallel to the road) and laid out in a straight line with a common a setback (distance from the lane's edge).

The mining companies' shared mental template led them to build single-family dwellings rather than more large boardinghouses. This sent the message that they wanted to employ more married men. Companies recognized an occupational hierarchy at the mine. At the location, not all employees were equal, nor was all housing. Cookie-cutter log houses, having three or four small rooms and a sleeping loft, stood in rows in neighborhoods. Married, skilled workers rented them, usually at a rate of one dollar per month per room. Bosses and managers lived somewhere else on the location, in bigger, more expensive houses that often flashed a bit of architectural style and elaboration (such as Gothic Revival touches) left off workers' dwellings. Meanwhile, single men or unskilled workers had to fend for themselves in getting housing. Perhaps a company still operated a boardinghouse or two. If not, the single men lodged with a family at the mine or trekked off to the nearest village to find a place to live.[9]

Besides erecting housing, companies engaged in other activities that made life on Lake Superior more comfortable and secure. Chief among these, they hired doctors, dispensed medicines, and built hospitals. In the first half of the nineteenth century, many

Many companies in the 1840s through the 1860s built worker houses similar to these shown at the Phoenix mine. Constructed of logs, their wooden sills often sat right on the ground, without benefit of a masonry foundation. They commonly had two rooms on the first floor, plus a kitchen off the back, with a sleeping loft. One cast-iron stove provided for cooking and also heated the dwelling. *(Michigan Technological University Archives and Copper Country Historical Collections)*

families throughout the United States nursed themselves when struck by illness or injury. Physicians did not have high status as medical providers, and they did not have many miracle cures or surgical arts to offer. Commonly, the first line of defense against suffering and illness was the wife or mother, who knew medical recipes and served as family nurse. But things were changing at about the time the mines started. Medicine was becoming more of a profession—and thanks to the industrial revolution and to the opening up of places like Lake Superior—many young, unmarried men were leaving home to strike out on their own. They left behind their traditional nurses—mothers or wives—and had no woman to take care of them. Enter the physician.[10] The mining companies recognized a need for doctoring on the mining frontier. As soon as a company employed two hundred or so workers, it hired a physician or at least shared one with a neighboring mine. Companies recruited doctors, treated them as important mine officials, and gave them good houses and good salaries.[11]

Even the early mines established company medical plans. Single men usually paid fifty cents into the plan per month; married men, a dollar. (Participation was mandatory for underground workers, but sometimes optional for surface workers, an acknowledgment of the different levels of occupational risk the two groups faced.) For this fee, men and their families received doctor's calls, all medicines, and outpatient treatments for no additional charge. Busy mine doctors treated all kinds of maladies, responded to trauma cases, set fractures, amputated limbs, and delivered babies. Sometimes mines pressed them into service as the company dentist or veterinarian. Within five years of going into production, if a company was still expanding and doing well, it not uncommonly erected a dwellinglike structure to serve as its hospital. Home care remained the order of the day for patients dying of disease or giving birth; company hospitals mainly treated men injured while working.[12]

Although the typical mine location lacked a full-blown commercial district, mining firms helped provision their employees with the stuff of life. Early on, especially if a range of commercial enterprises had not yet opened in some nearby village, many companies built and ran a general store for a while. Sometimes, under a contractual agreement, they allowed a merchant or two to open shop, with the understanding that they were not to gouge the men with high prices, which would only alienate workers and lead to agitation for higher wages. Many early companies ran farms, which produced feed for work animals and vegetables such as potatoes for human consumption. Their other option was to lease land to a farmer who sold his produce back to the company or its men. The Cliff mine and others set aside patches of cleared land as gardens for employees wanting to grow some of their own food. The sizable lots that came with company houses also gave families room for modest gardens or for raising fowl or a hog. And companies offered free or low-cost pasturage to employees fortunate enough to own a milk cow.[13]

As a location's population grew to include more women and children, more social needs became apparent, especially the provision of religion and education, whose histories at the mines were intertwined. Sometimes the first church and the first school were one in the same building, a rather nondescript wooden structure erected by a company. It served as a rudimentary school by day and then as a church on Bible nights or Sundays. Within a few years, though, each institution moved into its own space.[14]

Churchmen arrived on Lake Superior before miners did. The Methodists had occupied their Indian mission near L'Anse on Keweenaw Bay since 1834; the Catholics arrived on the opposite side of the bay, near Baraga, in 1843. The Methodists and Catholics competed against each other as they sought Ojibwa converts to Christianity. Once the rough mine camps began transforming into settlements, the Catholic missionary, Father Frederic Baraga, and the Methodist, Reverend John Pitezel, began tending to the religious needs of their populations. Reverend Pitezel preached his first sermon at a mine location, the Cliff, in 1846. Cornish settlers, mostly Methodists, made up his new flock. They first listened to Reverend Pitezel in the mine's cooper shop, pushing aside the tools of barrel making to make room for the service, but by 1848 they erected a church. Father Baraga made his first trips up to the mines in 1847. In private homes or in some secular building, he preached the word of God, offered the sacraments, and encouraged the Catholics, numerous among the German, Irish, and French Canadians in the area, to begin building churches.[15]

The Phoenix mine (first opened in the mid-1840s) stands in the foreground, while log houses stand along the curving road. In the middle distance, a pile of poor rock stands to the left. In the center stands the Phoenix store, school, and the Catholic church, moved there from the Cliff mine. Company houses cluster together off to the right. Phoenix reached a peak population of about one thousand in the late 1870s. *(Keweenaw National Historical Park Archives)*

Church construction proceeded slowly at first but boomed in the late 1850s as the population increased at the mines and commercial villages. Because both the Methodists and Catholics had missions on Keweenaw Bay, and because most settlers belonged to one or the other of those religions, Methodist and Catholic churches were usually the first two to stand in any community. (An Episcopal church was often the third one in.)[16] Sometimes the Protestants got there first, sometimes the Catholics. These early churches were modest structures—smallish, universally wood framed, and clapboarded on the exterior—but with some refinements, such as plastered interiors and the possession of a melodeon or organ. Modest or not, standing amid small workers' cabins on land cleared of trees, the church steeples stood out prominently on the landscape.

When Methodists, Catholics, or Episcopalians broached the idea of building a church, invariably the host mining company acquiesced. Companies did not favor one church over another; they supported church building in general. They stopped far short of paying for new churches and eschewed making large gifts to sustain churches, but they often contributed to a church's building fund and provided a building site, either freely donated or leased at nominal cost.[17] Once erected, a mine location church flourished

Methodists, mainly Cornish, erected the first church at the Cliff mine (or Clifton) in 1849. This church stood about midway between the mine and the bulk of the small log houses built by the company. While the mine thrived, Episcopalians added a church in 1856, followed by the Catholics in 1858. All three churches suffered when the mine declined in the late 1860s. Two went to rack and ruin on site, whereas the Catholics picked up their church and moved it down the road to the Phoenix mine. *(Michigan Technological University Archives and Copper Country Historical Collections)*

only if the company did. If a mine faltered and failed, workers and families abandoned the location, and the church lost its congregation. The fortunes of churches at commercial villages, too, rose or fell as nearby mines grew or declined. Up and down the mineral range, there was an ebb and flow of new church construction here and church abandonment there in early decades. By 1870, the copper district as a whole had thirty-three active churches. The Methodists and Catholics each had eleven; no other denomination had more than four.[18]

At the mines, churches allowed settlers to reclaim past traditions. They worshipped as they had in the old country, hopefully while being ministered to by someone sharing their ethnic identity and language. Churches helped immigrants self-select themselves out into smaller groups—for instance, Irish Catholics or Cornish Methodists—where they felt at home and comfortable being with like peoples. Public schools moved this largely immigrant society in the opposite direction. They brought diverse children together, where they sat with children unlike themselves. In school, they did not speak or listen to the dialect or language of a distant homeland but conversed and read in American English. The churches tended to perpetuate an old ethnic identity, whereas the schools helped forged a new one: American.

In schools, making new Americans out of the sons and daughters of immigrants was an activity that was rarely well orchestrated, well funded, or rapidly achieved. In commercial villages, students got the education the population wanted; at the mine locations, they got the schooling the mining company wanted. Nominally, schools were public institutions, but at a location where the company owned the land, provided the employment, paid the taxes, built the roads, and erected the houses, the company also set the course for the school. At a successful, growing mine, proper schooling was more likely achieved than at a struggling, declining mine, where survival was at stake and sustaining a school was a bothersome distraction. No early companies, however, even the most successful ones, funded or supported schools to the full satisfaction of teachers, administrators, or newspaper editors.[19]

The first instruction at the locations often occurred within houses or borrowed structures rather than in dedicated school buildings. The first teachers were paid tutors, priests or ministers, or volunteers from among the better-educated settlers. School buildings dedicated to public instruction appeared at the mines at the tip and at the base of the Keweenaw in the late 1840s and 1850s, after companies there had attracted a sizable numbers of families. They did not appear in the middle part of the peninsula until the 1860s. Starting as early as 1848, ministers or priests conducted the first schooling at the Cliff mine. By 1860 the mine had a permanent school of two rooms, supported by a two-mill tax on property in the district (mostly owned by the location's parent company, the Pittsburgh and Boston Mining Company). To the south, in 1852, when its location boasted thirty-three dwellings and a population of just over three hundred, including one hundred women and children, the Minesota mine erected "a new and comfortable building designed and regularly occupied as a church and school house." In 1861 an early Houghton County school went up in Franklin Township, where it enrolled children from the Franklin and Pewabic locations.[20]

Mine schools were ungraded and poorly equipped. A single teacher instructed fifty to one hundred students of all ages gathered in a single large room. At best, a mine school had two rooms and two teachers: a man who directed the whole school and instructed the older students (who sometimes could be fractious and difficult to deal with; hence the need for a male teacher), and a woman, paid about half of what the man received, who taught the younger children.

Class sizes and rosters constantly varied as working families moved from mine to mine. Students stayed home during the frequent epidemics that swept through the population. In the harshest days of winter, students did not venture out. Also, in this culture, settlers did not universally believe in or strictly enforce the notion of compulsory attendance. Michigan's education laws were themselves quite slack; they mandated as few as twelve weeks a year of schooling, and only six had to be consecutive. As a consequence, not all children enrolled in school. (Only half the children in Houghton County attended school in the early 1860s.) Those who did often came and went as they pleased.[21] Still,

The Central mine, incorporated in 1854 and operated until 1898, was one of the richest mass-copper mines in Keweenaw County. It produced about fifty-two million pounds of copper and paid out over two million dollars in dividends. While a poor mine might hold school in a room over a mine office or saloon, Central's fine school (shown here sixteen years after the mine closed) was done up in the Second Empire style, which was especially popular in the United States in the 1860s through the 1880s. *(Michigan Technological University Archives and Copper Country Historical Collections)*

this industrializing society on the frontier made a start at proper education, and the mining companies played a key role. Besides initiating some of the earliest schools, a few mines, such as the Cliff, Pewabic, and Franklin, started the region's first modest libraries (which sometimes consisted of only a bookcase or two of donated volumes).[22]

The mining firms could hardly have anticipated all the work needed to plant a heavy industry in the woods. They initially focused on prospecting, on finding a promising piece of property and discovering copper beneath it. As they moved along, they addressed more needs and tasks. Mining companies built roads and docks and invested in other companies that dredged out waterways and improved channels. They pieced together all the technologies needed for mining and milling copper rock; they worked out arrangements for smelting. Out of the same sense of necessity, they also engaged in a host of community-building activities as they shepherded the transformation of a piece of wilderness into a mine camp and then into a mine location. They decided what types of men they wanted, and tried to find and keep them. They operated farms or somehow supported early agriculture to the benefit of the population. They decided which commercial activities to allow right at the mines and which to exclude. Sometimes to help find a new home for commercial activities that they themselves did not wish to pursue, they created new towns adjacent to their mine works. They built houses, which not only provided shelter but a sense of order and even a social hierarchy. They hired doctors, built hospitals, supported churches, and set the directions, for good or bad, of many schools.

Much of this activity went beyond the bounds of the mental template the companies started with regarding what a mine was supposed to be, or include, or look like. But as locations matured, as companies found pragmatic solutions to social problems and needs, the mental template became more inclusive and clear. What the companies did at the start out of necessity (such as build houses), they continued to do out of habit. They learned in the first few decades that their mines were just that—*theirs*. They would control them in myriad ways, and at those mines, life and work would be inextricably linked. There was a word for this managerial approach and style, where a company's activities broadly affected the lives of employees and their families from cradle to grave. That word was *paternalism*.[23] The mining-company officers and bosses never used that word, paternalism, but their evolving mental template surely embraced the concept, and they practiced company paternalism for over a century on the Keweenaw.

5

❖ ❖ ❖

THE QUINCY MINE

Taking the Long Road to Success

Of the firms launched in the 1840s, the Quincy Mining Company survived the longest and produced the most, despite a difficult start. It acquired its mineral lands north of Portage Lake in September 1846.[1] A small crew immediately began prospecting at this site and found little copper—a result that did not substantially improve for years. While Quincy worked in obscurity, the Cliff and Minesota mines stole the spotlight, garnering fame as they found rich deposits of mass copper and paid handsome dividends. But the Cliff and Minesota mines, which profited so early, also faltered early. By the end of the Civil War, their glory days were over. Quincy, on the other hand, headed in the opposite direction. After struggling mightily for over a decade, the company began working a lode of amygdaloid copper that supported profitability well into the twentieth century.

Quincy was a well-managed company.[2] The lode it worked, the Pewabic amygdaloid, was not always easy to follow or easy to wring profits from. It was a good lode, one of the best, yet it was not so fabulously rich in copper that it assured high profits. Quincy managed its works carefully and never took success for granted. By producing a long run of dividend-paying years, the firm earned the nickname "Old Reliable." The company paid its first dividends in 1862–64 and then none in 1865–66. But starting in 1867, Quincy rewarded investors with dividend payments for fifty-four consecutive years, running until 1920.[3]

Quincy was exceptional because of its long string of dividends, because it mined its lode for so many years (1856 to 1931, and then 1937 to 1945), and because it was the deepest mine in the United States. By the 1920s, Quincy's No. 2 shaft reached a depth of over 9,200 feet, measured along the incline. Yet at the same time, Quincy was a typical Lake copper mine. It employed commonplace mine methods, tools, and technologies to mine and mill rock and transport materials. It engaged in house construction, doctoring, and other paternalistic practices. Like other companies, Quincy limited trade at its location, yet supported the growth of a nearby commercial village, Hancock. In its early years Quincy depended on key American settlers to direct its fortunes, yet immigrants did the hard work above and below ground. And like other Lake mines, it relied on the contract system of mining, borrowed in modified form from the copper and tin mines of Cornwall, as a way of putting men to work.

To conserve capital, Quincy employed only a handful of men in the late 1840s and early 1850s. The company first explored the hillside rising up from Portage Lake, rather than the hilltop. They found some mass copper, including one impressive specimen of 8,140 pounds, but failed to find a consistently rich lode or fissure vein.[4] A lack of capital slowed the search, and the lack of copper finds constricted the capital flow. After several years, prospecting shifted to the hilltop, where men explored near a run of Indian diggings. In 1854, while the directors were considering a sale of the property, Quincy opened the most promising lode yet found, which it dubbed the "Quincy lode." Quincy sank two or three shafts into its namesake lode, but it yielded little copper. By March 1855, the discouraged and poor company, which had sunk over $40,000 into the property, ceased mining.[5] Then, in July 1856, serendipity stepped in. The Pewabic mine, right next door, discovered a promising amygdaloid deposit—that crossed onto Quincy's property.

Between 1856 and 1858, three different, yet equally important, events reenergized the Quincy Mining Company. First came the discovery of the Pewabic lode. Then a new group of officers, directors, and major shareholders took charge of the company, and Thomas Fales Mason, who became president, provided a steady hand of leadership from 1858 all the way to 1899, save for a break from 1872 to 1875. Finally, Quincy relocated its headquarters from Philadelphia to New York City, much closer to sources of investment capital. With stable leadership forming, and with investment and geology both looking up, Quincy vigorously pursued its future. From 1856 to 1858 it worked the Quincy and Pewabic lodes simultaneously. To its shafts on the older lode, it added four new ones on the Pewabic. After 1858 Quincy developed the Pewabic lode exclusively, and it produced all the dividend-paying years to come.[6]

During its last years of prospecting in the mid-1850s, Quincy employed between 20 and 30 men. As the mine pushed development work harder, 110 to 115 men worked at Quincy in 1857–58, and 257 labored there in 1859. With the onset of real production on the Pewabic lode, employment jumped to 469 in 1860 and to 583 in 1861. Copper production rose similarly. Quincy produced 7 tons of copper in 1856, 61 tons in 1857, 179 tons in 1859, and then 970 and 1,283 tons in 1860 and 1861.[7] After a decade and a half of struggle, Quincy finally became a major copper producer.

The Pewabic lode, as it ran across Quincy's property, was about twelve feet thick near the surface. It descended on a dip of 54 degrees from the horizontal and was sandwiched between layers of traprock. Quincy's inclined shafts ran through the lode and followed it into the ground. The company sank its first shaft into the Pewabic lode in November 1856, its second and third in October and December 1857, a fourth in December 1858, and its fifth and sixth shafts in July and August 1859.[8] These numerous shafts, only two to four hundred feet apart, enabled miners to attack the lode at more places, and the shafts aided natural ventilation by serving as multiple chimneys for the exchange of fresh air for stale air.

As Quincy followed the Pewabic lode into the ground, mining captains told men to take out copper-bearing rock wherever they found it. The company did not systematically leave rock pillars alongside shafts or up in the stopes to hold up the hanging wall, and it did not timber the mine in any regular fashion. Timbermen propped up particularly loose parts of the hanging with stulls but let the rest go. This neglect of the hanging ultimately caused the mine serious structural problems, but in early decades Quincy put much confidence in the strength of its hanging wall and got away with it.[9] The company discovered that the Pewabic lode sometimes split, and that it was "bunchy" or "pockety."[10] Miners had trouble staying in good ground, and periodically they drove crosscuts to test whether good or better copper was lying to either side of the main lode. During the Civil War, Quincy extended its explorations in a major way by starting to drive an adit over eleven hundred feet long. Completed in 1868, the adit ran from the base of Quincy Hill toward the Pewabic lode, which it intersected at depth. The company hoped this expensive tunnel would reveal another lode to be worked, but it didn't.[11] The completed adit did serve later, however, as a useful drain for unwatering the upper reaches of the mine.

Going well past the Civil War era, miners still employed sledgehammers, hand drills, and black powder to blast out Quincy's openings. Initially, after miners broke the rock, men having *wheeling* contracts mucked it out and transported it underground in wheelbarrows; later, men performing this task were called *trammers*. By 1862 eleven tons of lightweight, narrow-gauge iron rails ran along Quincy's drifts, and tramming teams of two or three men loaded four-wheeled cars by hand and pushed them over the rails to the shaft for hoisting.[12]

When Quincy first opened the Pewabic lode, small contract teams raised the rock by using hand-powered windlasses or capstans mounted at the shaft collars.[13] When it moved into production, Quincy upgraded its hoisting technology. Some mines made extensive use of horse-whims, but not Quincy. Two credible sources, dated 1859 and 1861, mention a single horse-whim at the mine, which by the latter date was no longer in use.[14] Steam hoists arrived at the mine by the end of 1858, and an inventory of March 1859 listed two hoisthouses and two portable engines. Quincy erected more permanent structures and installed larger engines later in 1859 and in 1860. These engines were not engineered specifically to power hoists; they were general purpose, utility engines that could have driven all manner of machinery in a variety of industrial settings. At the mines, millwrights and mechanics connected them with gearing or friction drives to rotate hoisting drums.

In the summer of 1859 Quincy installed an engine built by George M. Bird and Company of East Boston, Massachusetts, between Shafts 1 and 2. By August 1860 Quincy had put a new high-pressure, horizontal engine built by J. B. Wayne and Company of Detroit between Shafts 3 and 4. The hoists were located between shafts, so the engines could lift from more than one simply by rerouting the hoisting chain or rope. This flexible arrangement meant that Quincy saved money by not needing an engine at each shaft. By the end of 1860, or shortly thereafter, Quincy's Bird engine hoisted from Shafts 1 and 2, as needed, and the Wayne engine pulled from Shafts 3, 4, and 5. By mid-1861 Quincy replaced its horse-whim at Shaft 6 with an unspecified "small engine." A few other engines were shuttled in and out of service in the early 1860s. By 1862 Quincy had used at least six different hoisting engines as it began mechanizing its mine works.[15]

With the arrival of steam power, Quincy modified other parts of its hoisting technology. Between 1861 and 1864 the company replaced all hoisting chains with wire rope, which was both lighter and less likely to fail catastrophically, and it replaced its kibbles with wheeled skips of two-ton capacity.[16] As the skips were filled underground and dumped at the shafthouse, their motion (up, down, or stopped) was controlled by the hoist, located one to three hundred feet from the shaft. Three men who could neither see nor speak to one another oversaw hoisting—the *filler* underground, the *lander* in the shafthouse, and the *engineer* in the hoisthouse. To communicate with one another and signal starts and stops, they rang bells in code. They pulled levers that moved ropes or wire connected to the bells that ran up the shaft and along the surface.

If a man looking for work walked from Hancock up to the Quincy mine in 1865, he discovered most of the company's major buildings standing beside the road at the top of the hill. Double-pitched roofs surmounted six board-and-batten shafthouses with unpainted plank walls. Three wood-frame hoisthouses stood over masonry foundations laid up of poor rock. By each hoisthouse a tall iron smokestack emitted wood smoke from the steam boiler inside. Nearby, large stacks of cordwood for feeding the boilers sat at the ready. Pulley stands supported on timber stilts connected hoisthouses to shafthouses, and carried the wire hoisting ropes from one to the other. Men sorted rock inside simple and rough sortinghouses attached to several of the shafthouses. Nearby, other men calcined copper rock in a couple of smoky kilnhouses. More men wheeled rock over light-rail tramroads from sortinghouse to kilnhouse, or from sortinghouse to waste-rock piles.[17]

The industrial landscape was rough hewn and showed plain, utilitarian, mostly unpainted buildings surrounded by stockpiled tools, stulls, and planking. In the vicinity of the shafts, virtually no vegetation grew. Scattered mine rock made up the landscape floor. Across the road from the shafts, large outcroppings of rock, fringed with weeds,

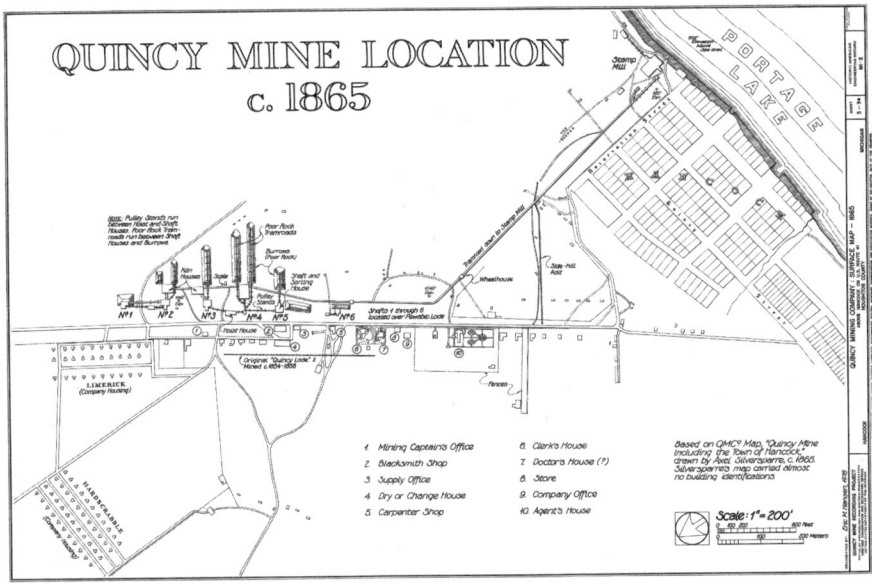

The Quincy Mining Company, after its fortunes picked up in the late 1850s, had no intentions of building or controlling its own town. On Quincy Hill, it erected a typical mine location, with many shafts, a full complement of mining structures, a range of company houses, and a single store. Quincy did, however, support the development of the village of Hancock. Quincy platted and sold lots to help create a town that offered a wide range of housing and services to Quincy employees and others in the area. *(Historic American Engineering Record Collection, Library of Congress)*

occupied the high ground. The company office building, mining captain's office, and a single store stood on this side of the road, as did the blacksmith and carpenter shops. After coming up from the underground, miners crossed over the road to wash up and change clothes in Quincy's dry house, a gray-black building made of poor rock from the mine.

Quincy's southernmost shafts, Nos. 5 and 6, proved disappointing, so the company soon shut them down. Past these closed shafts ran a tramroad that carried carload after carload of copper rock from the richer northern end of the mine to the head of a stamp-mill incline on the southern end. This double-tracked incline connected Quincy's hilltop mine with its mill on Portage Lake. While traversing 2,200 feet of hillside, the incline descended five hundred feet before arriving at Quincy's stamp mill, erected in 1859–60.[18] Loaded cars traveling down the incline pulled the empties back up. Portage Lake provided ample water for running the mill and also served as dump for Quincy's stamp sands or tailings.

The stamp mill represented one of Quincy's most expensive surface improvements. The mill contained sixteen stamp batteries, each with four stamp heads, for a total of six-

As the crow flies, the Quincy location and Hancock were not that far apart. But as this portion of an 1881 bird's-eye view shows, they were two distinct localities, with a tall and steep hill in between that reinforced their separation and differences. Quincy was the mine; Hancock, the town. The southern end of the mine is shown on the top right, with its tramroad leading down to Quincy's stamp mill, near the bridge to Houghton. *(Historic American Engineering Record Collection, Library of Congress)*

ty-four. A steam engine powered the Cornish-style drop stamps, while a steam pump sent Portage Lake water into each stamp battery's mortar box. In choosing Cornish stamps, Quincy ran counter to the new technological trend on the Keweenaw of installing more powerful steam stamps. Quincy argued for decades that steam stamps abraded amygdaloid copper too much and produced particles too fine to be captured by the rest of the mill's machinery. So Quincy stuck with Cornish stamps. The water, copper, and rock flowing from their mortar boxes passed to classifiers, Collum jigs, Evans Patent Rotating Slime Buddles, and other sorting and washing apparatus.[19] Captured mineral went into barrels, which lake boats delivered to the downstate Waterbury and Detroit Copper Company for smelting.

In the mid-1850s, Quincy had built some modest dwellings for its small labor force: a boardinghouse and several small houses, scattered on the hillside above Portage Lake, which was riddled with tree stumps, holes, and piles of rock. Quincy was prospecting

Quincy built sixty-eight of these small, T-shaped, wood-frame houses in 1864 and aligned them in orderly rows in the new neighborhoods of Limerick and Hardscrabble. For many decades, fences enclosed each of these unadorned, flat-faced houses. Fences may have given occupants a sense of proprietorship; they also kept animals in or out. *(Historic American Engineering Record Collection, Library of Congress)*

on the hillside and put its workers in the midst of it all. Between 1859 and 1861, as the company's works on the Pewabic lode took off and its employment climbed to nearly six hundred men, Quincy built about a hundred wood-frame dwellings for $30,000. Most of these single-family units stood on the hillside or close to the bottom of the hill, near the new stamp mill, on Quincy land that later became part of Hancock.[20] For a while, Quincy chose not to put houses on its hilltop right at the mine. It delayed building neighborhoods near its run of shafts until it was sure of where it might find copper or how much land it needed for industrial development. By the mid-1860s, Quincy apparently had resolved those concerns and started building neighborhoods atop Quincy Hill.

For its company agent, clerk, and doctor, Quincy built houses that stood alongside the dirt road that came up from Hancock and traversed the Quincy location. A bit off this road, the company constructed two housing clusters, dubbed Limerick and Hardscrabble.[21] Sixty-eight wood-frame houses erected in 1864 lined the lanes in these neighborhoods. The houses had a T-shape plan, with the top of the T paralleling the lane and carrying a centrally located door. The front of the house stood one-and-a-half stories tall, with two rooms on each floor; the single-story rear served as a kitchen. While some of Quincy's earliest dwellings might have been founded on cedar posts or on wooden sills laid right on the ground, these houses seem to have been built on foundation walls made of poor rock. Plain, without ornamentation or pleasantries, such as a front porch, these houses rented at first not even for a dollar per room per month (as would later become standard), but for just a dollar per house per month.[22]

Besides putting up Hardscrabble and Limerick, in 1864, at considerable distance from the mine proper, Quincy created another new neighborhood called Swedetown. This was part of a generally unsuccessful attempt on the part of the company to bring in a new breed of workers during the labor-short days of the Civil War. The housing at Swedetown was substandard: Quincy went back to small log houses whose sills sat right on the dirt. Relatively few imported men came to work at the mine or to live in Swedetown's thirty-seven log houses, and this neighborhood soon became a more or less forgotten part of the company's operation and history.[23] Failure or not, Swedetown, along with the long-term neighborhoods of Limerick and Hardscrabble, represented how a company's housing efforts intensified as copper production progressed.

As it matured as a company, Quincy engaged in several paternalistic practices. After sharing a physician with nearby mines in the early 1850s, in 1859 it hired its first physician, Dr. J. W. Robbins, and opened a hospital, which had thirty-five beds by 1862.[24] It deeded over lands for the site of a new Congregational church in 1862 and for a Catholic church in 1865. In 1864, during the high inflationary times of the Civil War, Quincy opened a store at the mine, where it made provisions available to workers at less cost than independent merchants on Portage Lake. It rented garden plots to workers for a nominal fee of three or four dollars a year so they could produce some of their own food. It established rules forbidding the sale of alcoholic beverages on company land.[25] And in 1867, the mining company's office hosted a meeting to organize School District No. 1 in Quincy Township. Quincy's company agent, clerk, and doctor became the school district's first board members.[26]

At the bottom of Quincy Hill, Hancock began to develop rapidly after Quincy platted lots there in 1859 and began selling them off. The growing town became home to most churches that Quincy employees worshipped in; these structures stood on building sites donated by the company. Many who worked at the mine lived not at the hilltop location, but down in Hancock. By 1862, some forty-one employees had built dwellings in Hancock on ground leased from the company. Many others, especially single men, lived in one of numerous boarding establishments that had recently been built.[27] By 1870, Hancock had several general stores; retailers of books, carriages, clothing, jewelry, hardware, liquor, and millinery. It had a bakery, a druggist, a couple of butchers, three barbers, several licensed hucksters, and many saloon keepers. It had a newspaper, several churches, and fraternal organizations. The growing village and the growing mine were separated from each other by a steep hill and nearly a mile, but they were much connected by the needs and activities of the people who moved back and forth from one to the other.[28]

Those who resided at the Quincy location and in Hancock were predominantly immigrants, with a smattering of Americans. Quincy's early managers were New Englanders or New Yorkers and sometimes kin to one another. Columbus Christopher Douglass served as Quincy's *agent,* the top man in charge of the mine, from 1856 until 1858. Douglass was born in Chautauqua County, New York. His interest in Lake Superior copper came naturally: he was the cousin of explorer-geologist Douglass Houghton. Ransom Shelden, who founded Houghton on the south side of Portage Lake, also put the first store on Quincy's property on the Hancock side of the lake in 1852. Shelden hailed from Essex County, New York, and was married to C. C. Douglass's sister. Samuel Worth Hill, Quincy's agent from 1858 until 1860, came from Starksboro, Vermont. Samuel S. Robinson, who succeeded Hill as agent and served until 1866, was born in Cornish, New Hampshire. James North Wright, who became Quincy's chief company clerk in 1863 and who later served as agent for Quincy and then Calumet and Hecla, was born in Haddam, Connecticut.[29]

Through the Civil War era, the principal immigrant groups on the Keweenaw were the Cornish, Germans, Irish, and French Canadians, along with a smattering of Scandinavians, plus "other" Englishmen (aside from the Cornishmen) and some Scots. Of the four main groups, the Cornish, Germans, and Irish worked across a wide range of mining occupations, while French Canadians tended to work at surface jobs in the forests or at stamp mills. Quincy acknowledged immigrant groups when it gave names like Limerick and Swedetown to its neighborhoods. The company perhaps intended for select blocks of houses to be occupied by members of a particular ethnic group, but such ethnic enclaves did not last long. The ebb and flow of groups into and out of the mines meant that neighborhoods came to have mixed populations.[30]

The U.S. census of 1870 demonstrates how a few groups dominated the ranks of miners in Houghton County. Of over eleven hundred men who identified themselves as miners (as opposed to trammers, stamp-mill hands, or other occupations), the Irish made up 41 percent of the total; Englishmen (mostly Cornish), 40 percent; and Germans, 8 percent. Just over 6 percent were Scandinavians, and about a third of those resided near the Quincy mine.[31]

In mid-1865, Quincy's labor force included 208 miners: 137 were Cornish or English; 58, Irish; and 13, German. Among Quincy's total employment of 594 men, 318 (54 percent) were Cornish or English; 138 (23 percent), were German; and 130 (22 percent), Irish. Besides filling blue-collar jobs, select immigrants moved into the top levels of the mine's hierarchy. In 1867, for instance, the agent at Quincy, George Hardie, was a Scot; the chief clerk, James North Wright, an American; the head mine captain, John Cliff, a Cornishman; the surface superintendent, John Duncan, a Scot; the master mechanic, Fred Labram, an Englishman; and the mill superintendent, Philip Scheuermann, a German.[32]

At Quincy the men in charge of different parts of the operation did their own hiring

Many mining companies periodically called their men together for group photographs, perhaps so that officers and directors back east could take a glimpse at what was happening in faraway Michigan. In ca. 1875, Quincy miners, including some boys, posed near the boiler- and hoisthouse between No. 3 and 4 shafts. The low board-and-batten structure is a snowshed built over a tramroad; the water barrels on the roof are for fire protection. *(Historic American Engineering Record Collection, Library of Congress)*

and firing, and they tended to look after their own. In 1853, when still in the prospecting stage, Quincy hired William Worminghaus, a German, as its first mine captain—and for a while the mining force was largely German.[33] In September 1863 an Irishman by the name of John McCormick got a place on a contract team paid to receive and sort all of Quincy's mine product at the shafts. At first, Germans surrounded McCormick, and Henry Obenhoff, German himself, headed the contract team. But McCormick somehow gained control of this operation, and by the end of the year, Irishmen the likes of Cuddihey, O'Brien, and O'Neil filled his crew.[34] The Germans were out; the Irish were in. But while the sortinghouses became an Irish stronghold, Germans still reigned at the stamp mill, where Phillip Scheuermann served as superintendent and where Jegel, Offenbacher, Holzbauer, and Wiedenhofer all labored. Underground, Quincy's many contract mining teams displayed ethnic alignments. The men selected their own coworkers, and they formed teams that were all Cornish, Irish, German, or Scandinavian.[35]

The contract system of mining defined relations between miners and managers, and between miners and other laborers. The system originated in Cornwall, starting about 1700, and immigrant Cornish miners carried it to Lake Superior.[36] Contract miners did not sell their time or work for a daily wage. They did not work under the close supervision of a boss. Instead, the men worked under performance-based mining. A contract team, if shaft sinking or drifting, received so much per lineal foot of advance; if stoping, they received so much per cubic fathom (six by six by six feet) removed from the lode over the length of the contract, which was usually a month.

The mine captain didn't put together teams or assign coworkers. Instead, the men picked their own teams. The boss didn't tell the mining teams where to work or exactly what work to do. Instead, the men negotiated a contract to work in a part of the mine of their choosing, and they had their say in whether they wanted to drift, stope, or shaft sink. They worked under minimal direct supervision and set their own work pace. Another feature of the contract system that helped men to feel more independent was that their employer did not provide tools and supplies for free. Contract miners paid for every pound of drill steel they used up and for their powder, their candles, ladders, and whatever else they needed. The company kept track of the supplies drawn by each team and deducted supply charges from earnings. Miners liked the status of working under contract: they were higher up the ladder than the unskilled men working for daily wages.[37]

Companies liked the contract system because it seemed an incentive system. Miners had to work to earn anything. What counted was not the clock, but the work completed: the number of feet drifted, the number of cubic fathoms stoped. Early in its production stage, Quincy put all kinds of workers, not just miners, under contracts. When production was small or intermittent, the use of contracts protected the company from paying out too much in wages when there was little to do. Contracts protected the company from paying men to stand around. So wheeling contractors trammed rock from the stopes to the shafts; windlass contractors lifted the rock out of the mine; a large contract team sorted the mine's product. Their pay was based on how much rock they actually pushed, lifted, or sorted. Others contracted to burn and dress the rock at the kilnhouse, and down at the stamp mill, a man, under contract, received so much for every car of rock he pushed from the foot of the stamp-mill incline to the storage bins serving the stamps.[38]

As copper production expanded, Quincy eliminated most contracts and put more men to work for daily wages, under the control of bosses who watched them. Wheeling contractors evolved into trammers, who worked where they were told and with whom they were told, for a daily wage. Technological change helped eliminate contractors, too. Steam hoists replaced windlass contractors, and, in the 1870s at Quincy, mechanical rock crushers eliminated the contractors who had burned and dressed copper at kilnhouses. Over time, most unskilled laborers belowground and aboveground ended up working for a daily wage. Quincy reserved the contract system for skilled miners, and their contracts set them apart from, and above, other workers.

After a frustrating decade of futile prospecting on Section 26, in the late 1850s and early 1860s Quincy went into production, built houses, hired doctors, and platted a new village, Hancock. All this was just in time for the Civil War, which proved to be a second frustrating era for the company. Quincy produced 2.6 million pounds of copper in 1861, but only 2.1 million pounds in 1865. Production fell during the war, despite high demand and high prices, because the company couldn't find enough workers.[39] Like some other Lake mines, Quincy tried to import men to increase the size of its wartime labor pool. Quincy arranged for a Swede, Axel Silversparre, to find fellow countrymen willing to relocate to Lake Superior, and he succeeded in recruiting some 140 Swedes from Canada who came to the Keweenaw in 1864.[40] Quincy had its new Swedetown neighborhood waiting for these men, but most Swedes did not stick. They had little ameliorative effect on the labor shortage during the war, and by the 1870s Quincy abandoned its Swedetown neighborhood.

In the labor-short years of the Civil War, Quincy and other companies paid higher contract rates and wages. Due to wartime inflation, they also paid more for mining supplies. What was true for companies was true for the general population. Residents had to weather a shortage of food and then pay more for what little they got. Social disorder sharply increased around the mines, as civil order seemed threatened by inordinate drunkenness, street brawls, and crime. Quincy's paternalistic, community-building activities briefly acquired a new wartime twist. The company received an arsenal of fifty muskets from the state of Michigan, and in the spring of 1864 Quincy established and drilled a volunteer militia company for the purpose of enforcing order, if necessary.[41]

Quincy had many difficulties on the road to profitable production, and the Civil War brought with it a new set of problems and challenges. Still, at the end of the war, Quincy was poised to be a very successful mine, perhaps the best on Lake Superior. But two things interfered with that. The first was the onset of a long and severe postwar depression in the copper industry. And the second was the rise of Calumet and Hecla, a new mine about ten miles north of Quincy. Calumet and Hecla soon grew to be the region's giant, and throughout the rest of the nineteenth century, all other mines, Quincy included, lived in its shadow.

6

❖ ❖ ❖

THE ERA OF MICHIGAN
DOMINATION

1865–1890

At the end of the Civil War, Michigan reigned as the nation's leading copper producer, and it kept that crown until Butte, Montana, surpassed Lake Superior production in the late 1880s. The 1865–90 era was not without economic problems, such as in the postwar years, when a severe depression hurt the Michigan mines. Over these years, copper prices generally fell and challenged the mines' profitability. On the whole, though, great growth and change marked this era, a time when the best mines and their attendant communities took on the trappings of permanent settlements.

The Lake mines never enjoyed only success and expansion. While new mines came on line at one place, older mines failed somewhere else. This proved very true in the quarter century after the Civil War. This era witnessed the decline of the once-rich Cliff and Minesota mines, and a general mining decline in Keweenaw County to the north and in Ontonagon County to the south. While mines faltered on the ends of the range, a surge of production came out of the middle in Houghton County. The Portage Lake mines became larger producers, and most importantly, the new Calumet and Hecla (C&H) mine, established in the late 1860s about twelve miles north of the Portage, quickly developed into one of the largest and richest mines in the world. By 1890 C&H alone employed nearly 3,500 men (more than a fourth of the district's total force of mine workers) and produced sixty million pounds of copper (60 percent of the region's total production).[1]

In Houghton County, one found the largest mines, mills, and smelters, and the most populated mine locations and towns. The early settlements of Eagle River, Eagle

Harbor, and Ontonagon were eclipsed by the rise of Houghton and Hancock on Portage Lake and by the rapid growth of Red Jacket village, next to C&H. In 1865, when total Keweenaw population ran a bit shy of 20,000 inhabitants, 8,500 (42 percent) resided in Houghton County. The Keweenaw's population doubled to 40,000 by 1890, when 35,400 residents (88 percent) lived in Houghton County.[2]

At the end of the Civil War, thirty-six Lake copper mines operated. That number dropped to twenty-four in 1870 and to fifteen in 1890. Although the number of mines declined, copper production increased dramatically. In 1865 the mines produced 14 million pounds of copper; in 1890 they produced over seven times as much: 101 million pounds. Fewer mines produced far more copper because they became larger, more mechanized, and more productive. While copper production increased sevenfold, total mine employment only doubled: from six thousand workers in 1865 to twelve thousand in 1890. Technological changes enabled each worker in the industry to account for greater production. In 1865, copper production per man-year was 2,300 pounds; by 1890, each man on the payroll accounted for nearly 15,000 pounds of copper.[3] Such productivity gains enabled companies to remain profitable in the face of generally declining copper prices. From a peak of thirty-six cents per pound in 1865 (a price reflective of Civil War inflation), copper fell to only twenty cents per pound in 1870 (a price reflective of the postwar recession). Copper prices rebounded to another high point of about thirty-three cents per pound in 1872, and in subsequent years steadily fell off to just fifteen cents per pound in 1890. During the Civil War the mines' dividend payments hit a then-record annual high of $1,150,000 in 1864. During the postwar recession, when copper prices fell precipitously, altogether the firms paid out only $100,000 to $210,000 annually in dividends. Profitability recovered in the 1870s and 1880s, in part because of the substantial productivity increases that countered the negative effects of declining prices. A few peak dividend-paying years were 1872 and 1890, both of which saw more than $3 million in dividends returned to investors.[4]

This post–Civil War era marked the true industrialization of mining. The first Lake mines exhibited many characteristics of a craft industry. Individual skill on the part of blue-collar workers still counted a great deal; the production of copper rock at individual shafts was small; and men did most labor by hand, from drilling shot holes to winding up rock. The technological changes that most dramatically altered mining between 1865 and 1890 were the introductions of power rock drills and high explosives, and the enhanced use of steam power.

Producing a machine that drilled rock, without breaking down, tested the engineering and metallurgical skills of the mid-nineteenth century. The Hoosac Mountain Tunnel became a showcase for testing early rock-drilling machines. Begun in 1851, the 25,000-foot-long tunnel was being driven through the Berkshires (a southern extension of the Green Mountains) to provide a shorter rail link between Boston and the Hudson River. This was the greatest tunnel that Americans had yet tried, and it proceeded too slowly, at too great a cost. They brought in new rock-drilling machines, hoping to speed the work and lessen its cost. Finally, in 1866–67, the Burleigh Rock Drill Company of Fitchburg, Massachusetts, manufactured a machine that won acclaim for driving the Hoosac Tunnel. Compressed air reciprocated a piston inside a cast-iron cylinder, and the piston drove the Burleigh's drill back and forth. With hopes it would reduce their costs, too, between 1868 and 1873 at least ten Lake mines tried the Burleigh drill.[5] They installed steam-powered air compressors on the surface and put in cast-iron pipes and flexible hoses to carry the compressed air underground and distribute it to mining teams. The companies discovered, to their disappointment, that the heralded new machine failed to lessen their mining costs. It was too cumbersome and heavy. It worked well in the flat Hoosac railroad tunnel, twenty-four feet wide, but proved hard to set up and maneuver in a mine's smaller, inclined spaces. Also, it broke too often and proved too complicated for ordinary miners, who had never run machines before. The mines shelved their Burleigh drills by the mid-1870s.[6]

After giving up on the Burleigh, the mines tested other machines, and in the late 1870s and early 1880s hit upon the machine produced by the Rand Drill Company of New York. In contrast to the Burleigh, the Rand was simpler and more reliable. Most importantly, the Rand drill was smaller, lighter, and had a simplified mounting system, so two men could carry this drilling machine anywhere in a mine and quickly set it up. The Rand drill proved a grand success. The mines switched from hand drilling to machine drilling and enjoyed large productivity increases. By 1882, with Rand drills, C&H produced 20 percent more copper with 20 percent fewer miners. Quincy produced 50 percent more copper while employing half as many miners.[7]

At the copper mines, machine drills and high explosives had interconnected histories. They came into use at about the same time, and just like the first drilling machines failed to earn a place in the mines, so did the first high explosives. These explosives burned much faster than traditional black powder, and their blasts had a shattering, rather than a heaving, effect on the rock. High explosives brought down more rock per charge, and in smaller pieces. The mines' initial flirtations with high explosives followed the work of Alfred Nobel, who in the late 1850s in Sweden developed liquid nitroglycerin into a commercial explosive. The first shipments of nitroglycerin blasting oil reached the United States in 1865 and the Lake Superior copper mines two years later.[8] But by 1867, something else had reached the mines: tragic tales of nitroglycerin oil accidentally detonating and claiming many lives. Many miners and mine managers feared the new explosive. Some companies eschewed trying it. The mines' early experiments with nitroglycerin blasting oil ended, quite literally, with a bang at the Huron mine in 1870. Huron miners wanted to stop testing the explosive in their mine and didn't want the company to convert to nitroglycerin oil, so they used their tried-and-true black powder to blow up all the stored blasting oil on the surface.[9]

Between 1870 and 1880, tragic accidents sometimes occurred as the mines experi-

The rock drill powered by compressed air was the most important machine introduced underground in the nineteenth century. After experimenting with several different power drills, in ca. 1880 the copper mines discovered the Rand drill, which proved a great success by increasing worker productivity and lowering mining costs. The Rand drill shown here is an illustration from the *Scientific American* of 25 December 1880, 402. *(Michigan Technological University Archives and Copper Country Historical Collections)*

mented with other high explosives. In 1874 a trial of a new explosive called Dualin ended at the Phoenix mine when the powder exploded in a mining captain's office, killing six men.[10] While the Lake mines remained skittish about adopting high explosives, other mining districts adopted them, especially Nobel's nitroglycerin dynamite. Nobel in 1867 had combined liquid nitroglycerin oil with an inert absorbent that held the explosive in a more stable form. Sticks of this nitroglycerin dynamite, which were detonated by blasting caps, proved safer to transport and handle than black powder because the sticks were less likely to explode prematurely or late and more likely to fire on cue. Just when the Lake mines adopted Rand drills, in the late 1870s and early 1880s they also adopted various nitroglycerin dynamites, such as Hercules powder and Excelsior powder.[11] The two new innovations worked together well. The Rand drills produced more and deeper shot holes, and dynamite brought down more rock per charge.

During this era, companies did not substantially alter tramming. Men still loaded cars and pushed them to a shaft for hoisting. At the shaft, though, some notable changes occurred. When their miners still drilled slowly by hand, to extract much product the companies had to attack their lodes in many places, reached from numerous shafts spaced only two to four hundred feet apart. Once miners were armed with faster Rand drills and

The mines sped up the transport of men to and from the underground by replacing man-engines with man-cars, which were attached to the hoisting rope in each shaft. They also sped up the hoisting of rock by installing a second skip track so that an empty skip was returning underground while a full one was raised. This ca. 1900 view of a Quincy shaft-rockhouse shows a man-car and a double skip track. *(Historic American Engineering Record Collection, Library of Congress)*

dynamite, each contract team produced more copper rock, so mines no longer needed as many different workstations or as many shafts. They hoisted from fewer shafts, but thanks to new drills and explosives, each shaft produced more rock per day. To handle the increased output, many shafts now carried two skip tracks instead of just one. They hoisted skips in balance. While one went down, another went up; the weight of the descending skip helped counterbalance and lift the ascending skip. Also, several companies stopped using their old man-engines and put man-cars in their shafts to transport workers up and down. At the beginning and end of a shift, men in the shafthouse removed a rock skip from the hoisting rope and replaced it with a *man-car*, a longer, specialized "skip" having a tier of bench seats. About thirty men at a time rode a man-car into the mine. In contrast to the single man-engine that most mines had used, the man-cars, running in multiple shafts, transported men underground faster and delivered them closer to their work.[12]

On the surface the most significant changes involved steam generation and the hoisting, breaking, and transporting of copper rock. In early decades the mines fueled their mine and mill boilers with local wood. The mines quickly consumed the timber right at their locations and then purchased additional woodlands as close to their mines as possible. Cutting, splitting, transporting, and stacking thousands of cords of wood was not inexpensive and became more costly when companies had to go farther out for their supply. By the 1880s companies generally abandoned wood as a fuel and adopted coal, shipped up from the Lower Great Lakes.[13]

As mines reached down another level or two each year, hoisting drums became larger to accommodate longer hoisting ropes, and the scale of hoisting engines also increased. It became common to have two engines drive a winding drum instead of just one. With *duplex engines,* the engines sat side by side in the engine house, with the hoisting drum mounted between them. Both engines' cranks connected to the drum's main shaft and helped turn it. By now some engines were specially designed to serve as hoists, but many remained general purpose American-style engines with horizontal cylinders, which just happened to be used for hoisting. Many engines on Lake Superior served long lives, often due to their versatility. Companies moved steam engines around as needed from mine to mill or from shaft to shaft.[14]

During industrialization, different industries often encountered similar problems. When one industry developed a machine to solve its problem, other industries often discovered that the same technology worked for them. The power rock drill, for instance, first developed for tunneling, found its way into quarrying, mining, and road and harbor building. Similarly, in the 1860s and 1870s a machine first developed by Eli Whitney Blake to crush rock for road surfacing in Connecticut wrought great changes in how the Lake mines dressed copper rock. Blake patented his *jaw crusher* in 1858.[15] Powered by a steam engine, the machine carried two heavy, cast-iron jaws, one stationary and one movable. The movable jaw alternately closed in on the other jaw and then backed off. The two jaws sat in nearly upright positions but were wider apart at the top than at the bottom. Rock loaded at the top was broken by the jaws as they moved together. When the movable jaw backed off, the rock moved lower into the machine and was crushed again. After several breaks, the reduced rock dropped out of the bottom.

The Blake jaw crusher allowed the Lake mines to abandon kilnhouses, with their high fuel and labor costs. Several mines acquired crushers by 1865, and others followed suit over the next decade. They installed the crushers in newly built *rockhouses,* substantial structures that often stood at the end of a line of shafts. A tracked, endless-rope tramroad ran across the mine site, picked up the copper rock from all shafts, and delivered it to the rockhouse for sorting and breaking. The rockhouses contained steam boilers and engines for driving the tramroad and for powering machinery. In addition to jaw crushers, rockhouses commonly included a drop hammer and a steam hammer. A large *drop hammer,* with a head weighing in excess of one thousand pounds, was mechanically lifted up and then dropped onto the largest pieces of copper rock. After this coarse break, the rock now

At Calumet and Hecla, tramcars running on an elevated trestle collected the copper rock from many shafts and then delivered it to the upper level of a rockhouse. Inside, jaw crushers and hammers broke the rock before it traveled to a stamp mill. The Hecla rockhouse is shown here in 1878. C&H, which often duplicated facilities to assure large and constant production, built another rockhouse to serve its Calumet branch to the north. *(Keweenaw National Historical Park Archives)*

fit into the jaw crushers. At a *steam hammer,* a smallish reciprocating head worked against pieces of mass copper, shedding them of adhering rock.[16]

In the earliest decades, companies moved their mine product via wagons or sleighs drawn by workhorses or oxen. In some instances, when mine and mill were closely situated, they used a tramroad. In the quarter century after the Civil War, many mines and mills stood five to fifteen miles apart, and steam locomotives and railroads transported most mine product. The Pewabic and Franklin mines, followed by the new Hecla mine, pioneered the use of steam locomotives in the mid to late 1860s. By the 1880s an expansive network of narrow-gauge, standard-gauge, and dual-gauge railroad lines connected mines, mills, and smelters up and down the Keweenaw.[17] Some were independent; several were owned and operated by parent mining companies. By the end of the 1880s a railroad bridge passed over Portage Lake, and Keweenaw rail service connected with lines running off to distant cities such as Chicago.

The mine of 1890 looked very different from the mine of the Civil War era. Mine locations that had been around for three or four decades looked more industrial. Tree stumps no longer littered the ground, and the forest line on the margin of the mine had

The Hecla and Torch Lake Railroad, starting in the late 1860s, delivered stamp rock from the mine to the company's mills on Torch Lake. Instead of returning empty, railcars picked up coal and other supplies shipped up to Lake Linden and delivered them to the mine. Two Hecla and Torch Lake locomotives are shown at the mine, ca. 1880. *(Michigan Technological University Archives and Copper Country Historical Collections)*

been pushed back. A poor-rock carpet obliterated natural ground covers and obscured original terrain. Poor rock filled ravines in some places and at others rose from the landscape as big burrow piles. Fewer animals and wagons traversed the ground because tram and rail lines now crossed the site, often elevated on wooden trestles. Stockpiled cordwood had been replaced by stockpiled coal. Commotion was more continuous. Mines hoisted from fewer shafts, but each shaft accounted for larger tonnages than before and by now had its own dedicated hoisting engine. Twin wire ropes ran from the hoists to the head sheaves in the shafthouses; the ropes hummed day and night as they crossed over wooden pulley stands and then down into the mine, where they raised and lowered skips. The shafthouses still had virtually no machinery in them but kept the head frame, skip tracks, and skip dumps out of the weather. Kilnhouses were gone, replaced by rockhouses. Air compressors for powering underground rock drills represented a new and important class of machinery on the surface. As the mines mechanized more, they often built larger machine, blacksmith, and drill-sharpening shops to perform essential maintenance and fabricating operations. The look of these buildings changed, too, as fewer of them were constructed of wood. Wooden structures lacked the cachet of permanence, and more than a few had burned to the ground. More and more, companies turned to poor rock and then to brick and dressed stone for construction.

At 1890 even the best mines' company houses hadn't been modernized yet; they still lacked utilities, such as running water, toilets and sewers, and gas and electricity. Differences among houses reflected that the companies built them as needed. The companies engaged in short bursts of construction, and each burst resulted in another neighborhood or two. Within each neighborhood, dwellings were often identical—but in their elevations and plans the houses tended to vary from one lane or neighborhood to the next. Generally, as the nineteenth century progressed, working-class families expected and wanted more: more rooms, more space and headroom, more windows, more stoves, and the like. Company housing at the mines exhibited this trait: later waves of housing were often better and more desirable than earlier ones. New houses at the mines did not replace old houses; they joined them. So, at a mine in 1890, some still lived in thirty-year-old log structures of just three or four rooms, with no cellars and a low sleeping loft. Others resided in new frame houses of five or six rooms, with clapboarding, cellars, a more commodious upper floor, and perhaps a second stove and more windows.

Rank and status also created housing diversity. One sign of the passing of the frontier era was a widening gulf between classes, evidenced by their lifestyles, dress, and housing. By 1890, more of a contrast existed between workers' and managers' houses. Workers' houses were repetitive in their design and carried no architectural embellishments, no "gingerbread," no porches. Meanwhile, mine agents, captains, doctors, and other key personnel lived in larger houses of individual design, with more amenities, that were spatially set apart from common workers' housing. In 1880–81 the Quincy Mining Company made a strong statement about status and role at its location. At a time when it spent several hundred dollars to build a miner's house, it spent $25,000 to erect and furnish a fine house for its agent, done up in the Italianate style.[18]

As the mine locations grew, so did the adjacent commercial villages that served them. They boasted more extensive downtowns replete with railroad stations, hotels, restaurants, general stores, saloons, specialty shops (such as millineries, drugstores, and tobacconists), plus service businesses, professionals, and some manufactories. A snapshot of Hancock in 1890 shows that it was home to some three banks, six churches, five hotels, five doctors, three dentists, three bakeries, three carriage dealers, six confectioners, four druggists, three dry-goods merchants, nine general merchants, nine grocers, six hardware retailers, four meat markets, three millineries, eight tobacconists, four barbers, three laundries, two jewelers, one florist, one photographer, and numerous saloons and billiard halls.

Red Jacket village, which was not even launched until the late 1860s, when the Calumet and Hecla mines started up, exhibited a similar diversity of businesses, trades, services, and professionals. By the late 1880s the village counted four churches, three hotels, four liveries, four tailors, two harness makers, one bowling alley and one coffin retailer, nine clothing stores, five dry-goods merchants, nine general stores, eight confectioners, three druggists, eight groceries, seven meat markets, three jewelers, three millineries, five barbers, and thirty-five saloons.[19] Additional churches and other institutions occupied parts of Calumet Township, right next to the village. Red Jacket village served as the primary retail, service, and entertainment center for the northern end of Houghton County and especially for employees and families from the C&H, Tamarack, and Osceola mines.

In the pioneer era, to accelerate the taming of a remote wilderness the mining companies helped fulfill a host of social needs. After the Civil War, when the region's population grew and businesses and institutions became more diverse, the mining companies withdrew from some of their earlier activities. Thanks to the presence of sheriffs, courts, and jails, they no longer had to concern themselves too directly with keeping law and order. Government rightfully conducted some of the public-works jobs that the mines had once assumed, such as road building. With better transportation links to the world below, and with a multitude of shops and stores in the villages, the mining companies concerned themselves far less with provisioning employees and their families. For the most part, they moved out of agricultural and commercial pursuits. By no means, however, did the companies retreat from all paternalism. They still built houses and ran medical programs. Within their townships, they encouraged managers to remain active in politics, which meant they retained considerable say in the building, staffing, and operations of schools, the setting of tax rates, and so forth. The mining companies did not seek to be all things to all people but did seek to maintain control over their workforce in many ways. They were successful at this, and in the metal-mining industry the companies earned a reputation for harmonious labor-management relations.

As conditions in the copper industry fluctuated, the mining companies profited better in some years than in others. They raised or lowered wages accordingly, and wages proved to be the one issue that most often created tension between companies and workers. When workers were plentiful, or when faced with a poor copper market, companies offered lower wages and resisted pressures to raise them. But when the mines boomed or when workers were in short supply, laborers often seized that time to press for an increase. Sometimes workers won these confrontations, but overall they manifested little collective power when standing up to the companies.[20]

Workers flirted with unionism during this era, but no labor organization proved effective or lasted long. In the early 1870s the International Workingmen's Association had a role in prompting a three-week-long strike at the Portage Lake mines and at Calumet and Hecla. This strike in 1872 resulted in workers achieving increased wages— but at a tremendous price. The strike so alienated the new president of C&H, Alexander Agassiz, that it engendered in him a staunch antiunionism that he took to his grave in 1910. In the intervening years, Agassiz, as leader of the dominant mining company in the region, effectively quashed unionism and other forms of worker unrest in and around his mine, even as unionism was gaining a foothold in many other industries around the Great Lakes.[21]

In the 1880s, the Knights of Labor, whose national membership peaked at 750,000 in 1886, arrived on the Keweenaw. The year 1886 was known as the "Great Upheaval" because of the great number of strikes (1,432) and lockouts (140). The rapidly industrializing, rapidly urbanizing American economy spawned deteriorating labor-management relations, and some of the worst outbreaks occurred in midwestern cities such as Chicago and Milwaukee. Under Agassiz's tutelage, C&H was careful not to import workers from places known for strident unions; careful not to hire any Knights of Labor members; and if any men on the payroll subsequently joined that union, they were to be discharged. Much of C&H's antiunionism during this era was conducted quietly. C&H did not fire men for being union members, an act that might have made them labor martyrs or fueled a labor backlash. Instead, the company fired them for some infraction of company rules. The copper companies, with C&H leading the way, succeeded in neutralizing unions, such as the Knights of Labor, in part by using guile, vigilance, and the steadfast resolve that they would not share their power with organized labor.[22] Paternalism played into their overall strategy here. By providing services such as housing and doctoring, they hoped to cement worker loyalty while controlling their workers' cost of living. A lower cost of living, companies believed, would lessen demands for higher wages.

People inside and outside the Lake Superior copper district often expressed the view that this was a special place, where a harmonious relationship existed between employer and worker. In 1882 a *Harper's Magazine* writer expressed the view that "two 'Molly Maguires' from the coal regions would make more noise than the two thousand employees of Calumet." Calumet, he wrote, was a mine village without peer, an exemplar of "the straightforward manly development of American civilization." It was a place where immigrants were "quietly and harmoniously developing into self-respecting American citizens." James North Wright was a longtime insider, an American from the east who came to Lake Superior and served as the agent, or superintendent, of both the Quincy and C&H mines. In 1899 he wrote, "From an experience of thirty years . . . , I can truthfully say that I know of no mining region where the relations between the companies and their employees have been more friendly and pleasant."[23]

During this era, when extolling the progress made by local society, praising company paternalism, describing the underground, or detailing the large and impressive facilities built on the surface, most observers paid particular attention, and often a kind of homage, to one mine in particular: Calumet and Hecla. After the Civil War it quickly became the best, the richest, and the most dominant Lake mine. During this era, C&H was never just a typical operation. Instead, it was in a league of its own.

7

❖ ❖ ❖

THE LARGEST AND BEST COPPER MINE
IN THE WORLD

Calumet and Hecla

Calumet and Hecla (C&H) got a late start because early prospectors missed finding its copper lode. But once mining began on the Calumet Conglomerate lode in the late 1860s, C&H quickly became the premier copper mine on Lake Superior. It produced the most copper, paid the greatest dividends, employed the biggest workforce, and built the most impressive physical plant. C&H, through much of the late nineteenth century, dwarfed not only its neighboring Michigan mines, but all other copper mines in the world.

In his *Report* for 1899, Michigan's Commissioner of Mineral Statistics wrote of C&H that "it sprang, fully panoplied, like Minerva from the head of Jove, to a commanding position among the wondrous treasure vaults of man."[1] In choosing the word "sprang," the commissioner made the rise of C&H seem rapid and easy. Even C&H encountered snags on its way to success. The discoverer of the Calumet Conglomerate, Edwin J. Hulbert, ran afoul of the Boston capitalists who invested in the property, and controversy and animosity marked the company's early years. Hulbert's first attempt to develop the lode left the mine in shambles and impeded subsequent work. He erred in thinking that the Calumet Conglomerate's rock was so rich that it could be smelted directly, without being concentrated. When the need for milling became obvious, the company invested in novel (and unsuccessful) roller crushers instead of tried-and-true stamps.[2] When the company built its railroad, the track gauge did not match up with its first locomotives. But following a fitful beginning, C&H soon became the dominant mine in the region. It could hardly fail to profit, because its lode was so rich. Two credit-ranking firms reported

This 1893 view looking southward from the Calumet branch toward the Hecla branch shows the linearity of the massive physical plant that C&H built over the Calumet Conglomerate lode. The industrial core was long, but narrow. All the shaft-rockhouses stood in a straight line, and close by stood the railroad roundhouse, machine shop, warehouses, hoisthouses, boilerhouses, and other facilities. *(Chuck Voelker/Copper Country Reflections website)*

that the Calumet and the Hecla mines (which started as two separate, but related, companies) were "A No. 1" and the "best on the Lake." By the mid-1870s, C&H was rated as "the largest and best copper mine in the world."[3]

C&H originated with Hulbert, a surveyor with considerable knowledge of geology. In the late 1850s the Michigan legislature provided funds to survey a road along the Keweenaw, and Hulbert received the contract to run the line from Copper Harbor south for sixty miles. Routing a wagon road in this country was not simple, and Hulbert made more than one pass at it, trying to avoid steep grades and swamps. During his survey, Hulbert kept a sharp lookout for copper-bearing rock. In 1858, in the forest between Portage Lake and the Cliff mine, he found promising geological specimens in pit, which he suspected was an ancient Indian digging situated atop a promising lode. But he found no hammerstones or other tools suggestive of ancient mining. Hulbert left the area "puzzled" by the pit he'd discovered. He kept it a secret and planned to return sometime for another look.[4]

Early in September 1859, Hulbert packed up tent, blanket, compass, tripod, and hatchet, and set off on foot from his office at Eagle River to revisit his secret pit. As he tramped on, whenever he saw anyone coming, he sat down until they passed, so they would not know which way he was heading. At his destination, he camped out and studied the geology around his pit, which he later called his "Sphinx of the Forest." To maintain secrecy, he recorded his field observations in cipher. Following this examination and subsequent ones, Hulbert determined that a great copper deposit indeed existed here. In 1860 he began purchasing mineral lands and started seeking the investment capital needed for development work. The coming of the Civil War interrupted Hulbert's efforts, and serious work to uncover the Calumet Conglomerate lode waited until 1864. Ironically, the mysterious ancient pit indeed sat right atop the lode, but was not deep enough to expose it. The pit that fascinated Hulbert proved not to be a place where Indians had taken copper, but a cache where they had hidden copper taken from elsewhere.[5]

Three companies became involved in testing and developing the new lode. The Hulbert Mining Company came first. Boston capitalists joined Hulbert in starting this firm. His key associates were Horatio Bigelow, James A. Dupee, James Beck, Henry Sayles, E. D. Brigham, Charles D. Head, T. H. Perkins, and J. W. Clarke. Their company obtained mineral land parcels all along the Keweenaw, including the site of the Calumet Conglomerate in Sections 13 and 14, Township 56, Range 33—land previously owned by the Metalline Land Company of Philadelphia and the St. Mary's Canal Mineral Land Company. To develop this particular site, in 1866 the Hulbert Company spun off the Calumet Mining Company. Later that same year the Calumet Mining Company created yet another company controlled by the same investors—the Hecla Mining Company— which acquired Section 23, Township 56, Range 33, next to the Calumet property. The two companies shared a boundary, had the same stockholders, the same administrative heads—and the same lode.[6]

After purchasing their respective properties in 1866, the Calumet and Hecla companies had little money left to pay for developing the mines. Later that year, the fledgling Calumet Mining Company entered an agreement with Hulbert to have him work its lands under tribute. Hulbert would take most of the risk and bear the expense of opening the lode. In return he and his associates would receive seven-eighths of the ingot copper yielded by the property. The Calumet Mining Company—which provided the mineral lands but avoided development costs—received the remaining eighth of the product.

Under this tribute contract Hulbert began losing his grip on the lode he had discovered. In 1866 he and his associates chose first to develop the Calumet property rather than the Hecla, which was wider and richer at its grass roots and less covered by overburden. On the Calumet end of the lode, instead of exploiting it through standard shafts, Hulbert mined it using large open pits that filled with water causing the pit walls to collapse. Hulbert sent glowing reports back to Boston about the richness of the lode, but his crew's work yielded little marketable copper.[7]

Other reports went to Boston penned by Henry D'Aligny, a French engineer who served as resident agent of the St. Mary's Canal Mineral Land Company, which owned a great deal of Keweenaw property, received as part of the land grant for building the Soo Locks and Canal. D'Aligny held Calumet Mining Company stock and knew many of the company's major investors. While Hulbert wrote optimistic reports, D'Aligny wrote letters "full of doubt and misgiving." The discouragement spread by D'Aligny caused some original Calumet Mining Company investors to pull out. One man who had more faith in Hulbert's reports was the Bostonian investor Quincy A. Shaw. Shaw and a group of investment friends bought up much of the Calumet stock and began directing operations in Michigan.[8] Shaw sent his brother-in-law, Alexander Agassiz, to the Keweenaw to assess the situation and Hulbert's performance. On this trip, and others to follow, Agassiz became a believer in the property, but not in the person. Hulbert's slow progress in developing the mine appalled Agassiz. In February 1867 he wrote, "The value of the mines, both Calumet and Hecla, is beyond the wildest dreams of copper men, but with the kind of management many of the mines have had, then even if the pits were full of gold, it would be of no use."[9]

Under Shaw's leadership, the new Hecla Mining Company was set up, and the Calumet Company halted its arrangement with Hulbert to work its land on tribute. Shaw made his first trip to the Keweenaw in June 1867 and became further convinced of the value of the properties. Both companies levied new assessments on their stock to fund development work, which in March of that year the companies' directors had entrusted not to Hulbert, but to Alexander Agassiz. Agassiz moved to Lake Superior to serve as resident agent of the Calumet and Hecla mines until they were well under way. Thus, in 1866–67 the Shaw-Agassiz team wrested control of the endeavor away from Hulbert, who from 1867 onward felt nothing but bitterness toward the Boston pair. In that year, Quincy Shaw wrote to Theodore Lyman, "I have the prospect of a general lawsuit with Mr. E. J. Hulbert, former Supt., who is confident that he is defrauded by me of all his property and that I have been scheming for his destruction, while he was slaving in that desolate region deprived of all social advantages to make me rich."[10]

Under the Shaw-Agassiz team the Calumet and Hecla mines improved rapidly. In 1867 the two mines together produced 1.35 million pounds of copper, or 7 percent of the Lake mines' total production. In 1868 the two companies totaled 5.10 million pounds, nearly a quarter of regional production. By 1870 production hit just over fourteen million pounds of ingot copper, and the Calumet Conglomerate lode's dominance was clearly established. Michigan in 1870 accounted for 87 percent of the nation's copper ingot. The Calumet and Hecla mines provided 57 percent of Michigan's product and a whisker under half the nation's. The Hecla Mining Company paid its first dividend in 1869; the Calumet mine, in 1870. The mines proved rich enough to pay for their own development through copper sales rather than through numerous stock assessments paid in by investors.[11]

In the spring of 1871 the two companies joined, forming the single Calumet and

Hecla Mining Company, which Quincy A. Shaw presided over briefly until Alexander Agassiz became president on 1 August 1871. Agassiz did not surrender that post until his death in 1910, so C&H prospered under consistent leadership for many years. Shaw and Agassiz stamped an indelible mark on their company, on its technologies, buildings, and relations with workers. They also left an indelible mark on the Lake Superior copper district as a whole.

The Calumet Conglomerate lode outcropped on top of the broad plateau that forms the spine of the Keweenaw Peninsula and stands about 640 feet above Lake Superior. The lode ran on a strike or line 39 degrees to the northeast; it dipped into the ground on an angle of 38 degrees from the horizontal. C&H was fortunate to own its particular portion of this lode. The Calumet Conglomerate ran for long distances along the Keweenaw, but over most of its length it existed as a thin sandstone or shale, in places only a few inches thick. But in Sections 14 and 23, the lode opened to a width of six to twenty feet and became much coarser, filled with pebbles and boulders. Long ago, the ground's porosity allowed the inflow of copper-bearing solutions. Meanwhile, impermeable margins of shale, which closed in against the lode, checked any outflow. Native copper concentrated in this lode to a degree not found in any other. On C&H property, the lode yielded 3 to 15 percent copper, with an average yield of 4.5 percent. This was well over twice the yield of Quincy's Pewabic amygdaloid lode.[12]

C&H had no way of knowing it at the start, but this lode's richness ran to great depths. In addition, the company's boundaries coincided with the lode's copper-rich portion. While C&H returned fabulous profits, competing investors failed to develop the same lode just to the north and the south. In 1881 a 4,200-foot-long stretch of the lode was considered prime; C&H owned all of it. North of C&H, the Schoolcraft and Centennial mines turned no profit on the Calumet Conglomerate lode. The Centennial extensively worked the lode over twenty-nine years. It sank seven inclined shafts, one extending down 3,200 hundred feet, but did not find commercial mineralization. To the south, the Osceola mine failed just the same. These companies survived by abandoning the Calumet Conglomerate and working amygdaloid deposits on their properties.[13]

The richness of its lode meant that C&H did not have to pinch pennies while developing its location. It was certainly less cost conscious than the Atlantic mine, south of Portage Lake, which squeezed profits from a lode charged with less than one-fourth the copper value of the Calumet Conglomerate. This is not to say, however, that all was perfect or ideal at C&H. Across its property, the Calumet Conglomerate ran richer in some parts than others. And while C&H had the richest lode, the Conglomerate lode was a much harder, tougher rock both to mine and mill than amygdaloid deposits. It offered great resistance to miners drilling shot holes by hand and to stamping or crushing machines on the surface. Also, the hanging wall over the Conglomerate lode was far less stable than the roofs of other mines, where overhead rock tended to stay put and required little support. At C&H the hanging demanded great attention: the company buried "a

Calumet and Hecla's rich Conglomerate lode provided great wealth but also posed special problems. The rock was harder to drill than most, and the hanging wall was more fragile, causing the company to plant a "forest" of timber supports or stulls underground each year. Stulls rode a timber skip underground, and then timbermen pushed them along tram tracks to the stopes. *(Michigan Technological University Archives and Copper Country Historical Collections)*

forest" of timbers in its underground levels each year. The timbers rendered C&H more secure from rockfalls, but more susceptible to underground fires.[14]

C&H did not have to invent its own business practices, technologies, or social and paternalistic activities. The Lake mines had been in business for a quarter century, so C&H had models to follow. Agassiz's company innovated in certain ways, to be sure. But what set it apart from other Lake mines was not so much what it did—but the way it did it, on a vaster scale.

Reconstructing the rise of C&H in detail in the nineteenth century is difficult. Extensive business records survive, but do not present great coverage of the early years.[15] Unlike virtually all other Lake mines, C&H published no annual reports until the 1890s, and those were hardly fact filled. Quincy Shaw and Alexander Agassiz put their mark on the company, and one trait C&H long exhibited was this: it did not leave itself open to scrutiny. It kept much economic, production, and employment data to itself. C&H volunteered little information to stockholders and dismissed outsiders' inquiries, whether they came from Michigan's Commissioner of Mineral Statistics or representatives of the Dun & Bradstreet investment firm. C&H was the most aloof, independent, and arrogant Lake mining company. Its attitude did not endear it to outsiders, and a trace of pique and censure can be discerned in Michigan's Commissioner of Mineral Statistics *Report* for 1899:

> The company does not give detailed figures to its stock-holders, but contents itself with an annual report of exceeding brevity. . . . The statements given shareholders are remarkably meager, but as the dividends are phenomenally large, the stock-holders content themselves with a minimum of information regarding their property, and the maximum of profits derived therefrom.[16]

The Calumet Conglomerate lode traversed C&H property over a total distance of ten thousand feet. Early on, the Shaw-Agassiz regime abandoned Hulbert's attempt to sink pits into the lode and explored where best to attack it using traditional shafts. Generally, the further north the lode ran, the less promising it appeared, so they chose not to push ahead, at first, with the northern end. This still left a long run of lode to open, and the Calumet and Hecla companies did so by following the common practice of sinking a line of closely spaced shafts, as little as four hundred feet apart. By 1871, when the Calumet and Hecla companies merged, plans called for sinking sixteen shafts, far more than any other mine had on its property. Only four of these were on the northern side of the old boundary line between the original two companies, ground now called the company's Calumet branch. Twelve shafts penetrated the richer southern ground of the Hecla branch.[17]

Companies often altered their underground works in response to changing copper values and technologies, and C&H was no exception. At least five of C&H's first sixteen shafts remained in service for decades and reached down to the mine's eighty-first level. The No. 4 and 5 shafts of the Hecla branch, on the other hand, never extended beneath the tenth level. Similarly, the company did not sink Calumet shafts Nos. 1 and 3 and Hecla shaft No. 1 below the twentieth level, because all three received severe fire damage in the late 1880s. Closing several shafts did not mean that C&H was declining or diminishing; it meant the company adjusted its plant to changing conditions. While closing some shafts, it also opened several new ones, so C&H still had twelve shafts operating along the Conglomerate lode in the late 1890s.[18]

At C&H, using sledgehammers to drill shot holes into the Conglomerate lode was laborious. Because the rock was so hard, some early C&H contracts paid miners on the basis of how many inches, rather than feet, they had driven shot holes.[19] The hardness of its rock prompted C&H to become one of the Lake mines most interested in adopting machine rock drills. In the early 1870s, before giving up the bulky machines, C&H ran more Burleigh rock drills than any other mine in the region. Then, like other Lake mines, in the late 1870s C&H put Rand Little Giant rock drills into service. By 1882 C&H ran sixty-five of the machines underground, and the Rand drills and dynamite significantly increased a miner's productivity.[20]

Given the production and employment data that survive, the best available measure of productivity at C&H is the number of tons of stamp rock produced per underground employee. From 1871 to 1875 C&H employed a low of 824 men underground, and a high of 1,114. During these years, miners drilled by hand and with the Burleigh drill. Output per underground worker (including miners, trammers, timbermen, and others) varied from 188 tons per man in 1873 to 224 tons in 1874 and 1875. This output more than doubled when Rand drills and dynamite appeared on the scene. Between 1880 and 1885 C&H employed a low of 692 and a high of 1,004 men underground. Tonnage per man varied from 407 in 1880 to a high of 564 in 1885.[21]

Since miners armed with Rand drills and dynamite worked faster and brought down more rock, C&H made several changes in how it exploited its lode. Each shaft could produce more copper rock, so the company needed fewer hoisting shafts to reach the desired amount of production. Because Rand drills could achieve more shaft sinking and drifting in a year, the company saved on development costs by spacing its levels ninety or one hundred feet apart instead of just sixty. The reduced costs of drilling encouraged C&H to open up more ground in advance. In 1880 the shafts and drifts already opened at C&H could support five years of stoping in the future. By the early 1890s C&H had sunk a total of 2 miles of shafts and driven 13.6 miles of drifts beneath its working levels. C&H opened ground ten to eleven years in advance of stoping, whereas most mines were opened only a year or two in advance. C&H's position enabled it to better assess its future stoping ground and plan for its exploitation.[22]

Where the Conglomerate lode carried commercial quantities of copper, its dip and strike were remarkably uniform, so drifts running through it were very regular. They did not turn or twist much to follow the lode. Also, the lode was richly charged in copper from the footwall to the hanging, so stoping miners took down the entire lode. The problem became, at the same time, keeping the hanging wall up. At the start, C&H used *stulls,* sections of tree trunks, to hold up the hanging. In the 1870s, woodsmen cut much of this timber from near the shores of Torch Lake. It was stockpiled at the lake near the stamp mills, hauled up an incline, and loaded onto cars on the Hecla and Torch Lake Railroad. The railroad carried the stulls to the mine shafts, where special skips lowered them underground. Timbermen struggled to handle the massive stulls, moving them along drifts, up stopes, and then putting them into place. By the mid-1880s C&H used square-set timbers, which were easier to handle. The company milled 12 × 12s, 14 × 14s, and 6 × 12s that had mortise-and-tenon connections. Timbermen interlocked the pieces underground, erecting cribwork that supported the hanging.[23] Sometimes, C&H used both square-sets and stulls simultaneously, using the stulls to serve as barricades to protect the more fragile square sets from rolling rock. Because of its need for timbering, C&H purchased a great deal of forested land along the Keweenaw. By 1885 the company owned an estimated eighty million feet of standing pine. By 1895, along with its three thousand acres of mineral lands, it owned twenty thousand acres of timberland in Houghton, Ontonagon, and Keweenaw counties.[24]

Over time, C&H relied less on cumbersome stulls and more on square-set timbering. The cribwork fashioned of smaller, milled timbers did a better job of supporting the hanging. This timbering reduced the risk of rockfalls but made fire a much greater hazard to the company and its underground workers. *(Michigan Technological University Archives and Copper Country Historical Collections)*

The timbering guarded against rockfalls, but exposed the company and its workers to the hazard of fire. Michigan's Commissioner of Mineral Statistics noted in his *Report* of 1887 that "the mine is a network of fine timber; millions of feet of dry pine; the best fuel for a great conflagration, and to such a calamity there is, inevitably, constant danger."[25] Indeed, fires broke out in 1887 and 1888 at C&H that greatly damaged the underground, caused lengthy shutdowns and production losses, and killed men. On 4 August 1887 a fire broke out along the sixteenth level near the No. 2 Hecla shaft. The mine closed for several weeks while fighting the fire with water, steam, and carbonic acid gas. C&H managed to reopen, but just a short time later, on 20 November, fire broke out again. The company sealed its shafts as tightly as possible to smother the blaze and again manufactured and injected carbonic acid gas into the mine. Despite vigilance and constant effort, C&H could not fully reopen the interconnected Calumet and Hecla branches of the mine until June 1888. These two fires cost C&H a production loss of about ten million pounds of copper. What saved the company from a total shutdown was that by this time it had three branches: the Calumet, the Hecla, and the South Hecla. The newest South Hecla branch, fortunately, did not connect through drifts or stopes to the original two branches. The South Hecla branch stayed open while the rest of the mine smoldered.[26]

On 29 November 1888 another fire broke out, this one in the No. 3 Calumet shaft. Eight men died of smoke inhalation in this fire, which again closed the Calumet and Hecla branches, but left South Hecla open. The company thought its attempts to smother the blaze had succeeded and started to reopen the mine in February 1889, but when fresh air poured into the mine, the blaze erupted again. Not until 1 May 1889 did the Hecla and Calumet branches reopen.[27]

The fires caused a number of changes at C&H. The company chose not to recover some extensively damaged parts of the mine. It closed three shafts, which left the company with three shafts on the Calumet branch, four shafts on the Hecla branch, and five shafts along the South Hecla branch. The fires drove C&H, in the 1890s, to open and equip a wholly new and expensive part of the mine: the Red Jacket shaft. C&H sank this shaft to open new ground that could stay in production even if other parts of the mine burned.[28]

In addition to using timbers to protect against rock falls, C&H used rock pillars to support the hanging wall. The company sought protection from collapses that might crush a shaft and shut down hoisting. Even at shallow depths, C&H left rock pillars twenty-five feet wide alongside its shafts. Miners did not stope within twenty-five feet of a shaft, in either direction, no matter how rich the rock was in copper. As the mine went deeper the company enlarged these pillars. By 1886 C&H left a fifty-foot pillar on each side a shaft.[29] Along each level of the mine a fair run of the Conglomerate lode remained intact, supporting the hanging. This rock was not a total loss to the company. C&H knew the shaft pillars constituted "a neat little mine" to be exploited later, when the works neared their end. Someday C&H would take down or "rob" the pillars of their copper, starting at the bottom of the mine and working back up.

Save for the introduction of power drills, little mechanization occurred underground at C&H prior to 1890. As early as 1871 it used diamond drills to take core samples, and around 1880, C&H initiated some mechanical ventilation to assist in moving air in the mine. This exhaust fan, thirty feet in diameter and turning fifty revolutions per minute, was probably mounted at the collar of an abandoned shaft. C&H ran the exhaust fan during hot weather, when increased air temperatures on the surface upset the natural flow of air into and out of the mine.[30]

Mucking rock into cars and pushing them to the shaft remained labor-intensive tasks done by hand. By the time C&H started up, the wheelbarrow era was over, and trammers pushed four-wheeled tramcars, loaded with about four thousand pounds of rock, over light-rails with about a four-foot gauge. They dumped the cars' contents directly into skips at the shaft. C&H's shafts carried a single skip road, and the skips' capacity was about two tons, the same as the tramcars'.

C&H relied briefly on ladders to transport men up and down, but by the early 1870s the company adopted the man-engine. Other mines operated a single man-engine. C&H operated two: one on the Calumet branch and one on the Hecla. By 1880

Because C&H failed to find a mechanical means of tramming rock, it continued to use men as "beasts of burden" well into the twentieth century. In 1915–16 a crew tips an end-dumping tramcar to fill a rock skip at the Hecla No. 7 shaft, seventieth level. Note the absence of any safety railings to protect a man from a deadly fall down the shaft. *(Michigan Technological University Archives and Copper Country Historical Collections)*

the C&H man-engine rods extended seventeen hundred feet into the mine; by 1885 they reached the mine's twenty-eighth level—and the man-engine could get a man there in twenty minutes.[31] The man-engines met their demise during the fires of the late 1880s. Instead of rebuilding them, in 1889 or 1890 C&H introduced man-cars that carried up to twenty-eight men at a time. The man-cars did not attach to the ends of the hoisting ropes used for rock skips, nor were they raised by the same large engines that lifted rock. Instead, they had their own ropes and hoists, which were smaller, slower, and safer to use in raising men.[32] Only the richest mine in the district built this level of safety into its man-cars. And only the wealthiest mine could have built the very capital-intensive aboveground works that miners walked through everyday after riding the man-engine or man-car to the surface and heading to the dry house.

8

❖ ❖ ❖

BEFITTING A COPPER KING

C&H's Visible Empire

In 1880, Michigan's Commissioner of Mineral Statistics opined that Calumet and Hecla's surface plant was "superior, it is believed, to that possessed by any other mine in the world."[1] Calumet and Hecla (C&H) made no attempt to hide its wealth, and any observer touring its facilities could easily tell that, among the Lake mines, C&H clearly was king. It erected more expensive and bigger facilities than any other company. In large measure, this followed from the simple fact that C&H had the richest lode, the largest production, and the highest profitability in the region. But something else was at work: C&H developed a style that incorporated *big* and *powerful* as key elements of its presentation of self to others. Two men in particular set this style: Alexander Agassiz, president, and Erasmus Darwin Leavitt, consulting engineer. The former conjured up the image of a richer and more powerful company than any other, and the latter executed that image in iron.

In 1880, if visitors wanted to get at the heart of things in the vicinity of the C&H mine and Red Jacket village, they didn't walk down a Main Street; they strolled down Mine Street. There, surrounded by a population of about 7,500 people, the C&H mine stretched out along the lode. A dozen shafthouses stood out on the landscape. Tramroads elevated on wooden trestles ran down the line of shafts. Cars received rock at the shafthouses and delivered it to one of two large rockhouses—one for the Hecla branch and one for the Calumet—where it was sorted and crushed. Next to the rockhouses ran rail lines. Hecla and Torch Lake Railroad locomotives pulled up with trains of rockcars, which constantly shuttled between the mine and the Lake Linden stamp mills, pulling full cars in one direction and hauling empties back in the other.

By 1880 each branch of the mine had its own large engine house replete with magnificent steam engines and massive boilers. Tucked in among the shafts were dry houses, a railroad roundhouse, pumphouses for unwatering the mine, a compressor building for powering new air drills, a couple of blacksmith shops, a machine shop, warehouses, man-engine houses, a carpenter shop, numerous tall smokestacks and, a bit to the north, the C&H waterworks. The surface plant was fully equipped to receive and ship copper rock; to forge and sharpen drill steels; and to maintain, repair, and fabricate all kinds of equipment. In terms of their materials, the structures evidenced an evolution at C&H. Many of the earliest buildings were framed and clad in wood; then followed a generation of buildings laid up of poor rock from the mine; more recently, dressed stone and brick gave form to the company's more important structures. In addition to its mine plant, C&H operated two large stamp mills at Lake Linden, a small industrial village on the shore of Torch Lake.

In the late 1860s, while still two separate companies, the Calumet and Hecla mines had faltered in launching their first mills. There was some question as to whether the Calumet Conglomerate rock needed milling at all—followed by questions about which machines to use and where to site the mills. The Calumet mine initially tried smelting the richest portion of its copper rock without concentrating it. In 1866 it sent this rock—15 percent copper by weight—to the Portage Lake smelter. At the time the poorest mineral being smelted was 40 percent copper. The attempt to smelt the Conglomerate rock in a cupola furnace failed completely, meaning it had to be milled.[2] It had to be reduced to three or four inches at the mine and then sent to a stamp mill.

The richness the Conglomerate lode largely eliminated the need to sort out poor rock from copper rock at the mine. By 1880 C&H sent only 2 percent of hoisted rock to poor-rock piles, whereas other mines rejected up to 25 percent.[3] C&H still had to size copper rock and break those pieces too large to go into a mill's stamps. Breaking Calumet Conglomerate rock was no mean feat because it included pebbles as hard as flint. Through the early 1870s the Calumet mine made limited use of fire, sledges, and picks to break oversized rock in kilnhouses. Most rock received a mechanical break delivered by steam hammers and two jaw crushers used in series.[4] Rock passed through a large Blake jaw crusher for a preliminary break and then went through a second, smaller crusher for a final break. With rock-breaking technology in place at the mine, the next step was to build a concentrating mill. In equipping this mill, Edwin Hulbert selected novel—and unsuccessful—rock-crushing machinery. Compounding his troubles, he sited the mill poorly, situating it just north of the mine alongside a small, dammed creek whose pond provided too little water and too little room to dispose of tailings.

Hulbert had once served as agent of the Huron mine above Houghton, where an earlier agent, John Collum, had used corrugated iron rollers to crush rock. Two parallel rollers rotated toward each other, drawing rock into the space between them and crushing it. In the mid-1860s, roller crushers seemed an attractive alternative to stamps because they promised higher capacity. Their rotary motion was continuous, whereas reciprocating stamps broke rock only on the downward stroke. Hulbert decided to equip the new Calumet mill with rollers, and his faith in roller crushers temporarily spread to Alexander Agassiz in 1866 and early 1867.[5] Agassiz ordered roller crushers from the nearby Portage Lake Foundry, which manufactured equipment for many of the mines. By the end of April 1867, Agassiz had already realized that the rollers, plagued by breakage, were poorly matched against the mine's hard rock. He searched for a replacement, choosing between Cornish drop stamps or steam stamps.[6] Agassiz opted not to use Cornish stamps because they lacked great force or high capacity. In 1867–68 he obtained two Ball steam stamps for the Calumet mill and two more for the first Hecla mill. This stamp, an adaptation of the Nasmyth steam hammer, was patented by William Ball of Chicopee, Massachusetts, in 1856. Chicopee's Ames Manufacturing Company, a premier American machinery builder, produced the first Ball stamps for a South Carolina mine. Shortly thereafter, several Lake copper mines turned to this technology, largely because one Ball stamp could do about the same amount of work as twenty to twenty-five drop stamps.[7]

At the first C&H mills, pairs of Ball stamps stood side by side and shared common mechanical devices, such as valve gearing. Timber framed the early machines. At the base, under the mortar box, sat an eight-ton anvil. The replaceable shoe on the vertical stem weighed three hundred pounds. The stamp stem rose up to connect directly to the piston rod entering the vertical steam cylinder. Within the cylinder, steam at about eighty pounds per square inch reciprocated a double-acting piston that pulled up and then drove down the stamp shoe, making about seventy-five strokes per minute. Stamp tenders shoveled rock from storage bins into the machines' mortar boxes, and water, rock, and copper constantly flowed outward through screens, passing to the next steps in the milling process. The first Ball stamps treated about eighty tons of rock per day each.[8]

Besides changing from rollers to stamps, Agassiz changed mill locations. He knew the original Calumet mill was poorly sited, so when locating the first Hecla mill he selected Torch Lake, about five miles from the mine. The lake provided all the water needed, as much as thirty tons of water per ton of rock stamped, and its deep bottom could receive enormous tonnages of stamp sands for decades. The lake was not land-locked but connected with Portage Lake, which in turn connected with Keweenaw Bay and Lake Superior. C&H dredged the Torch Lake–Portage Lake connection and kept it clear. This enabled the largest freighters on the Great Lakes to deliver and receive shipments at company docks on Torch Lake.

The Hecla mill started up at Torch Lake in 1868; a Calumet mill, replacing the one near the mine, followed in 1871. The Hecla and Torch Lake Railroad, initially wholly owned by the Hecla Mining Company, connected the mines and mills. Built in 1867, the railroad's four-and-three-quarter-mile-long, four-foot-gauge main line stopped about one-half mile from the mills. It terminated at the top of a hill running down to Torch Lake that was too steep for locomotives to handle. A tramroad or incline delivered rock down to the mills and dumped into storage bins above the Ball stamps.[9]

Water carried the discharge from the Ball stamps to machines that separated the

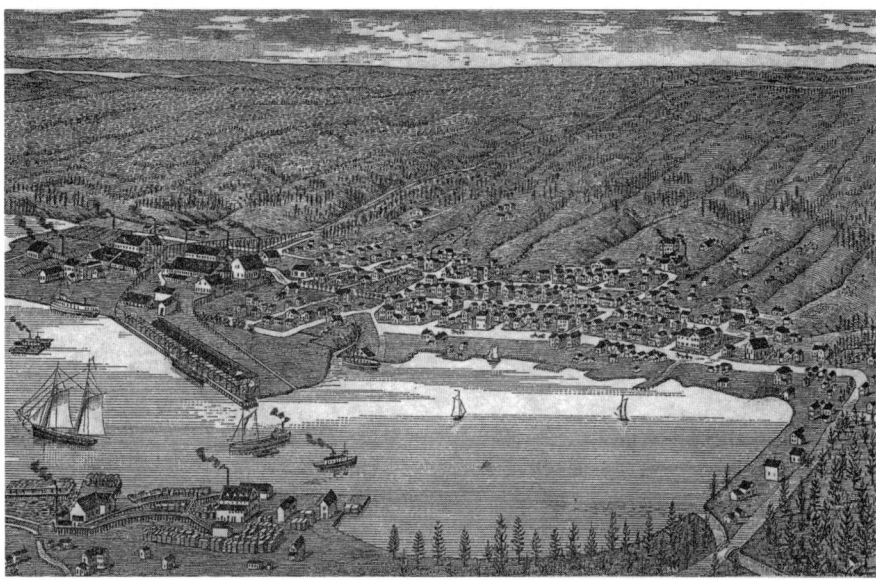

Major stamp mills, like mines, fostered the growth of new settlements. As depicted in this early 1880s bird's-eye view, the Hecla and Torch Lake rail line runs down the hill to deliver stamp rock to the Calumet and Hecla mills. A coal dock stands along the water, and launders or troughs running out from the mills dump tailings into Torch Lake. Next to the mills, the substantial village of Lake Linden provided housing, goods and services. *(Keweenaw National Historical Park Archives)*

copper and gangue by gravity methods. Copper has a specific gravity of about 9; the specific gravity of Calumet Conglomerate rock ran from 2.67 to 3.25. This difference was great enough to allow the copper to be recovered with considerable efficiency, but this recovery was never complete or perfect. The coarsest copper caused problems because, when subjected to the flow of water, it did not move. It sat in the bottom of the stamp's mortar box, for example, and tenders periodically picked it out by hand. The finest copper presented the opposite problem. When traveling with water and rock, it could not be stopped. Instead of becoming part of the captured mineral shipped off for smelting, it flowed out the mill as waste and ended up in Torch Lake.[10]

The Calumet Conglomerate lode contained a higher proportion of fine copper particles than rock from amygdaloid lodes. In 1879–80, 67 percent of C&H's copper took the form of flat scales less than a millimeter (one-twenty-fifth of an inch) in size. Only 42 percent of the Atlantic mine's copper, and only 10 percent of Quincy's, was so small. This fine copper, if liberated from the rock, was hard to capture within the mill, and the smallness of the particles meant that much copper was never liberated from the rock at all. It remained bound up within grains of stamp sand and went out with the tailings.[11] The rock that C&H milled in the late 1860s and 1870s yielded 4 to 5 percent copper, extremely high percentages for the district, yet C&H failed to recover 25 to 30 percent

Boys who worked in the C&H mills went to a studio in 1896 for a group photograph. They posed with their lunch pails and with tools (such as hoes, brooms, wrenches, and oil cans) that would help them move stamp sands along or keep machinery in good repair. The boys appear to range from about thirteen to eighteen years of age. Across the copper industry, companies employed more boys from 1880 to 1910 than they did in earlier or later decades. *(Keweenaw National Historical Park Archives)*

of all the copper hoisted out of the mine. Assays done in 1879–80 showed that 1,000 tons of rock delivered to the mills contained 130,848 pounds of copper. The mill recovered 90,000 pounds of copper, and the other 40,848 pounds flowed out into Torch Lake.[12] Only decades later, armed with new technologies, could C&H reclaim the copper once lost.

C&H employed about ten locomotives and hundreds of railcars to feed stamp rock to the mills. A rerouting of its rail line in the mid-1880s eliminated the old stamp-mill incline, and rockcars thereafter discharged their loads directly into rock bins at the backs of the mills. C&H elevated these bins so that gravity delivered rock into the stamps' mortar boxes; men no longer shoveled it all in. The rock slid out of the bin and along a shaking tray that fed the stamps. A stamp tender, freed of shovel work, stood alongside the tray, regulating the flow of rock and picking out barrel copper.[13]

Through 1890 the biggest changes at the mills involved the stamps. By 1880 each Ball machine, reciprocating at ninety strokes per minute, stamped as much as 150 tons of Conglomerate rock daily. Then the company switched to stamps engineered by Erasmus

Darwin Leavitt, one of the foremost machinery designers in the United States, the second president of the Society of Mechanical Engineers, and C&H's primary engineering consultant. The first Leavitt stamps showed a 25 percent gain in capacity over the Ball and a 10 percent gain in fuel economy. Second-generation Leavitt machines achieved a 50 percent increase in tonnage, and a 35 to 40 percent savings in fuel.[14] Ultimately, C&H claimed a capacity of 325 tons per day for each Leavitt stamp. By 1886 the Calumet mill ran seven of the stamps, and the Hecla mill ran five. By the mid-1890s eleven Leavitt stamps worked in each of the mills.[15]

For two decades, C&H shipped its mineral to the Detroit and Lake Superior Copper Company's smelter on Portage Lake. By 1885, C&H believed it could do its own smelting at less cost. At the mine, C&H ran two divisions: the Hecla and Calumet branches; at Lake Linden, it operated two mills. C&H was known for duplicating facilities in order to avoid accidental shutdowns and production snags. So it was not too surprising that, when C&H decided to build a smelter, it built two. It put the first (opened in 1887) on Torch Lake at Hubbell, less than a mile from its mills; it sited the second smelter (opened 1891–92) on the Niagara River at Black Rock, New York, near Buffalo.

Near the end of 1885, C&H and the Detroit and Lake Superior Copper Company entered an agreement to build a new smelter. They put up equal capital and jointly formed the Calumet and Hecla Smelting Corporation. The agreement stipulated smelting charges—and also gave C&H the option of buying out Detroit and Lake Superior's interest in the Hubbell smelter after five years of operation. In 1892, C&H exercised its option, bought out Detroit and Lake Superior, dissolved the separate smelting corporation, and ran the works under the banner of the Calumet and Hecla Mining Company.[16]

C&H's smelter at Hubbell occupied thirty acres and originally contained four reverberatory furnace buildings. As was common in the region, each building contained four furnaces, one in each corner. Numerous access doors for mineral and coal pierced the walls, and above each of the four furnaces stood a tall brick chimney, surmounted by a large damper plate. The site also housed a cupola-furnace building (for recovering copper from slag), a blister furnace, and numerous ancillary structures: cooper, blacksmith, and carpenter shops, an office, and storage buildings for mineral, coal, and other supplies.[17] The plant was not novel, except in size. It embodied standard smelting practices that had changed little on Lake Superior since 1860. The sixteen furnaces at the C&H smelter were larger, though, than earlier ones and could refine a thirty-thousand-pound batch of copper per melt, up from the original sixteen thousand pounds.[18] After refining the metal, furnacemen cast it into molds of different sizes and shapes: flat, square cakes for copper-rolling mills; notched ingots for brass founders; and bars for wire-drawing firms. Men ladled fifteen to twenty pounds of copper out of the furnace, carried the copper to molds, and poured it. When pouring a large copper cake of 150 to 300 pounds, as many as four to eight men ladled at a time. Later, to enable faster casting and to reduce labor, C&H installed larger ladles, which a man handled with the aid of an overhead trolley.

To smelt copper on Lake Superior, C&H had to pay to ship metallurgical coal up from the Lower Great Lakes. To smelt copper near Buffalo, C&H paid less for fuel but had to pay to ship mineral down to Lake Erie. Part of that mineral was not copper, but waste sands that would emerge from a smelter as slag. C&H smelted lower-grade mineral at Hubbell and shipped higher-grade mineral to Buffalo—so that each ton of mineral shipped yielded the most copper possible. At Buffalo the company paid reduced prices not only for coal, but for virtually all supplies. The economics of smelting in Buffalo worked in the company's favor. By the mid-1890s it smelted more product there (fifty-one million pounds annually) than it did at Hubbell (about thirty-five million pounds).[19]

From the mid-1870s onward, C&H's corporate style distinguished it from other mines. While other companies built their plants to meet short-term needs, C&H overbuilt to assure that facilities served for decades. Michigan's Commissioner of Mineral Statistics repeatedly made the point that C&H had a one-of-a-kind physical plant. In 1881 he noted, "It is claimed that the machinery at the Calumet and Hecla excels that found at any other mine in the world, particularly the Hecla and Calumet [steam] engines." The following year, he even implied that C&H's plant was too lavish: "It may be with companies as with individuals, [that] an immense and assured income begets a seeming extravagance." In 1885 the commissioner sent out another similar message: "Nowhere else on this continent, if indeed in the world, is there so much powerful and costly machinery employed in mining work."[20]

The chief architect of C&H's corporate style was Alexander Agassiz, a rare individual and atypical mining company president. Born in Neuchatel, Switzerland, in 1835, Agassiz grew up in a remarkable family. His father, Louis, well educated as a naturalist in Europe, became well known for his work in ichthyology, geology, and paleontology, and for being a great popularizer of science. His mother, Cecile, was a fine artist and illustrator. Alexander emigrated to America with his family in 1849. They settled in Cambridge, Massachusetts, where his father enjoyed a professorship at Harvard. Like father, like son: Alexander moved into scientific studies and pursued several different fields. Harvard awarded him a bachelor of arts degree in 1855; he received a bachelor of science degree in engineering from the Lawrence Scientific School in 1857, and Lawrence awarded him another degree, in zoology, in 1862. Some of his early positions were as diverse as his education. Agassiz taught at a school for girls; worked on the U.S. Coast Survey, which allowed him to travel up and down the Pacific coast of the Americas; worked some in Pennsylvania coal mining; and wrote and illustrated scientific articles. He was a well-traveled man, thirty-one years of age, when his brother-in-law, Quincy Shaw, sent him out to the Keweenaw Peninsula in 1866 to check on two fledgling copper mines.[21]

His first visit to the mines convinced Agassiz that the Calumet Conglomerate lode was one of the world's great mineral deposits. As mining on that lode took off, so did Agassiz's career as an industrialist. From his mid-thirties until his death in his mid-seven-ties, Alexander Agassiz juggled the demands of two different worlds. Residing most of the time in Cambridge, he served as a prominent Harvard professor and eminent zoologist. At the same time, for nearly forty years he presided over C&H.

Agassiz exhibited a technological style that only a rich mine could afford. One hallmark of this style was the acquisition of truly monumental steam engines. Agassiz liked large pieces of technology that impressed. Instead of installing small steam engines all over the mining landscape, each powering one finite task, he preferred large engines that often drove multiple kinds of machinery. Agassiz also believed in planning, in identifying potential problems that might interrupt production, and in using money and engineering to keep them at bay.

If Agassiz, figuratively speaking, was the chief architect of C&H's technological style, Erasmus Darwin Leavitt, quite literally, was its engineer. These two men, who did so much together to shape C&H's style and technologies, came from very different backgrounds. They were nearly the same age. Leavitt was born in 1836, just a year after Agassiz. Age aside, their differences predominated. Agassiz was European, well bred, and well educated at universities. After coming to America as a child, he came of age in Cambridge, alongside Harvard. Leavitt was born and came of age only twenty-five miles away from Cambridge, but in an altogether different place and culture: Lowell, Massachusetts. Lowell was one of the premier industrial cities in America—a city dedicated to textile production. Agassiz grew up with academies, science, and the elite. Leavitt grew up with textile mills, water-power canals, and machines.[22] Agassiz went to college more than once; Leavitt never attended college. Leavitt graduated from the "shop culture" school of engineering.[23] He learned his engineering on the job, in the shops. He worked his way up by gaining more responsibilities at work until he became experienced and skilled enough to do his own design work. He became so successful as a machinery designer and consulting engineer that the university world, which he'd never attended, sometimes asked him to teach there and paid him respect: in 1884 he received the first honorary doctorate in engineering awarded by the Stevens Institute of Technology in Hoboken, New Jersey. Just a year earlier, he had become the second president of the American Society of Mechanical Engineers. No small measure of the accolades earned by Leavitt followed from his work done for C&H over thirty years, starting in the 1870s.

A future shop-culture engineer could not have picked a better birthplace than Lowell. Ten cotton textile mills practiced new-style factory production, and the Lowell Machine Shop built textile machinery, steam locomotives, and water turbines. At age sixteen, Leavitt apprenticed as a machinist in Lowell. At age nineteen, he started moving from shop to shop, advancing himself with every change. Leavitt moved from Lowell to the Corliss and Nightingale firm in Providence, Rhode Island, where he worked with steam-engine valving. Then he went to the City Point Works in South Boston, where he helped construct engines for the *USS Hartford*. Next he served as chief draftsman for Gardner and Company of Providence and then entered the U.S. Navy as an assistant engineer during the Civil War. He served on a gunboat and then worked in Navy yards in Baltimore, Boston, and Brooklyn. In 1865 the Navy detailed him to Annapolis, Maryland, as an instructor in engineering. Two years later he struck off on his own as a consulting mechanical engineer, specializing in the application of steam power. He achieved success and recognition early, thanks to his execution of large steam pumping engines for Lynn and Lawrence, Massachusetts. In the mid-1870s, C&H needed large steam pumps, too, for its stamp mills. Upon the recommendation of James B. Francis, famed hydraulic engineer for the Locks and Canals Company in Lowell, C&H turned to Erasmus Darwin Leavitt.[24]

Leavitt's first job for C&H, in 1874, was a stamp-mill pumping engine. His second job, in 1876, was another pump for the mills, having twice the capacity of the first.[25] Leavitt designs came to dominate much of the mills' "big ticket" engineering. Besides the pumps and his steam stamps, in the mid-1880s he designed large sandwheels (fifty feet in diameter) for the mills.[26] By then, stamp sands already filled the lake bottom close to the mills. To deposit additional tailings, they had to flow out of the mills in *launders* (troughs) that started higher up and had a steeper pitch. Leavitt's sandwheels worked rather like waterwheels running in reverse. As they rotated, buckets at the bottom picked up water and waste sands, which they carried up fifty feet and then discharged. The water and sands flowed into an elevated launder that dumped them farther out in the lake.

The most celebrated machines Leavitt designed for C&H stood at the mine. The predilection he and Agassiz shared for large prime movers resulted in C&H operating perhaps the most impressive array of steam engines found at any one American company. In the last quarter of the nineteenth century, many Lake copper mines acquired large engines specially designed for hoisting. Typically these engines sat right near the shaft they served, and all they did was hoist rock. Leavitt's engines not only powered hoists, but other machinery at the same time. C&H's first Leavitt engine, named the Hecla, went into service at the mine in 1877. The Hecla was a compound engine: it used the same steam first in high-pressure and then in low-pressure cylinders. The engine produced about 1,000 horsepower. It drove four hoisting drums twenty-four feet in diameter, each of which served a different shaft. The Hecla also powered new air compressors for the Hecla branch of the mine, plus rock breakers and other machinery.[27]

In 1879 C&H contracted with Leavitt to design an even bigger engine. This engine, aptly named the Superior, stood in a large brick engine house along Mine Street on the Calumet branch of the mine. The Superior became a signature piece for Leavitt and for C&H. The engineering press paid considerable attention to the acclaimed Superior engine, built at a time when monumental engines served as important cultural symbols. The centerpiece of America's Centennial Exposition in Philadelphia's Fairmount Park had been a massive Corliss engine. The Superior, although working in a far grittier environment, functioned similarly as the centerpiece of the C&H mine.

When started up in the early 1880s the Superior was said to be "the largest stationary

Installed 1881, the Superior engine, designed by Erasmus Darwin Leavitt, reflected C&H's penchant for erecting equipment large enough to remain in service for decades. Weighing in at 700,000 pounds, the 4,700-horsepower Superior incorporated German steel from the Krupp Works in its main shafts. The engines that Leavitt designed for C&H made the company a showcase of steam technology. *(Michigan Technological University Archives and Copper Country Historical Collections)*

engine in the world." I. P. Morris, a well-established engine firm from Philadelphia, built the engine to Leavitt's plans. The compound engine weighed nearly three-quarters of a million pounds. It could produce 4,700 horsepower, although its regular working load was about 2,700. The engine drove winding drums that hoisted from four separate shafts: Nos. 2, 4, and 5 on the Calumet branch, and Hecla No. 3. The winding drums, twenty-and-a-half feet in diameter, could raise rock from over four thousand feet deep. (The mine was not that deep when the Superior was installed, but it was engineered to last for decades.) Running at fifty to sixty revolutions per minute, the engine turned two 45-ton flywheels, whose outer surfaces served as belt pulleys. Thirty-inch-wide leather belts took power from these wheels for myriad purposes. Using belts, gears, and/or a wire-rope power transmission system, the Superior engine drove two large air compressors, two mine pumps, two man-engines, and the rockhouse tramroad. The wire ropes transmitted power as far as two thousand feet from the engine before that power was applied.[28]

As C&H grew it added more Leavitt engines. In the 1880s and through the early 1890s, C&H added the Baraga and Rockland engines to the Calumet branch; the Frontenac, Gratiot, Houghton, and Seneca engines (all about 2,000 horsepower) to the Hecla branch; and the Hancock, Pewabic, Detroit, and Onota engines to its newer South Hecla branch. These engines could reach depths of 4,000 to 5,500 feet and hoist five- or six-ton skips at 1,000 feet per minute. By the end of the 1890s, C&H had some fifty steam

Finnish Lutherans erected the Bethlehem Church in 1891–92, and in 1897 they added the parsonage to this small wedge of land at Calumet. Shown in 1911, a surrounding fence created a religious island within a complex landscape that mingled life and work, where church steeples, smokestacks, and shafthouses all pointed toward the smoky heavens. Services in English were first delivered in this church (six times a year) in 1934. *(Michigan Technological University Archives and Copper Country Historical Collections)*

engines in service at the mine, plus more at its mills and smelters. The engines produced about 50,000 horsepower, or about as much power as was then used by a manufacturing city of two hundred thousand people.[29]

By 1890 the greatest concentration of population in the Copper Country lived near C&H in a landscape that was primarily industrial, but with important commercial and residential elements. C&H, like earlier mining companies, did not build a complete town. It built its mine, and, on the margins of the mine, it built neighborhoods of company housing. On company property, it allowed little mercantile development. In the formative years of C&H, only two general stores (one each near the mine's Hecla and Calumet branches), plus a bank, meat market, and hardware store stood on company-owned ground.[30] Like other companies, C&H depended on an adjacent village, Red Jacket, to provide a downtown commercial district with a full range of shops and services.

The landscape around C&H and Red Jacket village differed from the one found in the vicinity of the Quincy mine and Hancock. The Quincy mine promoted the concurrent development of Hancock as its attending village. Hancock rounded out local society by providing entertainment, housing options, all kinds of goods and services, plus culture and churches. Quincy sat atop its hill as a mine location, relatively isolated and unto itself. Hancock was downhill from the mine and out of sight. Quincy and Hancock, though closely related both socially and economically, were two very different places. One was the mine; one was the town.

This mid-1890s photo shows the village of Red Jacket (later called Calumet) nestled between two major mines. C&H's blacksmith and drill-sharpening shops occupy the foreground, along with several large stockpiles of timbers to be used as underground supports. In the middle distance, stores, houses, and several church steeples are visible within Red Jacket. The Tamarack mine location stands beyond the village in the far distance. *(Michigan Technological University Archives and Copper Country Historical Collections)*

That was not the case by C&H, which stood on a level plateau of ground. The mine and its attendant village of Red Jacket sat close together, within view of one another. Over time, as C&H and other nearby mines such as Tamarack and Osceola grew, so did the village of Red Jacket and another neighboring village, Laurium. This assemblage of mines, mine locations, and villages became unique in the copper district. No other mines on the Keweenaw had as much town right next to them as did C&H. Similarly, villages like Houghton, Hancock, or Ontonagon never had as much mine right next door as did Red Jacket. Up around C&H, one found clearly identifiable zones of commerce, transportation, housing and heavy industry, but these zones were more or less contiguous and integrated. True, various governmental boundaries divided this space up, but in a real sense, separate governmental units of village and township existed only on maps. Out on the land itself, this was one place, a populous community built around an industrial core, a place where life and work were hardly separated at all; where the spires of churches competed for attention with smokestacks; where railroad lines intersected streets; and where a school had boilerhouses and engine houses for neighbors.

Edwin Hulbert, the man who discovered the Calumet Conglomerate lode, established the boundaries of Red Jacket village, which he platted in 1868 as an L-shaped piece of land encompassing ninety acres. A grid of north-south and east-west streets divided this ground into twenty-six rectangular blocks.[31] After the separate Calumet and Hecla companies merged in 1871, president Agassiz and other company leaders did not launch their own rival commercial village. Instead, they encouraged the development of Hulbert's Red Jacket plan even though Hulbert and company officials were estranged from one another. Indeed, C&H literally bought into the Red Jacket plan by purchasing property there, which it later sold off for select uses that met with company approval.

Some structures stood in what would become Red Jacket village even before it was platted, but fire destroyed the earliest part of the town, or most of it, in 1870.[32] This was not an uncommon occurrence on the mining frontier. Other villages, including Hancock, suffered catastrophic fires. As they rebuilt and expanded Red Jacket, small commercial buildings constructed of wood remained the norm for about two decades. By 1890 Red Jacket's Fifth Street was the village's principal run of shops and stores. One- and two-story storefronts, still built largely of wood, lined this unpaved street. Some Red Jacket streets were residential, but rather than housing people, the village's major role was to offer the surrounding population goods, services, and entertainment. More people lived in Calumet Township than in Red Jacket village. In 1890 the village claimed 3,073 residents, while 12,529 lived in the township.[33] Within the township, C&H owned much of the land and housing, and exercised its will in determining the run of streets, rail lines, and the location of neighborhoods. The village offered a mix of commercial, residential, and cultural buildings, whereas the township presented a different mix of mining, railroad, and residential structures.

Unlike the village's streets, which were aligned by compass direction, many streets at C&H, within Calumet Township, were aligned with the copper: they ran parallel or perpendicular to the strike of the Calumet Conglomerate lode. Along its streets, C&H erected its housing in a manner consistent with other mines in the region. Initially, it erected boardinghouses. Then, as it went into production, it moved toward single-family dwellings. It first erected log houses, then switched to wood frame, usually one-and-a-half stories tall.[34] It tended to batch build houses, as needed, creating new blocks or neighborhoods. Like some other firms, the company named its neighborhoods, starting with Albion location and over the years adding many others, such as the Blue Jacket, Yellow Jacket, Hecla, Newtown, Raymbaultown, Swedetown, Red Jacket Shaft, and Waterworks locations.[35]

On some streets, C&H erected and owned full blocks of houses. Along other streets, C&H employees built their own houses on company land. C&H was not alone in allowing workers to build and own houses on company ground, leased for a nominal annual

As the mines neighboring Red Jacket grew, so did the village. Numerous shops, stores, bars, and hotels lined Fifth and Sixth, its main commercial streets. A portion of Fifth Street is shown here, ca. 1900. At this time, the streetscape included a mix of two- and three-story wooden structures, while an increasing number of brick or sandstone buildings were being added. The village boomed starting about 1890, but the boom ended around 1910. *(Michigan Technological University Archives and Copper Country Historical Collections)*

fee.[36] Other companies had done it, thinking that if workers built their own houses, then they could get away with building and maintaining fewer houses themselves. Companies also believed this practice cemented worker loyalty. When a man put a house on company property, he looked to work for his company for a long time. C&H boasted more worker-built houses on company land than any other mine on Lake Superior. That was not surprising, because C&H was the biggest and wealthiest of companies, with an assured future that no other company could match, including Quincy, which was a distant second to C&H in nearly every aspect of the business.

9

❖ ❖ ❖

A FAR MORE TYPICAL MINE

Quincy, 1865–1890

While Calumet and Hecla (C&H) invested lavishly in its physical plant, other Lake mines, working more modest deposits, had to limit their capital expenditures. Of these companies, Quincy was among the best at making the most of what it had. Michigan's Commissioner of Mineral Statistics, in his *Report* for 1881, noted,

> If any one mine were to be selected as an example from which to derive important lessons, undoubtedly the Quincy deserves the preference. Its management may be characterized as, on the whole, a fortunate medium between the conservative and progressive; it has ever held to that which has stood the test of experience, and availed itself of whatever was new that proved to be of value.[1]

In the quarter century after the Civil War, Quincy invested in new technologies that raised worker productivity and total copper production. In some important ways, however, the company did not grow. Quincy did not expand its mineral lands but continued to exploit its original stretch of the Pewabic lode. The company erected few new houses for workers at the mine, because Quincy produced more copper with fewer people, and employment declined. Overall, through 1890, Quincy did not look like a booming company, busily making itself over, reinvesting capital at every turn. Instead, it looked like a workaday company, living within its means, changing whatever needed changing, while leaving other things be.

In 1866, when the mine was about six hundred feet deep, Quincy put in a man-engine between its No. 2 and No. 4 shafts.[2] Requiring its own steam engine and shaft, this mechanical ladder was expensive to erect, and it had to be extended as the mine reached ever-deeper levels. Still, Quincy felt the benefits of the mechanical ladder outweighed its costs. Men traveled faster than when climbing or descending ladders, so they spent less time transporting themselves and more time working. Also, men could work more vigorously until the end of their shift and not have to save up energy for a long ladder climb. A man-engine also enabled a deeper mine to compete for laborers with a newer, shallower mine; without this device, Quincy could lose workers to mines that were easier to climb up and down. Finally, like other companies using man-engines, Quincy recognized that relieving men of arduous ladder climbs was simply the right thing to do.[3]

Miners who did Quincy's drilling and blasting continued to work under contracts. Compared to trammers or timbermen, who worked for daily wages, contract miners enjoyed higher pay, higher status, and greater perquisites, such as access to company housing and other favors. Miners even enjoyed a slightly shorter six-day workweek because they worked a shorter shift on Saturday than others did.[4] While the organization of contract mining changed little, the way the men won ground changed a great deal. Quincy, like other Lake mines, failed with the Burleigh drill and later succeeded with the Rand; it avoided the use of nitroglycerin oil and then introduced, considerably later, nitroglycerin dynamite.

In Quincy's 1871 *Annual Report,* the president and agent urged the procurement of "suitable machinery . . . for the introduction of power drills into use in the mine." In April 1872, stockholders appropriated $100,000 for improvements to the mine's plant, and by the end of that year, Quincy's air compressor and drill account amounted to $25,093. Quincy purchased an air compressor, installed the requisite pipes, and ordered five Burleigh drills, plus an assortment of spare parts, mounting clamps, and two *mining carriages,* or low, wheeled trucks having vertical screw posts to carry the drills. By October 1872 two Burleighs worked in Quincy drifts twelve hundred feet underground.[5]

The Burleigh drills weighed up to 550 pounds and were up to sixty-seven inches long. To give the drills room to work, Quincy enlarged its drifts from five by six feet up to ten by ten feet. While miners pushed the larger, machine-drilled openings a little faster than smaller, hand-drilled drifts, it cost the company "considerably more than it would have [if] done by hand-power." Quincy's agent, A. J. Corey, noted that "we cannot afford to carry a 10 × 10 drift through poor ground."[6] Indeed, when the first two machine-driven drifts ran into long stretches of barren ground, Quincy stopped using the Burleighs for a time.

Quincy's Burleighs drilled shot holes fast once in position but took too long to set up and align. In mid-1873 the company received two smaller Burleigh stoping drills, which were thirty-eight inches long and weighed 206 pounds. The original Burleighs were too long, bulky, and hard to maneuver; the stoping drills were easier to set up, but underpowered. They did not break "as much ground" as had been hoped. Near the end of 1873 Quincy deactivated its compressor and shelved all its Burleighs. That same year, the *Engineering and Mining Journal* reported that, at the Michigan copper mines, "the introduction of drilling machinery is at a stand-still."[7]

Quincy and neighboring mines did not give up entirely; they kept searching for a workable machine. The mines tested—and found fault with—drills made by Wood, Winchester, Ingersoll, Horton, Brown, and Duncan.[8] Then they discovered the Rand Little Giant drills, which proved a phenomenon in this mining district. At least fifteen local mines turned to Rand drills by 1883, including all major producers. Quincy acquired its first Rand in 1879 and ran twenty-two of them by 1882. With the new machines, from 1880 to 1881 Quincy's production jumped nearly 50 percent—from 3.7 to 5.5 million pounds of ingot copper. At the same time, the mine's running expenses increased by only 3.5 percent, and the number of contract miners declined from 192 to 167. In 1882, Quincy employed only 90 contract miners, and production reached 5.7 million pounds.[9] The Rand drills caused over half the company's miners to lose their jobs, but because Quincy eased them into place and offered men incentives to use them, miners accepted the machines. The use of Rand drills provoked no strikes or protests.

In 1879 and 1880, Quincy's first Rand drills went to regular miners who volunteered to try them. After becoming proficient on the machines, those men taught other miners how to use them. Quincy encouraged contract miners to try the drills by temporarily abandoning the usual manner of settling contracts. Quincy gave them "Rand drill contracts" that guaranteed them a monthly payment higher than the average earnings of hand-drilling miners. In July 1881 Quincy ran four Rand drills. It paid one team $57 per man; one team, $55; and two teams, $52. During this same month, hand-drilling miners earned an average of $47.92. As miners became more proficient with the machines, Quincy once again paid them on the basis of the number of feet driven or by the number of fathoms stoped. Rand contractors received less per foot or fathom than did hand drillers, but the machine broke more ground and earned them more money. In 1878 and 1879, Quincy contract miners averaged monthly earnings of $40.80 and $38.34, respectively. By 1882 and 1883, with over twenty machines in use, average monthly earnings stood at $53.15 and $50.10. An increase in earnings of 25 percent attended the introduction of the machine drill.[10]

Men at Quincy, C&H, and other mines accepted the Rand drill because it took two men to set up and then run its six hundred pounds of post, clamp, and machine. One miner fed the drill steel forward with a hand crank while a second man at the front of the machine stood ready to rap the drill steel with a sledge if it stuck in a hole. He also poured water in the holes drilled downward (to clear chips for faster cutting and to keep down the dust) and chucked new drills into the machine when necessary. The need for two men preserved the social nature of contract work and the tradition of men working together on teams. Every miner still had his "buddy" to share the work with, and nobody worked alone in the dangerous underground. To take full advantage of the Rand drill as a laborsaving device, Quincy did make an important change in its contract system. A traditional stoping team consisted of six hand-drilling miners, three working each shift. Quincy sought to run the Rand drill with four-man stoping teams, or only two men per shift. The company instituted this change throughout the mine in February 1881. In 1882, Quincy and C&H both recognized they had a problem with the rule of having two men per machine. The smaller teams needed some help, some extra hands. In Quincy's case, its twenty-five new hands were *drill boys.*

Prior to the introduction of the Rand drill, few boys worked underground. The 1880 U.S. census reported only twenty boys underground in the entire Michigan copper district. Quincy and C&H started using more boys underground as the mines sought to maximize the machines' economic benefits by keeping them in use more constantly. The two miners tied up in running a drill could not attend to other chores. The machines dulled many drill steels that had to be taken to the shafts for hoisting to the surface, where blacksmiths applied new edges. Then the resharpened drills had to be returned to contract teams. Quincy turned to boys to run such errands. These boys may have been as young as twelve, but most were probably about fifteen to seventeen years old. The boys typically did not work with strangers, but with mining teams containing relatives—a father, brother, or uncle. They ran for drill steels, water, tamping clay, and other supplies, while the men stayed at their machines. The boys provided miners with extra help and the company with cheap labor. In 1882 most drill boys received one dollar per day, or about half what an adult miner earned; some, perhaps the youngest ones, earned only eighty cents per day. The *Portage Lake Mining Gazette* faulted parents for pulling boys from school and putting them to work, but the protest fell on deaf ears among mining families, happy to receive the extra income that drill boys provided.[11]

Miners lost jobs at Quincy and other mines when Rand drills were introduced, but the machines did not cause a drop in overall employment in the region. If that had been the case, one would have expected protests from the labor force. Quincy more than halved its number of contract miners between 1880 and 1882, dropping to 90 from 192.[12] But its total employment declined by only 53 men (11 percent). For the copper district as a whole, employment *rose* as the drills were introduced, from 5,000 in 1880 to 6,300 in 1884. Because they lowered mining costs, the drills protected jobs at some mines, such as the Atlantic, which worked a marginal lode. And they created jobs by making possible the opening of the Tamarack mine, which became an important producer. To reach and begin working the Calumet Conglomerate lode, located far beneath the surface of its property, the Tamarack mine had to sink a shaft 2,270 feet deep through poor rock. Using Rand drills exclusively, Tamarack accomplished the feat between February 1881 and June 1885. It was "the most rapid sinking in hard rock, that has anywhere been done," and, without the machine drill, it would not have been tried.[13]

In Quincy's *Annual Report* for 1879, agent A. J. Corey wrote, "It is becoming more and more evident that for the future cheapening of our mining costs, we must place more reliance upon the use of power drills and high explosives." Quincy linked these two innovations. From 1878 to 1882, just when it converted to machine rock drilling, it tried several high explosives for the first time. Quincy only cautiously accepted high explosives. The company had not flirted with nitroglycerin oil in the 1860s, and even after Nobel introduced the more stable, nitroglycerine dynamite, Quincy shied away from it for years. The men running the mine in Michigan were sensitive to the fact that, for miners, the choice of powder was important and personal because their lives were at stake. In the 1860s and 1870s, blasting accidents accounted for fully a third of all mining deaths in the district.[14] While the profit-driven company might want to switch to new, more powerful explosives, a miner did not want to be blinded or killed by a premature blast while charging a hole, or by a late blast if he had to go back to check on a misfire, only to have the powder then explode in his face. Miners, above all else, wanted a dependable powder that detonated only when it was supposed to. It would have to perform properly and be unaffected by temperature changes, dampness, or age. Its detonation could not produce overly noxious gases that burned eyes, seared lungs, or caused headaches.

In 1878, Quincy purchased at least 25,000 kegs of its traditional explosive, a DuPont saltpeter powder, an explosive that mine agent A. J. Corey had long preferred. Quincy's treasurer, William Rogers Todd, had at times purchased some less expensive "soda" powders for the mine, but Corey always rejected them, writing to Todd, "The Dupont has proved without exception the best powder in use on the Lake and gives universal satisfaction among the men, and by actual test breaks from 25 to 33 percent more ground than any *soda* powder."[15] By 1878 other mining districts used high explosives extensively, including the Marquette iron range, a hundred miles east of the copper mines. Dynamite did not seem as dangerous as before. Corey still relied on the DuPont powder, but he was finally ready to seek out an effective high explosive that his miners would accept.

Quincy began its experiments with high explosives, beginning with just four hundred pounds of No. 2 Hercules powder, bought from the California Powder Works in 1878; it purchased 21,750 pounds of that explosive in 1880, along with some traditional black powder, but then in 1881 it switched from Hercules to No. 2 Excelsior powder. In 1882 it also tried Diamond J powder from J. H. King and Company in California. After a few years the powder experiments were over, and Quincy had made its choice. By 1884, in an average month Quincy consumed twenty-six kegs of black powder and 9,933 pounds of dynamite.[16] Mine agent S. B. Harris then noted, "The high explosive used in the Quincy for some time past is the No. 2 'Excelsior,' manufactured at Marquette. . . . The grade used here is 50 percent [nitroglycerine], price 31 cents per pound. They [the miners] like this powder here better than any other they have used."[17]

Not all parts of a mine's operation changed at once. While Quincy changed drilling and blasting technologies, others stood pat. Men still worked by candlelight, breathing air circulated by natural ventilation. Underground sanitation was as primitive as always, and the company timbered sparingly and not systematically. Trammers still mucked up rock by hand and pushed their cars from stope to shaft. Until 1890 each Quincy shaft had only one skip track, and single-tracked shafts raised all the copper rock that the company's stamp mill on Portage Lake could handle. That mill had become a bottleneck to the mine's production, limiting it in the 1880s to six million pounds of ingot per year. In 1890, when Quincy opened up a higher-capacity mill on Torch Lake that used steam stamps, it widened its shafts and added a second track in each one.[18] Two skips hoisting in balance lifted twice as much rock per day from each shaft.

To save on mining and operating costs, Quincy trimmed the number of shafts it operated. It abandoned four or five early hoisting shafts atop the Pewabic lode, and for most of this era brought up its entire mine product through two shafts less than six hundred feet apart: Nos. 2 and 4.[19] In 1865, these shafts reached depths of 600 to 700 feet, or the mine's eleventh level; by 1890, the shafts descended forty levels, or down to 2,400 feet, measured along the incline. Quincy periodically replaced the shafthouses at No. 2 and No. 4, and in 1890, when the company started double-tracking the shafts, the company modified the shafthouse interiors, putting a second skip road and skip dump into each one. The shafthouses remained simple and contained facilities only for receiving and dumping rock; all sorting and breaking operations took place elsewhere.[20] When it came to the hoisthouses and engines at Nos. 2 and 4, Quincy cut quite a contrast to C&H's monumental structures and Leavitt-designed engines.

In 1882 Quincy erected a new hoisthouse with poor-rock walls at No. 2 and in 1885–86 built a similar structure to service No. 4. Building two hoisthouses presented Quincy with a seemingly good opportunity to equip both shafts with new, fully modern hoists—yet the mine agent decided not to do it. Sometimes this successful company spent very conservatively and stuck with the tried and true instead of the new. In 1867, Quincy had traded for a used, upright Hodge and Christie engine that it first put to work at No. 2 and other nearby shafts. As No. 2 ran to deeper levels, Quincy periodically *lagged* (or built up) the hoist's winding drum or replaced it with a larger one that could carry a longer length of wire rope.[21] In 1882, Quincy put that same old engine in the new No. 2 hoisthouse. It installed a friction gear to transmit power to a cylindrical drum, fourteen feet in diameter, which could carry four thousand feet of rope.[22] The new hoisthouse was big enough to enable a second drum to be added in 1890–91, when the mine double-tracked the shaft and converted to balanced hoisting. Quincy got is money's worth from the Hodge and Christie engine; it stayed in service at No. 2 all the way until 1894.

In 1872, at No. 4, Quincy's machinery riggers had installed a horizontal Jackson & Wiley engine that had a twenty-six-inch bore and a five-foot stroke. To demonstrate its penchant for squeezing out all the usefulness it could from steam engines, Quincy took the old No. 4 hoist engine and set it up to run the mine's pump and man-engine, and it took the old engine from the pump and put it to work powering new rock-breaking machinery.[23] By the mid-1880s, mine agent S. B. Harris contemplated replacing the Jackson & Wiley engine at No. 4, largely because of recommendations made by Nathan Daniels, an eastern stockholder and unofficial steam expert. In January 1884 Daniels wrote Harris to complain of the "dilapidated condition of the hoisting apparatus" at No. 4. "You want," he wrote, "a modern engine with a variable cut off and dispense with the old slide valve you now have—and new Drum and all that goes with it." Over the next half year, Daniels, the New Englander, chided Harris, the Cornishman, and pressed for a new engine. In May he wrote, "The power required at the shafts and the stamp mill is large, and the engines at each place, would not in our Eastern mills be allowed to remain in place longer than the time required to replace them with others of modern construction."[24]

"Modern construction," for Daniels, meant an engine with Corliss valves and a variable cutoff, which shut off steam flow into the cylinder as soon as possible in the piston's stroke—which saved on steam and thus on fuel. Harris argued back that "the old is equal to the requirements for years to come." Daniels agreed that the Jackson & Wiley engine at No. 4 was "big" enough, but size was not the issue:

> The real question is not whether the old slide valve Engine now there is big enough . . . , but rather can the corporation afford to literally throw away the extra fuel required to furnish steam, and hold on to an old-fashioned engine, entirely behind the requirements or economics of the present age.[25]

Daniels shopped around for a new engine at No. 4 and found one that could be had for $6,900; he also suggested that the mine investigate the Reynolds-Corliss engines built by the E. P. Allis Company in Wisconsin. In the next decade, Quincy would do just that. But in the 1880s, S. B. Harris dug in his heels. He had the old Jackson & Wiley engine relocated to the new No. 4 hoisthouse, where it drove a new sixty-ton cylindrical drum, eighteen feet in diameter. Nathan Daniels had wanted a hoist of thoroughly "modern construction." S. B. Harris happily settled for one more to his liking that was "plain, strong, easily operated and durable."[26] After being moved into the new No. 4 hoisthouse, the Jackson & Wiley engine saw service for another quarter century there.

Prior to the 1890s, largely to avoid the expense of new equipment, Quincy chose not to purchase more efficient hoists. In other ways, however, the company paid more attention to energy consumption and generation. Sitting on the plateau almost six hundred feet above Portage Lake, Quincy was not endowed with an abundance of water for steam generation. During some winters, it even melted snow to augment the water collected and stored in cisterns near the boiler plants—cisterns that provided not only boiler water, but water for firefighting. Starting in 1872, Quincy pumped water from its abandoned shafts on the Quincy lode to its cisterns. This arrangement did not work as well as had been hoped. In 1881 the company constructed a pump works at the mill on

Portage Lake and sent water four-fifths of a mile uphill to the mine. Quincy continued to rely on Portage Lake water until the mine closed.[27] In 1882 a shortage of steam, the danger of old boilers exploding, and the inefficiency of its old steam plants prompted Quincy to build a central boilerhouse east of the No. 4 shaft. The stone structure contained eight new tubular boilers and all necessary feed-water pumps and connections. This single plant provided steam—through cast-iron pipes laid in stone-lined, insulated trenches—to the No. 2 and No. 4 hoists, to the mine pump, the machine shop, and the air compressors.[28]

Quincy changed its water source; it also changed its fuel. Originally, wood harvested near the mine fired all Quincy's boilers. In 1862, "to provide against the gradually increasing costs of timber and fuel," the company purchased surface rights to Sections 15 and 22, adjoining the mine, which it used for woodlots. These properties fueled the mine for a quarter century before Quincy started converting to coal shipped north on lake freighters from the Lower Great Lakes.[29] The switch started around 1886, when the Mineral Range Railroad Company built a branchline to Quincy's central boilerhouse near No. 4. The Mineral Range line made it more convenient and cheaper to transport coal and other freight from docks to the mine.

Coal, like blasting powder, sometimes spurred debate between the Michigan and New York ends of the company. In 1887 S. B. Harris selected Mansfield coal for Quincy. He deemed it 20 percent more efficient than two other coals that had been tried. For the next several years, Harris happily stuck with Mansfield coal. But the New York officers sometimes felt that the mine agent was too prone to stick with the "tried and true" instead of shopping around for something better or cheaper. The officers saw it as their duty to help select supplies consumed in large quantities. They occasionally ordered shipments of new coal that the agent didn't want, and he stubbornly defended his choice of one coal over another.[30]

Aboveground, Quincy substituted mechanical crushers for kilnhouse work. The company anticipated making this change as early as 1863, but high prices and busy machine shops during the Civil War delayed things. After the war a depressed copper market caused Quincy to again hold off on acquiring crushers. By 1871, with the arrival of higher copper prices, James North Wright, Quincy's agent, foresaw coming demands for higher wages. To reduce labor costs at the sortinghouses and kilnhouses, Wright recommended "that a rock house be built near the head of [the stamp mill's] tramway incline, to be furnished with a full set of Blake's rock-breakers, and connected by a railway with the working shafts of the mine."[31]

Quincy completed its first rockhouse in 1873; it was the most expensive structure Quincy had ever erected at the mine. The rockhouse contained boilers and two steam engines. One engine powered the endless-rope tramroad that pulled full cars from the shafts to the rockhouse and then returned them empty. The cars dumped their contents into the rockhouse's upper level so gravity feeds could be used to move the rock. A sec-

Between 1865 and 1890 the adoption of mechanical rock crushers was the most important technological change Quincy made to its surface plant, and its 1873 rockhouse was the most costly structure yet built at the mine. It contained boilers, two steam engines, five jaw crushers, a drop hammer, and a steam hammer. Despite all this mechanization, many men were still needed to sort and move rock within the structure. *(Historic American Engineering Record Collection, Library of Congress)*

ond steam engine drove five Blake jaw crushers, one large and four small. The engine also lifted up a two-thousand-pound drop hammer, which was then dropped on pieces of mass copper to clean them of adhering rock. Initially, the rockhouse also contained a small steam hammer that broke rock away from smaller pieces of barrel copper. All this machinery sat in the midst of a series of iron screens (called *grizzlies*), slides, hoppers, and bins that sorted, moved, or stored material. The machinery replaced the arduous hand labor conducted at the old kilnhouses, but men still pushed, dragged, carried, or picked over a great deal of material by hand. When rock passed down the structure, gravity moved it; but when it moved across the rockhouse, men moved it.

Upon arrival at the upper level of the rockhouse, the stamp rock tipped out onto the inclined grizzly. The rock small enough (three inches or less) to fall through the grizzly's iron bars was small enough to enter the Cornish drop stamps at Quincy's mill. Needing no further reduction, this material fell into a storage bin, and from there laborers chuted it into cars for transporting to the stamp-mill tramroad. The oversized rock slid to the bottom of the grizzly. Men pulled the rock down onto a heavy floor and sorted it by hand. They sent rock less than ten inches in diameter to one of the small crushers; the bigger pieces went first to the large crusher for an initial break and then on to a smaller crusher for a final break. Laborers fed the crushers by hand while keeping an eye out for barrel copper, which they picked out to assure that gummy pieces of copper did not jam the machine's jaws. Once material passed the crushers, it fell into to storage bins and later moved on to the mill.[32]

The rockhouse and its equipment proved to be a new mechanical technology well worth its investment. Quincy's 1873 *Annual Report* noted that they handled "all the rock we can hoist, and at much less cost than by the old method of calcining, and breaking by hand."[33] The cost reduction was substantial. In 1872, the last full year for kilnhouses, Quincy sorted and handled 60,628 tons of rock at a cost of forty-eight cents per ton. In 1874 and 1875, using the rockhouse alone, Quincy handled 67,112 and then 71,441 tons of rock. The costs fell to thirty-seven and then thirty cents per ton. In 1872, kilnhouse labor charges amounted to $26,735; in 1874, rockhouse labor amounted to $16,450.[34] The biggest problem with Quincy's rockhouse was its habit of burning down. Fires destroyed the rockhouse 1879 and again in 1887. Quincy used each rebuilding as an opportunity to introduce some changes into the structure and the arrangement of its equipment. Even in ordinary times Quincy occasionally modified parts of the structure to try to make rock breaking more efficient and less costly. In 1884, for example, Quincy adjusted the grizzlies for screening rock several times until fourteen men could do the work that previously required twenty-two.[35]

Copper rock sorted and crushed in the rockhouse traveled in cars down the gravity incline to Quincy's stamp mill beside Portage Lake. Within its mill, Quincy again showed its propensity for sometimes holding on to older ways of doing things while other companies adopted new ways. Through the 1870s and 1880s, Quincy eschewed adopting steam stamps and stayed with its lower-capacity Cornish stamps.[36] But in the late 1880s, when the new electrical industry was demanding more copper, Quincy planned to purchase a longer length of the Pewabic lode and to increase its production significantly. To do so, it needed a higher-capacity mill. It also needed a new mill site to resolve its first environmental issue that brought it into conflict with the federal government. The government was pressuring companies with stamp mills on Portage Lake to stop depositing their tailings into that navigable waterway.

In 1891 the federal government acquired the waterway for $350,000 from two companies: the improvement company that, starting in 1859, had made the Portage River navigable to lake boats and had built the lower entry; and the canal company that in 1868–74 had dug its three-mile-long channel to Lake Superior and built the upper entry. But even before it acquired the waterway, the government set about stopping the use of the Portage as a tailings dump, particularly at Houghton-Hancock. There, the waterway was quite narrow to begin with, and the tailings from many mills ran hundreds of feet out into the water.[37] Quincy decided to site its new, higher-capacity mill away from Portage Lake and build it on a three-hundred-acre site at Torch Lake, about seven miles from the mine.

Quincy tried to negotiate good freight rates to deliver its stamp rock to its new mill

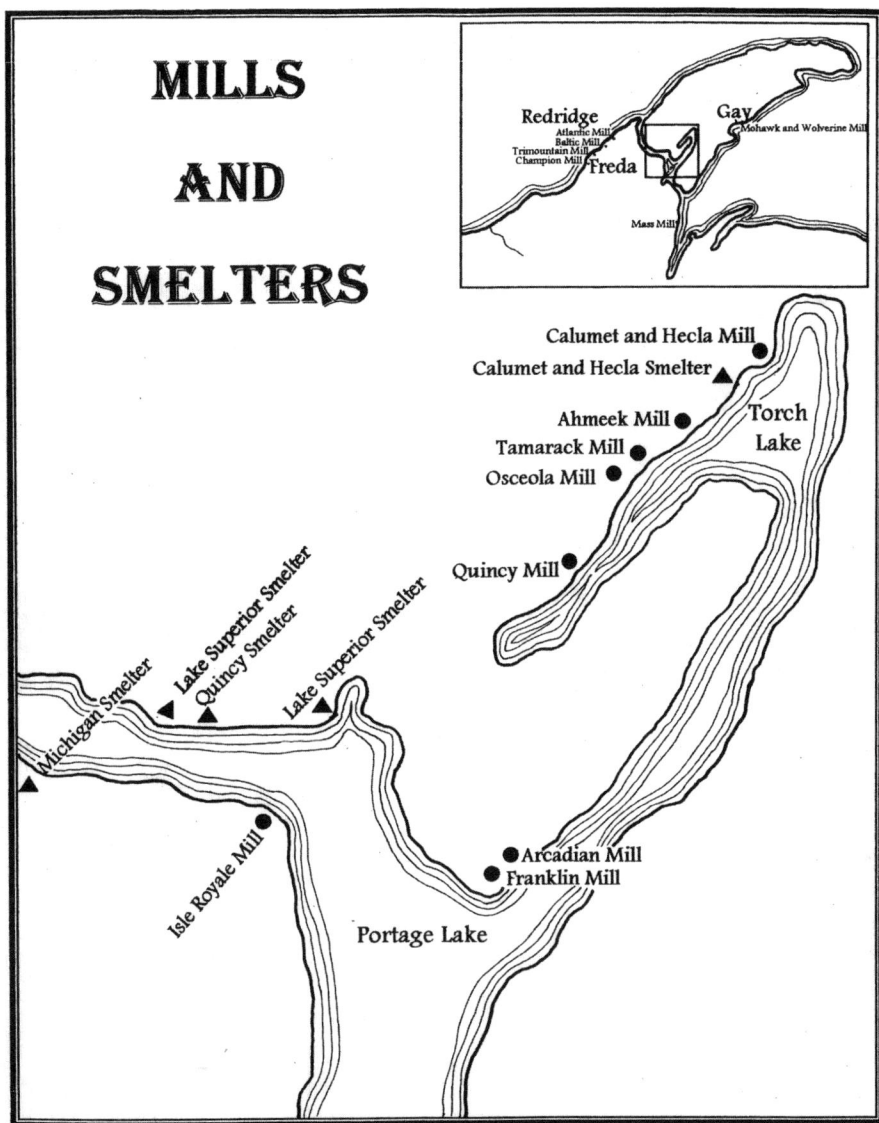

MILLS AND SMELTERS

The shorelines of both Portage Lake and Torch Lake were heavily industrialized, with stamp mills and smelters intermingled. Several major mills also stood along the Lake Superior shoreline, especially near Redridge and Freda. After erecting its first mill on Portage Lake, the Quincy mine later erected two mills and a reclamation plant on Torch Lake. (*Society for Industrial Archeology*)

site but could not come to terms with the Mineral Range or the Hancock and Calumet railroads. So it created its own line, the Quincy and Torch Lake Railroad, and proceeded to lay track, erect support structures, and obtain rolling stock. Using poor rock from the mine as construction material, Quincy in 1889 built a locomotive engine house on the southern end of the mine, attended by a turntable and covered water tower. The engine house

Quincy's mill No. 1, built on Torch Lake in 1890, exhibited the step-down construction common in the region, which allowed gravity and water to carry rock and copper from one milling process down to the next. The Quincy and Torch Lake railcars at the top of the structure delivered copper rock to bins that fed the steam stamps inside. The masonry structure with the stack is a boiler and pumphouse, which provided essential water to the mill. *(Historic American Engineering Record Collection, Library of Congress)*

originally had two engine stalls; later the company added a third stall and attached a small machine shop.[38] Quincy in the same year obtained its first narrow-gauge steam locomotives from the Brooks Locomotive Works of Dunkirk, New York, and laid three-foot-gauge track across the mine site, along the brow of Quincy Hill, and on to Torch Lake.

Quincy's new mill, which opened in 1890, followed the step-down design common in the Copper Country. The highest part of the structure, abutting a hillside, held the stamp-rock storage bins, which were supplied directly from the Quincy and Torch Lake Railroad's rock cars. Gravity fed the rock into the stamps, and water and gravity carried the discharge of the stamps to other milling machinery lower in the structure. Quincy's old mill on the Portage contained eighty heads of Cornish drop stamps and had a maximum capacity of 118,000 tons of rock per year, or a daily capacity of about 400 tons. The new mill, with just three E. P. Allis steam stamps, could treat 1,800 to 2,400 tons of rock daily.[39]

Off to one side of the mill, in a tract it called "Bunker Hill," Quincy built a cluster of six or seven new single-family dwellings for mill bosses or managers. The houses were

Within Quincy's new mill No. 1, the massive steam stamps were nearly three stories tall. Large foundations, mortar boxes, and the stamps' discharges occupied the lowest level; the steam cylinders occupied the highest. In the middle, men on the feed floor shoveled copper rock from bins into the mortar box. *(Historic American Engineering Record Collection, Library of Congress)*

generally L-shaped, with a total of nine rooms on two floors. These houses had a covered porch and another nice amenity—an enclosed walkway to the privy out back. Also, these houses stood along their own lane, which curved away from the more trafficked main road running along the industrialized shoreline of Torch Lake. On the opposite side of the stamp mill, Quincy erected small houses for ordinary mill hands. Thirty-five to forty houses stood in rows in a settlement dubbed "Mason," after the company's long-standing president, Thomas Fales Mason. No quiet settlement was this. Between the rows of houses ran Torch Lake's main wagon road, as well as a busy rail line. Two other rail lines ran nearby. The T-shaped houses carried no porches, and no covered walkways to the privies. They typically had four rooms on the first floor and two or three bedrooms on the second. A worker's house had about two-thirds as many rooms as a boss's house, yet was much smaller overall. In Bunker Hill, the first floor footprint of a manager's house covered 1,468 square feet; a worker's house in Mason had a footprint of only 596 square feet.[40]

By 1890, underground technology had changed markedly with the introduction of rock drills and high explosives. Aboveground, the mine location had changed with the building of new masonry hoisthouses, the installation of a compressorhouse, the abandonment of kilnhouses and the erection of a rockhouse, and the introduction of tram-

In the quarter century after the Civil War, Quincy built few worker houses at the mine. In 1880–81 the company did, however, build and furnish this handsome Italianate structure to serve as its agent's house. Those who administered paternalism at the mines also lived under it and received the best perquisites. Behind the house, a barn sheltered the agent's company-provided carriage and horses. *(Historic American Engineering Record Collection, Library of Congress)*

ways and rail lines, including the company's own Quincy and Torch Lake Railroad. The domestic side of the location did not witness such important changes, in large measure because the mine's labor force did not increase over this time. The company abandoned its Swedetown neighborhood, whose twenty-six houses had been built during the Civil War era, about a mile and a quarter from the mine.[41] Closer to the mine, the company erected only eighteen to thirty dwellings over this span. Some went up in the early 1870s and others in the late 1880s. Most were single units, but Quincy erected some double houses at the mine. Also, the company bought up a small number of residences that employees had earlier built on lots leased from the company.[42]

Quincy had three discernible neighborhoods atop the hill (Hardscrabble, Limerick, and Frenchtown), as well as numerous houses on the hillside running down toward Portage Lake. The grandest house in this mix, by far, was the agent's Italianate house, located near the company's office.[43] Quincy hoped the house would help it attract and keep the best agents it could get. Quincy had lost some top men to C&H, which had lured away James North Wright and others. At C&H they received higher salaries, more perquisites, and the bump in status that attended managing the greatest mine in the district. Quincy hoped the grandeur of its new agent's house new agent's house would help to slow or stop its loss of key personnel to its giant neighbor to the north.

Quincy's overall labor force, besides remaining quite steady in terms of numbers over this time, also remained relatively constant in terms of its ethnic mix. Most of the early-to-arrive ethnic groups from the Civil War era still remained at Quincy in the mid to late 1880s, and the tremendous influx of new immigrant groups, a hallmark of the post-1890 period, was only just beginning. In 1865 the Cornish, the Germans, and the Irish made up the three largest ethnic groups employed at Quincy; taken together, Englishmen (not from Cornwall) and Scots were a distant fourth. In 1885 the top three ethnic groups remained the same, but the Finns—indicative of a wave to come—had grown to be the fourth largest group.[44]

By 1890 the rising electrical industry created a booming market for American copper producers, and Quincy participated in that boom. It had just double-tracked its shafts and built a higher-capacity mill and a railroad. Soon it would acquire additional mineral lands along the Pewabic lode. For Quincy, tremendous growth lay just ahead in terms of its physical plant, production, and employment.

10

❖ ❖ ❖

QUINCY MAKES ITSELF OVER

1890–1912

With the advent of the new electrical age, the size of the U.S. copper industry expanded tremendously after 1880. In that year, American mines produced 30,200 tons of the red metal, a figure that increased more than fourfold over the next decade: U.S. mines produced 130,000 tons in 1890. Production continued to climb, hitting 303,000 tons in 1900 and 544,000 tons in 1910.[1] Quincy wanted and needed to get in on this expansion, and it did just that. It acquired the neighboring Pewabic mine, just to its north, in 1891. It also acquired mineral rights to land lying in the path of the No. 2 and No. 4 shafts, so it could sink those shafts still deeper for years to come. In 1896, Quincy purchased property to the northeast along the strike of the Pewabic lode, land on the far side of the Franklin mine that had once belonged to the Pontiac and Mesnard mines. Quincy's land purchases hemmed in the smaller Franklin mine, throttled its growth, and forced a sale to Quincy in 1909. Now one company, Quincy, held the property once worked by five.[2]

From the 1870s to 1891, Quincy hoisted from two shafts, Nos. 2 and 4, which stood six hundred feet apart. By 1900, Quincy pulled copper rock from five shafts. It opened a No. 7 shaft on the southern end of its original works. It hoisted from a rehabilitated Pewabic mine shaft, called Quincy No. 6, and from a No. 8 shaft on the old Mesnard property. The distance between Quincy's southernmost shaft, No. 7, and its northernmost, No. 8, was 7,500 feet. In 1909, Quincy finally closed the No. 4 shaft, one of its earliest, because the ground tributary to that shaft could be exploited by neighboring shafts. At the same time, Quincy opened a No. 9 shaft to the north, beyond the No. 8 shaft, on the site of the old Pontiac mine. No. 9 never proved a major asset to the company but did extend Quincy's works to a total run of ten thousand feet along the Pewabic lode.[3]

Through the 1880s, Quincy remained cautious and frugal when investing in its surface plant. All this changed in the next few decades, as Quincy upgraded existing shafts and outfitted its new ones: Nos. 6, 7, and 8. Of particular import, Quincy erected new shaft-rockhouses and finally installed modern steam hoists. Quincy also added support facilities such as new machine and blacksmith shops. The made-over mine wore all the trappings of a very successful and growing company.

While Quincy hoisted all its copper rock from two single-tracked shafts, it needed only one rockhouse to receive the entire product of the mine and ready it for shipment to the stamp mill. Then Quincy double-tracked its shafts and increased them to five, so it needed much more sorting and breaking capacity. It started placing the required new machinery right at the shaft. It grafted a rockhouse onto a shafthouse, creating a new hybrid structure at the mine: the *shaft-rockhouse*. Quincy was not the first to do this; several other Lake mines had made the combination earlier. But after Quincy erected its first two shaft-rockhouses at No. 6 in 1892 and at No. 2 in 1894, the firm never again lagged behind in developing efficient means of handling copper rock.

The No. 2 and No. 6 shaft-rockhouses dominated the landscape at the Quincy mine: they were impressive landmarks that could be seen for miles from the opposite side of Portage Lake. The structures had heavy-timber frames and were clapboarded on the outside. They had several sections rising up to shelter the climbing skip tracks inside, and, because of their myriad rooflines, they became known as the "many gabled" shaft-rockhouses.[4] Within the tall buildings, rock skips rose to near the top before dumping their contents; the height of the structures allowed the rock to be gravity-fed downward. Each shaft-rockhouse contained a grizzly to screen rock by size, a steam engine, jaw crushers, a drop hammer, and a small steam hammer. When sorted and crushed, copper rock small enough to send to the stamp mill fell into bins elevated over railroad lines running through the structure. The bins delivered the rock into railcars.[5]

The rock-handling system used in the many-gabled shaft-rockhouses of the 1890s was not perfect. Improvements were to come, principally to enlarge bin capacities so that scheduling rail service to the structures was less critical, and to eliminate the manual labor involved in moving rock along, in picking it up, and in feeding it into crushers. The pieces of Quincy's rock-handling, cleaning, and breaking technology did not change much from the 1870s into the twentieth century, but their arrangement or spatial organization did. In the new century, Quincy committed to scientific management, to streamlining processes, eliminating labor, and increasing worker productivity. The company made this commitment evident in the partial reconstructions or additions it made to extant shaft-rockhouses standing over Nos. 6, 7, and 8. It best embodied its new ideas, however, in the design of a new shaft-rockhouse erected at No. 2. In 1908, Quincy replaced the many-gabled wooden structure at No. 2 with an even taller shaft-rockhouse

A rock-strewn landscape, scrubby vegetation, and smoky skies characterized Quincy in the mid-1890s. The No. 4 hoisthouse, constructed of poor rock in 1882, and the No. 4 shafthouse (1895) stand in the foreground. The No. 4 boiler (1882) belches smoke in the middle distance. Further away stand the tall, many-gabled shaft-rock-houses erected over No. 2 in 1894 and over No. 6 in 1892. The diagonal running through the picture is not a sidewalk, but a planked-over, insulated utility trench carrying steam pipes. These trenches crisscrossed the mine, connecting boilers to steam engines. *(Historic American Engineering Record Collection, Library of Congress)*

that differed considerably from its predecessors in form (no double-pitched roofs), materials (steel frame with sheet-metal cladding), and arrangement (which eliminated labor and streamlined materials handling). The No. 2 shaft-rockhouse built in 1908 was one of the company's—and one of the region's—finest pieces of engineering.[6]

The American Bridge Company erected the nearly 150-foot-tall No. 2 shaft-rock-house. Its double skip road led up almost to the head sheaves (pulleys carrying the hoist ropes) mounted at the structure's top. Along the way, the side-by-side skip roads had dumps (rather like railway sidings or switches) at three levels. These three dumps, plus the landing station right at the shaft collar, facilitated the handling of all the different things arriving at the surface from underground. Underground workers got off the man-car at the landing station. At the first elevated dump, a rock skip could be tipped to discharge extremely large pieces of mass copper or, more frequently, tools, such as bundles of dull drill steels, or drilling machines in need of repair. These materials would slide down a rail-clad ramp to a platform just above ground level, where men loaded them onto flatbed railcars pulled into the building. If the rock skip contained poor rock, the lander switched it off the skip road at the second dump. The car tipped, and the rock

In 1908 laborers dismantled the wooden shaft-rockhouse at No. 2 while simultane-ously erecting the steel-frame structure that replaced it. Tearing down and building up at the same time kept the shutdown period and production losses at No. 2 to a minimum. Note the workers standing atop the pulley stand. *(Historic American Engineering Record Collection, Library of Congress)*

fell into a storage bin, again elevated over a rail line. Bin chutes sent the poor rock into railcars, which carried it away and disposed of it. But if Quincy needed railroad ballast or concrete aggregate, the poor rock could first be routed through a crusher before going into the bin or railroad car.

Skips filled with eight tons of copper rock each carried their loads to the top of the shaft-rockhouse and then tipped. The rock spilled out onto an inclined double grizzly—one that had two screens, one over the other. The uppermost grizzly was made up of long steel bars, six inches in diameter and set twenty inches apart. Rock larger than twenty inches couldn't pass through, so it slid off the end of the grizzly and landed next to a three-thousand-pound drop hammer. Rock smaller than twenty inches fell through the first screen and onto the second, which had bars 2.75 inches apart. Rock passing through those bars was ready for the mill. It fell through the grizzly and into the stamp-rock storage bin, which held up to two thousand tons of rock awaiting rail shipment to the mill. The oversized rock tailed off the 2.75-inch grizzly and fell into a separate, high bin. This bin fed the rock down inclined chutes into one of two jaw crushers. On this, the main crusher floor, the drop hammer, a steam hammer, and the crushers all worked in close proximity. The rock that fed into the crushers received a break and fell into the main stamp-rock bin. After being cleaned, pieces of mass copper were lowered down the outside of the building by a crane installed just for that purpose. Men tossed pieces of barrel copper down a chute into a mass-copper tube, which, when filled, could be emptied into a railcar bound for the smelter. All products of the mine had a place to go, and men moved very little rock by hand. Just three men working on the crusher floor treated upward of a thousand tons of rock per shift. All the material ended up in bins elevated over rail lines running under or alongside the shaft-rockhouse. Rock spilling from the bins filled a car bound for the stamp mill in only ten seconds.[7]

Besides being well engineered for handling copper rock, mass copper, poor rock, men, and tools, the shaft-rockhouse's lower level contained overhead cranes for holding and switching the various cars used in the shaft—the man-car, water-bailing skips, rock cars, and others designed for tools or explosives. Shaft workers using the cranes switched from one shaft vehicle to the next. For instance, at the start of the shift, they took a rock skip off the end of a hoist rope and replaced it with a man-car. Then, when the full shift of men had been sent underground, off came the man-car and back on went the rock skip. Men made these switches in a matter of minutes.

In the 1870s and 1880s Quincy relied on old hoist engines that were not fuel efficient but were powerful, steady performers, easy to maintain and operate. In the 1890s things changed dramatically. The mine no longer purchased general purpose steam engines, which mechanics connected to drums via friction drives or gears to serve as mine hoists. The design and construction of hoisting engines had become a specialty. When installing modern hoists, Quincy became dependent on the knowledge and skills of outside manufacturing experts. Large hoists were not "off-the-shelf" items. While a manufacturer's engines and drums shared certain design features from one to the next, each large engine was in part custom designed to meet the requirements of the purchasing mine. Quincy did not draw up detailed specifications on its own but told bidding firms what it wanted in general. The engine builders then decided most of the details and submitted their bids.

In 1892, 1894, and 1900, Quincy purchased modern hoists from E. P. Allis and Company of Milwaukee. These were some of the largest hoists Allis had ever built; they were installed at the mine's No. 6, No. 2, and No. 7 shafts.[8] Each hoist cost between $43,000 and $60,000, and each was bigger than the last. All three were *duplex, direct-acting hoists*—meaning that the hoisting drum sat between two identical steam engines, and the engines' cranks rotated the drum, with no intermediary drive or gearing. These hoists were more fuel efficient than Quincy's old engines because their steam cylinders had Corliss valves with automatic cutoffs rather than old-fashioned slide valves. These cutoff valves meant that the engines consumed no more steam than they really had to in order to do their job. The engines' cylinders were large: 40, 48, and 52 inches in diameter. They all had a seven-foot stroke. The hoisting drums were 21, 26, and 28 feet in

Quincy expanded dramatically to the north after acquiring the Pewabic Mining Company's land in 1891. In less than a decade, Quincy erected its first combined shaft-rockhouse at No. 6 (shown in the distance), plus a supply office, carpenter shop, machine shop, blacksmith and drill shop, compressorhouse, and boilerhouse. *(Historic American Engineering Record Collection, Library of Congress)*

diameter. The largest hoist could carry eight thousand feet of inch-and-a-half wire rope in a single layer on its drum; it could lift six-ton rock skips in balance at three thousand feet per minute. In 1905, Quincy added yet another new hoist to its plant; this smaller duplex hoist, put in at No. 8, was built by Nordberg Manufacturing Company, also of Milwaukee.[9] For the first time at Quincy, this hoist did not have a straight or cylindrical drum. Instead, to give it a mechanical advantage when initiating a lift, it had a smaller conical section on each end, making it a *cylindro-conical* drum. The drum diameter was 12.5 feet on the ends and 18.5 feet across the middle. The Nordberg hoist could lift skips—now rated at eight-ton capacity—from a depth of five thousand feet.

Quincy's wholesale transformation of its surface plant did not end with the shaft-rockhouses and hoisthouses. During the company's greatest building era ever, it erected compressorhouses, holding some of "the finest air compressing machines ever built" and boiler facilities to power its new hoists. Quincy better equipped itself to fabricate and/or maintain its works and machinery by erecting a new carpenter shop (1893), blacksmith and drill shops (1900), and machine shop (1900). It erected other storage or work buildings: a paint shop (1895), supply office (1893), warehouse (1900), pipe house (1895), oilhouse, for lubricants (1893), and an assay office (1897). Under an agreement with Quincy, the North families operated a new store at the mine built in 1895. Company managers took care of themselves, too. In 1896–97, Quincy built a company office building just up the road from the agent's house.[10]

Quincy looked bigger, more modern, and well-to-do. Until 1890, Quincy constructed most structures of wood or poor rock. The company definitely upgraded its look during this construction boom. The firm still erected some structures of wood, especially its shaft-rockhouses through 1900, and some other buildings such as (fittingly enough) its new carpenter shop. But more often Quincy constructed masonry buildings by using red sandstone quarried locally, called Portage Entry or Jacobsville sandstone. Masons laid up numerous buildings, large and small, of this attractive material: the new hoisthouses, the blacksmith shop, the supply and oilhouses, North's store, and the main office building. The new machine shop was done in brick, with a steel-truss roof—materials little used before. And the No. 2 shaft-rockhouse built in 1908, with its concrete foundation and floors and steel-frame superstructure, represented a definite design and materials transformation from the nineteenth-century Quincy mine to the twentieth.

With property acquisitions, more shafts, double skip tracks, larger skips, new hoists, new shaft-rockhouses, and a new stamp mill, Quincy achieved a much larger production. Besides exploiting the lode in new places, it extracted lower-grade rock that prior to 1890 it would have passed over. Because it could process far greater tonnages, Quincy instructed miners to take rock charged with smaller percentages of copper. And still, as agent S. B. Harris noted, the company profited: "We now, both from choice and necessity, mine larger quantities of 'low grade rock' and thus make money in many ways too numerous to mention."[11]

Quincy's contract miners still drilled shot holes by using large, two-man machines. Beginning as early as 1906, Charles Lawton, the company's new mine agent, began experimenting with different drills. Some were like jackhammers but mounted on posts; some were smaller versions of the piston drills Quincy had been using for nearly thirty years. The goal was to find, ultimately, a machine that could achieve as much hole drilling in one shift as the current machines and yet require only half as many men, one per machine, to operate.

Lawton through 1912 remained frustrated in this quest.[12] The drills he had tested weren't much to his liking, and the miners weren't enthusiastic about making this change. They did not want to break up contract teams, they did not want to work alone, and they did not want to see half of their fellow miners lose their jobs. Lawton and other mine bosses looked for a drilling machine that cut costs and increased productivity. The men, in their turn, wanted tradition, security, and a buddy to work with. Through 1912 the men were winning. The older, heavier two-man drill remained the standard. When smaller new drills were introduced, the men saw to it that the new machines didn't drill holes as fast as they were supposed to, and the drills suffered reliability problems and often broke. A little sabotage and soldiering on the part of miners kept the new technology at bay and created an uneasy standoff between miners and managers regarding the drills.[13]

At Quincy the company's choice of explosives also created a bit of a stir now and then. During most of this era, miners still charged and fired shot holes with nitroglycerin dynamite. But company officers in the East seemed bent on finding cheaper alternatives, so they sometimes purchased ammonia and gelatin powders, which had different chemistries.[14] Interestingly, with all the changes swirling about the mine, Quincy's agent often stuck with established dynamite that miners had long used and liked. If the men liked a given powder, he deemed it worth keeping.

The workaday world of the underground laborer saw some changes for the better, some for the worse—and some things changed little at all. By now, miners in some other districts could avail themselves of latrine cars moved from spot to spot on tram tracks; at Quincy, men still squatted over powder boxes. Quincy still relied on natural ventilation to circulate air, but by now the air temperature at the bottom of the mine approached 80 degrees, making workers less comfortable. Men still illuminated their work by using lighting devices they carried with them. By the 1890s, instead of candles, they used sunshine lamps that burned paraffin-based fuels. At the very end of this era, 1912, they started switching to calcium carbide lamps.[15] Burning acetylene gas, and equipped with reflectors, the carbide lamps produced more light and aimed it in the direction a man was looking.

For the miner, the tools and organization of work changed little. The trammer, however, saw some significant changes at the turn of the century, when Quincy became the first mine in the district to mechanize haulage successfully. In the 1890s, tramming not only remained the most arduous task underground; in some ways it became even harder for the men. As Quincy stretched out along the Pewabic lode, some tramming runs from the stopes to the shafts grew longer. After filling a tramcar, men pushed it farther before dumping its load. In 1896, Quincy's trammers struck to ease their burden. To resolve the strike, the company (temporarily, at least) increased the size of tramming teams from two men to three. For the men, this seemed a victory: it lessened the burden on each man to fill and push cars throughout a nine-hour shift.[16] Managers, however, saw this as an unwanted burden being shifted onto the company. They had long felt that tramming was too labor intensive, too costly. With three men per car, things seemed worse. To better the cost sheet, Quincy's managers sought a new means of tramming.

Mines had tried and failed to change tramming before, either through the use of animals or by some kind of mechanized system operated like cable cars. In 1901, Quincy became the first local mine to make a notable improvement in underground haulage. In that year, working with General Electric, it installed electric haulage locomotives underground. By the end of 1903, Quincy operated fifteen G.E. locomotives.[17] The power to drive the locomotives came from the Peninsula Electric Light and Power Company. Its alternating current, taken underground, powered a motor-generator set; the direct-current generator produced the power that the locomotives picked up from overhead trolley wires, much like the way streetcars operated. Electricity had been used minimally underground prior to this—a bit had been used for lighting or driving pumps. Quincy's G.E. locomotives represented a large step forward into the electrical age.

Quincy took pride in beating C&H to anything new. In 1901 it beat C&H in the race to mechanize tramming when it successfully deployed electric haulage locomotives. In this ca. 1920 photo, a trolley locomotive is stopped, along with its automatic side-dumping tramcar, next to an underground storage bin. *(Historic American Engineering Record Collection, Library of Congress)*

Quincy used the locomotives on long haulage runs of two to three thousand feet. Each 5,500-pound locomotive had nine cars in its stable. While three 3-ton cars were being transported along a drift at six to eight miles per hour, three others were being filled and the other three, dumped. The electric locomotives considerably reduced the cost of tramming by reducing the number of men required for this essential work. In 1902, Quincy's agent, J. L. Harris, figured that, with a locomotive (driven by a motorboy) six men (four loaders and two dumpers) could transport 132 tons of rock across the longest level in a single shift. To move the same amount by hand required twelve men.[18]

Tramming became less labor-intensive because men now had fewer tasks. They still filled cars with rock, but no longer had to push them, because the electric haulage locomotives pulled them. And they no longer had to accompany the tramcars to the shaft to dump their contents. Quincy cut underground storage bins of five-hundred-ton capacity over the shafts. Locomotives pulled tramcars by the mouths of these bins, and *automatic side-dumping* cars discharged their rock into the bins without ever stopping. Later a filler chuted that rock into a skip. Thanks to these devices, in 1903 Harris figured the cost of electric tramming at twelve to thirteen cents per ton; the cost of hand tramming was about twenty cents. The number of trammers employed declined, and Quincy even attempted to trim back their wages, arguing that their jobs were now easier, thanks to the new technology.[19] Meanwhile the number of locomotives went up, reaching twenty in service by 1910.

By the mid-1890s, new vehicles rode up and down the skip tracks at Quincy. Quincy's man-engine, which had taken men to and from the underground for nearly thirty years, was finally put out of service in 1895. Starting with the opening of No. 6 in 1892–93, Quincy switched to man-cars at each shaft.[20] These carried thirty men per trip, up and down. Man-cars rode the regular skip tracks and were moved under the control of the hoist engineers. After 1895 the same engineers also, upon occasion, controlled the passage into and out of the mine of water-bailing skips, which collected up to thirteen hundred gallons of mine water per trip at an underground sump cut in the mine rock and then hoisted it to surface.

While reconfiguring its mine, Quincy made some major modifications at its Torch Lake mill built in 1890. Quincy enlarged it as the mine added shafts and upped their output. In 1892, as Quincy equipped its new No. 6 shaft, it add two more Allis steam stamps to the mill, giving it a total of five. In 1900, Quincy added a second and separate mill, containing an additional three stamps. By 1905 the two Torch Lake mills stamped 1.1 million tons of rock per year—about ten times the greatest annual production achieved by Quincy's old Cornish stamps on Portage Lake.[21]

From the inception of the company, Quincy had always sent its mineral and mass off to a custom smelter, such as the Lake Superior Smelting Company. Quincy believed that it did not produce enough copper to justify the capital expense of building a smelter. That condition changed in the late 1890s because of mine and mill expansions. Quincy opted to build its own smelter in 1898 on the northern shore of Portage Lake. The smelter stood on the site of the defunct Pewabic stamp mill, a property that Quincy had acquired in 1891 when it assumed control of the old Pewabic mine. The site was not far from the mine and was connected by rail to the company's mills, which supplied it with mineral. (Quincy would also accept mineral from other local mines and smelt their product.) The waterfront location facilitated the shipping out of ingot copper to market, as well as the receipt of waterborne supplies such as metallurgical coal for the furnaces.[22]

The most important buildings at the smelter, both erected of Jacobsville sandstone, were the reverberatory-furnace building and the cupola-furnace building. Mineral that had traveled six miles via rail from Torch Lake went to the reverberatory building, which had four furnaces in it, one in each corner. Smelting was a batch process rather than a continuous one. Each furnace was charged, fired, tapped—and then the process repeated itself. Furnacemen still did much of the work of tending the furnaces by hand. They still rabbled it, poled it, ladled the molten copper into molds, and pushed molten slag out of the building in slag buggies. The slag was not pure waste; it included some copper that had not been separated out. Recapturing that copper was the purpose of the cupola building, which stood close to reverberatory building.

Workers charged the cupola furnace, a vertical blast furnace, from the top with hard coal, slag, and limestone flux. The charges passed down the stack into the combustion zone, where the slag melted. The remelted copper from within the slag settled to the bot-

Four tall brick stacks marked the reverberatory furnace building at the smelter Quincy built in 1898 on the site of the abandoned Pewabic stamp mill. Shown in ca. 1915, the hill rising from the smelter up to the Quincy mine was sparsely vegetated. *(Historic American Engineering Record Collection, Library of Congress)*

tom of the furnace's hearth and was tapped out. The molten slag was then transported by a narrow-gauge industrial rail line out to a slag pile and dumped. It solidified, and made the slag pile a key element of the smelter landscape. With the continued expansion of production into the twentieth century, Quincy added several new structures to the smelter in 1900–1906, including a No. 5 reverberatory building, which provided an additional furnace.

So much looked new and promising on the surface, but underground Quincy was deep and, for an American mine, old. The underground suddenly became less stable. After 1906, structural problems in stoped out parts of the mine started to plague the company and scare the men. Quincy had been hollowing out the Pewabic lode since 1856 without systematically supporting the hanging wall. It had timbered sparingly; it had not left regularly spaced rock pillars as supports. Instead, random pillars of poor rock left in place supported the hanging wall over the footwall. Finally, rock pressures grew great enough to shatter the poor-rock pillars, causing stoped-out portions of the mine to collapse. When portions of the mine fell, the collapsing rock compressed nearby air and shot it through the mine: hence the local term for the rock bursts—*air blasts.* They were known after their effect, not their cause. These collapses were often felt on the surface, too, and when they rumbled houses on Quincy Hill or down in Hancock, they earned monikers like "old rousers" or "radiator rockers."[23]

The rock bursts in abandoned stopes would have been of small consequence if their damage had stopped there, but the collapses sometimes extended to the shafts. In 1906, as the air blasts were beginning, rockfalls commencing at the mine's fortieth level crushed stretches of the No. 2, 6, and 7 shafts, putting them out of business for up to ten days. Besides interrupting production, burying shafts, and shattering compressed-air lines, the air blasts posed a potential disaster. Large numbers of men sometimes occupied the shafts, particularly at the beginning and end of a shift, and a collapse at the wrong time might kill many. Or, instead of being localized, a rock burst might collapse a huge section of the mine at once, jeopardizing an entire shift of workers. These blasts did not occur just infrequently. No count exists for the first years of the frightening phenomenon, but between 1914 and 1920, Quincy recorded over four hundred rock bursts.[24] The air blasts frightened the mine's agent, Charles Lawton, as well as the men who went underground every day.

Quincy, over many years, studied means of protecting life and property from air blasts. The mine flirted with switching from its established advancing system to a retreating system. Under the proposed retreating system, miners would first drive a long drift to the end of the ground tributary to a shaft. Then other miners would begin stoping at the end of the drift and retreat back toward the shaft. This was deemed safer because miners and trammers did not have to pass regularly through stoped-out ground that might collapse. But the system was also expensive to initiate because it called for driving a host of long, new drifts while suspending much stoping. Quincy through 1910 or 1912 experimented with the new system without broadly adopting it. Instead, it started leaving rock pillars alongside each shaft to provide a more secure hanging wall and also had laborers lay up rib packs or rib walls of poor rock.[25] As stoping progressed along a level, men put up a rib wall that ran alongside the drift and helped support the hanging.

By adopting some new underground mining techniques and busily making itself over on the surface, Quincy achieved some record highs. In five-year intervals, starting in 1890 and running through 1910, it produced 8, 16, 14, 19, and then nearly 23 million pounds of copper ingot. Over those twenty years, Quincy's share of overall Michigan production continued to run in the range of 8 to 10 percent annually. The number of mine, mill, and smelter workers at Quincy mirrored the jump in production. Employment ranks climbed from 484 in 1890 to 1,349 in 1900 and to 2,019 in 1910.[26] Yet despite its high production levels and all the visible new construction on the surface, Quincy's pesky and persistent air blasts reminded the company that it operated a deep and fragile mine.

11

❖ ❖ ❖

CALUMET AND HECLA

Profits Now, Problems Later

After a quarter century of operation, the mighty Calumet and Hecla (C&H) mine faced more problems than in earlier times. It needed to support the hanging wall of the Calumet Conglomerate lode more systematically. C&H had to reduce the great underground fire hazard posed by its labyrinth of timbering. By the 1890s the mine managers and the company president, Alexander Agassiz, were clearly vexed by new immigrant workers, especially trammers, who seemed bent on causing trouble. By the early 1900s the copper yield per ton of rock hoisted from the Calumet Conglomerate lode rapidly declined. C&H recognized that to secure its future, it had to seek out other sources of copper. It also needed new technologies for drilling and tramming. Two-man drills remained the norm at C&H and throughout the district. The switch to smaller one-man machines was in the works but was not happening very fast. Men still hand trammed rock from stope to shaft, while C&H struggled to find a mechanized method that would cut costs and reduce the company's reliance on fractious immigrants.

By the 1890s the ethnic mix at the mines began changing dramatically, as Finns, Italians, and eastern Europeans arrived in large numbers. These men usually started work as trammers. In its 1891 *Annual Report,* C&H noted that "it is becoming more difficult, from the number of men who do not speak English, to deal directly with our employees." The gulf between employer and employee widened in May 1893, when trammers working the Calumet branch refused to go down for their night shift on Sunday. By Tuesday, nearly all of C&H's trammers had joined the strike. They tried to keep other men from going to work and briefly seized control of the Superior engine house. To oust them,

C&H called in the county sheriff and a large force of deputies. In its *Annual Report* for 1893, the company noted, "It was a great disappointment to find that there were among the employees of the Company so many men ready to forget the friendly relations which had always existed between the men and the officers of the company." If the men were the first to forget "friendly relations," the mine bosses were the second, and in short order. By the end of 1893, Agassiz wrote his mine superintendent that "we must be prepared to do our tramming in some other way than man power very soon, for the men, judging from scraps in the paper, are beginning to talk about the use of men as beasts of burden." Agassiz, never forgiving of transgressors, made his dislike of trammers clear: "I shall be very glad not to have any of them."[1]

C&H could never eliminate all trammers; that was just wishful thinking on Agassiz's part. But to save on labor costs and to reduce the ranks of trammers, in 1894–95 C&H experimented with mechanized underground haulage by using a cable-car kind of system with motive power provided by engines driven by compressed air.[2] The drums on these engines wound and unwound wire ropes that ran past the shafts and down along the drifts. Men still loaded and unloaded the two-and-a-half-ton tramcars but no longer pushed them. The endless cable pulled full cars to the shaft and returned empties to the stopes. But the system didn't work as planned, and C&H never put it into practice throughout the mine. Next, the company tried electric locomotives such as those used at Quincy after 1901. But the great mine failed to figure out how to deploy electric haulage locomotives in a manner that saved money. When quizzed once on why C&H hadn't succeeded in mechanizing tramming, James MacNaughton replied, "It had been sort of an impossibility."[3] In trying to revolutionize underground work at his mine through 1910–12, MacNaughton was twice stymied. Drilling and tramming remained unchanged.

The instability of the mine's hanging wall prompted C&H to alter its mining method. By 1895 it used the retreating system. Stoping miners started taking copper from the end of a drift, and as stoping progressed, they retreated back toward the shaft. The object was always "to have solid ground between the miners and the shaft," so they could not be trapped by falls of the hanging wall in stoped-out ground.[4] C&H also protected its shafts better from collapses of the hanging. C&H had left modest shaft pillars in place even at shallow depths. By 1895 to 1900 these pillars extended seventy-five feet on both sides of a shaft. Shaft pillars and the retreating system did not free C&H from timbering. Supports remained necessary to secure the hanging wall until stoping in a given area was completed. By 1900, C&H installed thirteen million board feet of square-set timbers underground per year and twelve million feet of stulls.[5] To reduce the fire hazard posed by all that timber, starting in the 1890s the company treated underground timbers with zinc chloride and whitewash to make them more fire resistant. Men regularly sprinkled shaft timbers with water. Once a timbered level was stoped out, laborers built walls of rock to close it off from the shaft. To respond faster to a fire, C&H installed an electric fire alarm and telephone system underground. It also installed water pipes, hydrants, hoses, and chemical extinguishers.[6]

The disastrous fires of 1887 and 1888 also prompted C&H to sink a major new shaft into the Conglomerate lode. During the fires, while two interconnected branches of the mine were closed and sealed off at the surface, the separate third branch, South Hecla, remained in service. C&H wanted to extend its ability to mine significant product even if hit by additional fires. So in 1889 miners started sinking the Red Jacket shaft, located north and west of its main works. Red Jacket shaft was not inclined, nor did it run through the lode. Instead, it passed straight down through poor rock. Red Jacket shaft sinkers intersected the dipping Conglomerate lode 3,260 feet below the surface. Then they continued on, bottoming out at 4,900 feet in 1896. At this depth, Red Jacket was reportedly the deepest mine shaft in the world.[7]

C&H outfitted Red Jacket shaft in a manner unlike all others. The company published no cost figures for this work, surely to avoid controversy and criticism over whether the investment was worth it. Sinking and equipping a shaft, 15.5 by 25 feet, to a depth of 4,900 feet was immensely expensive. C&H divided this shaft into six compartments. Two hoisted men, who rode in double-decker, elevator-like *cages*. The other four compartments hoisted rock. The four rock skips were paired, and each pair operated in balance, one going down, one going up.

Because the shaft was vertical and intersected the lode at only one point, miners drove numerous crosscuts from the shaft to the lode to open it up for drifting and stoping. Atop the shaft, C&H built an impressive surface plant with a new steel-frame shafthouse, engines, boilers, and other support buildings.[8] Even president Agassiz entertained doubts, during the economic hard times of the early 1890s, about the efficacy of the endeavor.[9] Agassiz believed the Red Jacket shaft was essential for mining the northern end of the Conglomerate lode to great depths. Being independent of the mine's other branches, it could stay in operation even if they went up in flames. Or so it was thought. Originally, the Red Jacket shaft was not supposed to connect underground to the other works. But a safety issue was raised—could they have only one way out of that part of the mine, up Red Jacket shaft? Also, C&H encountered a ventilation and air temperature problem at Red Jacket. Rock temperatures at its bottom hit 87 degrees. To offer a second way out, lower air temperatures, and improve ventilation, C&H holed through openings at Red Jacket with ground tributary to the Calumet branch's No. 4 shaft and, where they met, installed fire doors.[10]

Ultimately, Red Jacket shaft proved its worth and became the company's major production unit on the Calumet Conglomerate lode. But this had not happened by the end of the 1890s, and in 1900 the shaft failed the test for which it had been created. In May, workers discovered a fire near Hecla shaft No. 3. They were unable to douse it with water or chemicals. Men ran to close fire doors along the various branches—including fire doors where the Red Jacket works connected to the rest of the mine. The doors kept

In 1889 C&H started sinking Red Jacket shaft to open up a new branch of the mine that could remain in production even if fire closed older branches. This large vertical shaft, sunk to a depth of 4,900 feet, represented an extremely expensive bit of fire insurance—so expensive, in fact, that C&H never released its cost to stockholders or the public. *(Michigan Technological University Archives and Copper Country Historical Collections)*

the fire from spreading toward Red Jacket shaft but did not seal that ground off from smoke and gases. The gases sank to the bottom of the older works, then infiltrated the Red Jacket openings, fire doors or not. The Red Jacket works suffered no damage, but like the rest of the mine they remained shut and sealed for three weeks before the fire died out. During those weeks, only the South Hecla branch remained in production.[11]

Above the mine, C&H's surface plant in the 1890s was second to none in the United States. President Alexander Agassiz's penchant for large systems and for building big to meet tomorrow's needs—coupled with Erasmus Darwin Leavitt's engineering—had resulted in a plant that could handle great production levels. When the rise of the electrical industry stimulated a greater demand for copper, C&H—unlike Quincy—did not have to remake itself to augment production. Still, the company made some significant changes to its physical plant, some of which mirrored what Quincy was doing. C&H discovered that having just a few rockhouses presented a bottleneck to production. C&H put sorting and crushing equipment right over the collars of many shafts. It erected eight combined shaft-rockhouses along the Conglomerate lode in 1891–93 and later added more.[12]

C&H moved more into the electrical age in the 1890s. As early as 1878 the company lighted its stamp mills with arc lamps, and by 1881 a Brush arc-light dynamo illuminated the mine's surface plant. After 1891 C&H used electricity to drive machinery. Fittingly, as a symbol of its movement from mechanical toward electrical power, C&H installed its enlarged electrical generating facilities at the mine in its old *gearhouse*.[13] This structure had been part of C&H's earlier, cumbersome system of transmitting mechanical power along the mine by using ropes and reciprocating rods. The 1891 generating station housed a 400-horsepower Porter-Allen steam engine and a compound Westinghouse engine of 740 horsepower. The engines drove two electric light dynamos, each sufficient to light a thousand lamps of 16 candlepower, three arc-light dynamos of seventy-four lights, and five Brush generators. From the gearhouse, reborn as a modern powerhouse, a line of electric poles ran the full length of the surface plant, and in 1892 electric lines for the first time ran underground to power the mine's new motor-driven pumps.

While using more electricity, C&H remained very much in the steam age. To an already impressive complement of engines, in the mid-1890s C&H added the Mackinac engine, a triple-expansion "steel giant" of 7,000 horsepower, to its Calumet branch.[14] Even more important and monumental were the new facilities at Red Jacket shaft, where C&H installed ten new boilers of 1,000 horsepower each. It built a 220- by 70-foot engine house, which sheltered steam engines producing about 8,000 horsepower. Like so many of the company's engines, Red Jacket's Minong and Siscowit engines were designed by Leavitt and built by I. P. Morris. These vertical, triple-expansion, beam-type engines ran with condensers to make them more fuel efficient. The twin engines had a six-foot stroke and cylinder bores of 20, 32, and 50 inches. The engine house, as originally configured, also contained the secondary engines, Mesnard and Pontiac. Raising 7.5-ton skips at 2,700 feet per minute, the two engines lifted two thousand tons of rock per day over two shifts.[15]

Red Jacket's hoisting machinery was notable not just for its size, but for its mode of operation. The engines did not provide 4,900 feet of hoist rope for each skip in Red Jacket shaft, nor did they wind hoisting rope on an individual drum for each skip. The Red Jacket hoists used the endless-rope, or Whiting, hoisting system, designed by C&H's general superintendent at the time, S. B. Whiting. The machinery occupied a 412- by 32-foot *tailhouse* near the engine house. Using this system, each pair of skips required about 6,500 feet of hoisting rope. A rope connected to one skip came out of the ground at the shafthouse and ran to the hoisthouse. There it made only a few turns around a hoist drum before leaving the engine house and running to the long tailhouse. The rope passed around a *head sheave* (pulley) mounted on a tension carriage that moved over rails. The rope, after passing around the sheave, returned to the shaft, where it connected to the second skip. To further enhance the operation of these two skips in balance, C&H equipped them with tail ropes in their vertical hoisting compartments. A wire rope connected to the bottom of one skip ran down to and around a sheave mounted at the bottom of the shaft. It then passed up to connect with the second skip. Thus a long length of rope always hung beneath an empty skip at the surface, and that rope's weight helped lift the heavier weight of the paired, loaded skip, when its lift started.[16]

The Red Jacket shaft complex did nothing to hurt C&H's reputation of having one of the nation's most impressive collections of steam engines. The engine house contained four engines, including two 3-cylinder (triple expansion) engines designed by Erasmus Darwin Leavitt. Note the traveling overhead crane needed to install or move heavy engine parts inside the structure. *(Michigan Technological University Archives and Copper Country Historical Collections)*

The machinery in place by 1900 at C&H's various branches basically handled the output of the Calumet Conglomerate lode for as long as it remained in production. Right at the turn of the century, with major capital investments in new machinery at the mine completed and with copper prices at recent highs, C&H enjoyed its three most profitable years ever. But the Conglomerate's diminishing yields clouded the company's future. One way or another, the company needed additional sources of copper.

In the October 1931 issue of *Mining Congress Journal,* which was devoted to the past, present, and future of C&H, James MacNaughton wrote an article that set forth some of his company's problems and opportunities in the twentieth century. Of particular importance, as a problem, was that in 1874 the Calumet Conglomerate lode yielded a high of 96.8 pounds of copper per ton of stamp rock, or 4.8 percent. In 1900 the rock yielded only 52.15 pounds per ton. The lode had increased in thickness at depth, but the concentration of copper fell off. By 1907 the yield had declined still further, to about forty pounds of copper per ton of rock hoisted (just 2 percent). In the face of such decline, C&H needed to develop new sources of copper.[17]

C&H looked for copper on the mineral lands it already owned in the vicinity of Red

Jacket and Laurium. It went on the eastern side of its once-great Calumet Conglomerate lode to open works on two amygdaloid lodes on its property. This activity began in the last years of the nineteenth century, when C&H began sinking shafts into the Osceola lode. C&H's No. 13 shaft opened there in 1897, followed shortly thereafter by Nos. 14 and 15. The company continued to expand on the Osceola, adding shafts No. 16 and 17 in 1899 and a No. 18 shaft in 1906.[18] Some of these shafts were more exploratory than productive, and C&H sometimes suspended operations on the Osceola when copper prices dipped. Still, the company now had an eleven-hundred-foot-long run of the Osceola amygdaloid ready for possible exploitation, whose rock had yielded as much as twenty-two pounds of copper per ton. When copper prices ran high enough, that yield could be made to pay. In similar fashion, but in a less intensive way, in 1903 C&H opened yet another mine, east of the Osceola lode, on the Kearsarge lode. The company did not push this development with great vigor, but by 1906 it did put down three shafts (Nos. 19, 20, and 21), sinking two of them to the sixth level. These works represented a test, and the test showed that this ground did not merit full-fledged production.[19]

C&H purchased large tracts of land in Keweenaw County, where it hoped to find new copper beyond its old boundaries. But two or three years of digging test pits and diamond drilling failed to turn up anything of value. By about 1906, C&H tried another approach. It started buying up stock in other copper-mining companies. In a 1910 letter "To the Stockholders," C&H management explained the move: "It was with the purpose of assuring the continuance of the life of this company and the profitable use of its very valuable plant after the exhaustion of its own mineral deposits."[20]

C&H looked to connect its interests with those of other companies and properties. Some companies that interested C&H were active producers. Others had property, perhaps, but lacked the capital to develop it. C&H had sunk shafts into its run of the Kearsarge lode and acquired interests (at a cost of more than $7 million) in the Osceola, Allouez, and Centennial mining companies, which owned other parts of the Kearsarge lode. It also purchased controlling or major interests in other companies, including the Isle Royale, Tamarack, and Ahmeek mines. It organized new companies on the northern and southern ends of the mineral range, such as the White Pine Copper Company, located at the Nonesuch lode in Ontonagon County. C&H's empire now reached along much of the Keweenaw.[21]

C&H managed of all these mines, but they continued to operate as separate, individual companies. By 1910–11, C&H wanted to consolidate many of these mines, but not all, into one company. (Those companies to be folded in to C&H included Ahmeek, Allouez, Centennial, Osceola, Tamarack, Seneca, Laurium, LaSalle, and Superior.) In his piece in the *Mining Congress Journal* of 1931, MacNaughton gave the prime reason: "Greater economies could be secured by a consolidation into one company, by abandoning intervening boundary lines, making common use of more favorably located plants, operating fewer shafts, centralizing management, accounting, purchasing, milling, smelting; and many other minor operations."[22]

After 1900 the copper content of the Calumet Conglomerate lode fell off sharply, causing C&H to look for additional sources of copper. Besides hunting for new deposits, in 1910–11 it bought up controlling or major interests in older, active mines, including Tamarack. Tamarack's No. 2 shaft-rockhouse is shown here, flanked by some very large stulls. *(Michigan Technological University Archives and Copper Country Historical Collections)*

The consolidated firm would eliminate unproductive shafts, trim the number of railroads transporting materials, reduce the number of mills and smelter furnaces in operation, and run surviving units to full capacity. That became C&H's dream of the future, but the future did not arrive in 1911, when the consolidation was proposed to take place. A lawsuit filed by non-C&H stockholders in one of the other companies stopped it for years to come.[23]

In the mid-1890s, nine to eleven steam locomotives and four hundred rock cars delivered stamp rock to C&H's two mills on Torch Lake. The mills, like the mine, were bastions of impressive steam power. Inside them, split equally between the mills, twenty-two large Leavitt steam stamps pounded away, and some of the largest pumps in the nation, having a combined daily capacity of over 110 million gallons, delivered the water needed to process the copper and rock. By 1899 the mills poured out five thousand tons of stamp sands per day.[24] Another way for C&H to find more copper was to capture more of it at the mills and wash less of it out with the tailings. Fortunately for C&H, while the yield of copper per ton of rock declined at the mine, the recovery of copper at its stamp mills improved. Early C&H milling technologies captured only three-fourths of the copper included in the rock. The last fourth of the copper—over 400 million pounds of it—was contained in C&H's 120-foot-deep deposit of stamp sands in Torch Lake. At 1900, C&H's mills still sent out more than twenty-two million pounds of copper per year with the tailings.[25] The company looked to reduce this ongoing loss of copper, and by 1912 it also planned to recover copper lost once already, by reclaiming it from the lake bottom.

Starting in the late 1890s, over the next fifteen to sixteen years C&H improved its milling technology. Instead of capturing 75 percent of the copper, it captured 90 percent. C&H took a major step forward in 1898, when it became the first Lake company to put *Wilfley tables* in its mills. This technology, first developed in Colorado by Arthur R. Wilfley, captured finer particles of copper. These tables, with their slightly inclined tops, were surfaced with linoleum. They carried parallel rows of low riffles across part of the surface. Water and small particles of copper and rock flowed onto the high side of the table, which was given a vibrating or shaking motion. The water and rock flowed down, across, and over the riffles while the heavier copper collected behind them. The table's back-and-forth motion moved the copper along the riffles until it reached the end of the table, flowed off, and was collected. Because Wilfley tables captured fine copper, C&H started using *Chilean mills* to grind its mineral to a smaller size. Set upright on their edges, these rotating wheels, about six feet in diameter, broke and abraded the rock against a die at the base of the machine. This grinding liberated more copper. Before, that copper had remained bound up inside coarser grains of sand, which had flowed through the mill and out into the lake.[26]

With Wilfley tables and other new milling technologies in hand, C&H for the first time in thirty years effected a wholesale change in its milling operations. Over a multi-year period ending in 1906, the company phased out its two original wood-frame mills and employed the Wisconsin Bridge and Iron Company to replace them with thoroughly modern steel-and-concrete mills clad in corrugated metal. The mills had a total of twenty-eight steam stamps and all new washing and separating machinery, mostly driven by electric motors.[27] C&H built the mills to capture a higher percentage of copper and to increase milling capacity so it could treat the rock coming from new properties and not just the Calumet Conglomerate lode.

Milling technologies that captured more copper from mine rock could also be applied to waste sands. Those sands, once retrieved, could be ground more finely and treated to capture copper missed the first time around. Starting in 1901, C&H and other companies, including Quincy, started experimenting with the idea of *reclamation,* of pumping stamp sands back to shore and re-treating them. C&H entered an agreement with the Metals Recovery Company to investigate a profitable means of reclamation. In 1904 it ran a limited reclamation test on stamp sands deposited in Torch Lake in 1866–70. This test indicated that reclamation might well pay. In 1912, C&H initiated engineering plans for a new plant at Torch Lake, which began reclaiming tailings in 1915.[28]

Besides revamping its mills, C&H made substantial changes to its refining and

C&H often doubled up its facilities. It essentially had two mines, and the Calumet and the Hecla branches had many duplicated facilities. The company had two stamp mills at Torch Lake. It also erected and operated two large smelters. This C&H smelter stood along the shore of Torch Lake, near the mine and mill. A second company smelter operated on the outskirts of Buffalo, New York. *(Michigan Technological University Archives and Copper Country Historical Collections)*

smelting facilities. Early in the 1890s the company followed the standard practice of having men smelt copper in small reverberatory furnaces that produced one batch of ingot copper at a time. Sixteen furnaces sat in four separate buildings at Hubbell, and a second smelter embodying traditional technology operated in distant Buffalo, New York. By the end of the 1890s, however, C&H began introducing some major technological changes, which were in full swing by the early 1900s. These involved new furnaces, new casting techniques, and C&H's first use of electrolytic refining.

C&H was not always the innovator. It watched other mining companies try something new, and if it proved successful, C&H adopted the change, too. In 1888 the Tamarack and Osceola mining companies jointly built a smelter at Dollar Bay on the Keweenaw that followed established designs and practices. But in 1898 the Dollar Bay smelter introduced a new furnace concept that C&H rapidly borrowed. They abandoned the small furnaces that did both melting and refining and replaced them with a larger reverberatory furnace where the copper was melted. Then at intervals they drew the molten copper from this furnace and delivered it to other furnaces for refining.[29] This idea of melting and refining in separate furnaces, coupled with an increase in furnace size, was an important innovation that saved on labor and fuel costs.

C&H built its first specialized melting and refining furnaces at Buffalo. By 1900 its Buffalo furnaces had a charging capacity raised from thirty thousand to one hundred thousand pounds of mineral. C&H eliminated much human shoveling, and reduced labor costs, by loading the melting furnace with a motorized conveyor, and mechanization also reached the casting end of the operation. At Buffalo, a circular, rotating Walker casting machine stood at the tapping end of a refining furnace. Men still ladled the copper out, using trolley ladles of 100- to 150-pound capacity, but they didn't walk over to the molds; the casting machine brought empty molds to them. After the filled molds passed the ladlers, a water spray helped cool and solidify the copper faster. Then the machine dumped the ingots into a water bath for further cooling, and a conveyor carried them away. The Walker machine reportedly enabled men to pour a furnace charge in less than one-fourth the time required by the old method of ladling. In January 1911, C&H started rebuilding its smelting plant at Hubbell. There it erected a new steel-frame smelter containing two 150-ton Jumbo reverberatory furnaces.[30]

In 1894–95, C&H experimented with a wholly different means of refining some of its copper. It built an experimental electrolytic refining plant at Buffalo, and in 1899–1900 it built a full-scale production plant there. This plant made use of the principle of *electrolysis,* discovered by Michael Faraday in 1835: metals can be separated and then selectively deposited by using a continuous electric current. A Newark, New Jersey, smelting works first applied electrolysis to copper refining on a commercial scale in 1881. In the 1890s this technology rapidly evolved, and between 1892 and 1902 the production of electrolytic copper rose from 25,000 to 250,000 tons. By early in the new century the bulk of the world's copper was electrolytically refined to remove those impurities that lowered its workability or electrical conductivity.[31]

Copper companies in the western United States led the way in the large-scale adoption of electrolytic refining. C&H became the first Lake mine to use the process. In its case, C&H wanted to reduce the level of two impurities: arsenic and silver. As C&H mined deeper levels, more arsenic turned up in its copper, which hurt its workability. By removing or at least reducing that element, more of the company's product could go into wire for the electrical industry. And if traces of silver could be removed and recovered, the value of this precious metal would help pay for the construction and operation of the electrolytic refining plant.

The plant consisted principally of a dynamo to provide continuous current, large tanks filled with a current-carrying (electrolytic) solution, and racks submerged in the solution, which held copper anodes and copper cathodes. C&H cast its anodes from copper made from mineral that assayed high in silver content. These copper anodes were the ones broken down by the electric current. Copper migrated from the anodes, through the solution, and across to the cathode plates, where it was deposited. By manipulating the electric current and the electrolyte, C&H adjusted the process to ensure that only copper reached the cathodes. Silver and other impurities, once liberated from the anodes, precipitated out as sludge on the bottom of the tank. By remelting the high-purity cathodes with its richest grades of mill concentrates, C&H produced a more desirable, lower-arsenic copper of higher electrical conductivity. And by collecting, drying, and smelting the tank sludge, it recovered valuable silver.[32]

The Calumet and Hecla Mining Company of 1910 differed greatly from the company of 1890. It had substantially expanded its physical plant at its original mine site, at its mills, and at its two smelters in Hubbell and Buffalo. It no longer mined the Calumet Conglomerate exclusively but also mined amygdaloid deposits. It had expanded territorially up and down the Keweenaw, acquiring new property that it explored in hopes of tapping into a heretofore undiscovered major lode. It had created new mining ventures and purchased large and usually controlling interests in other companies, which it now managed. It had adopted important new technologies at its mills to capture more copper per ton of rock stamped and was primed to launch a new reclamation plant to reprocess tailings and thus "mine" over 400 million pounds of copper sitting at the bottom of Torch Lake. New technologies at the smelters cut production costs there while enabling C&H to market a more desirable product.

C&H employed 1,576 workers in 1880, 3,490 in 1890, and 5,090 in 1900. That figure dipped just a bit to 4,940 in 1910. In terms of production, looked at every five years from 1890 to 1910, C&H produced 60, 77, 78, 95, and then 72 million pounds of copper.[33] C&H created new opportunities for itself during this era because its old source of easy money, the once fabulously copper-rich Calumet Conglomerate lode, was letting the company down. Standing pat would have meant certain contraction and, ultimately, corporate suicide. The 1910–11 edition of Horace Stevens's *Copper Handbook* went on page after page describing the vast industrial empire that C&H had built for itself. The review placed that growth and change in the context of the single greatest problem confronting C&H:

> The conglomerate mine, which, until a few years ago, was the entire Calumet & Hecla, has a life, at the present rate of production, of between 10 and 15 years, followed by 5 to 10 years of scramming [robbing pillars, searching abandoned stopes for overlooked copper], with greatly decreased output. The conglomerate is deteriorating rapidly in average copper contents with depth.[34]

C&H faced another challenge in terms of maintaining its leadership role and hegemony among the Lake copper mines. At the turn of the century, three new mines—the Baltic, Trimountain, and Champion—all owned entirely or partly by the new Copper Range Consolidated Mining Company—opened up about five to seven miles south of Portage Lake. The well-capitalized Copper Range mines developed very rapidly. They quickly surpassed Quincy's production and aimed to challenge C&H. Baltic, Trimountain, and Champion had some key advantages because they were new and shallow, whereas the long-established producers were among the deepest mines in the United States. Many of their lodes at great depth became harder to drill and yielded less copper per ton. The older mines suffered high production costs, in part because they now worked the nation's lowest-grade copper deposits.[35] While Quincy coped with air blasts and C&H with a diminishing copper lode, the Copper Range mines coped with the "problems" of new development and rapid growth.

12

❖ ❖ ❖

AN IMPORTANT FIND

The Copper Range Mines Open the Baltic Lode

Throughout the history of the Lake Superior copper district, "families" of investors tended to dominate production. Sometimes the capitalists literally were family through marriage or kinship—and sometimes they were just from the same city or had long done business with one another. They often invested in numerous companies, hoping that at least one would become highly profitable. This pattern started early, when an investment group led by Thomas Howe and Curtis Hussey struck it rich at the Cliff mine and formed another seven mines by the 1860s. Thomas F. Mason headed another investment family that controlled Quincy and another eight companies during the Civil War era. The Quincy A. Shaw and Alexander Agassiz group controlled Calumet and Hecla (C&H), while a different cluster of Boston investors, led by Horatio and Albert Bigelow, controlled the Tamarack and Osceola mines and several others.[1]

In the late 1890s a new investment family, headed by John Stanton and William A. Paine, arrived on the scene. Stanton, the son of a mining engineer, came to the United States two years after he was born in England in 1830. Growing up in the east, he kicked around coal and iron mines where his father worked. About the time the Keweenaw opened up, he became interested in copper deposits in several states to the east and south, and in the early 1850s he managed copper mines near Ducktown, Tennessee, and built a copper smelter there. The Civil War interrupted that phase of his career, and he relocated to New York City. In 1864, while still in New York, Stanton became secretary of Keweenaw County's Central mine. He then went on to invest in and serve as secretary, treasurer, or president of several other Lake mines: the Atlantic and Winona mines, south of Houghton; the Wolverine, Mohawk, Allouez, and Phoenix Consolidated mines, north of Calumet; and the Michigan mine at Rockland in Ontonagon County. Stanton partnered with fellow investor and fellow New York City resident, John Gay, to form the Stanton-Gay investment family that controlled many of these mines.[2] In the late 1890s, Stanton paired with an even more important partner, William A. Paine.

A Bostonian, Paine started his business career by working at a shoe store. A fire forced him to move out of that business, and he took a job at a bank, where he worked for eight years at several different positions. Believing that banking had grounded him in matters of money and business, Paine left his bank and joined with Wallace Webber to form the investment firm of Paine, Webber, and Company. Webber soon left the firm, leaving Paine to develop it into one of the nation's most important brokerage houses. When once asked what his business philosophy was, Paine answered, "My philosophy is simple: The great thing in business is to know what you want to do. Then do it." In the 1890s Paine wanted to invest in the Keweenaw, and he did it. By the 1920s Paine-Webber and Michigan copper made William Paine one of the richest men in New England.[3]

Two related developments on the Keweenaw in the late 1890s brought Stanton and Paine together: the opening of the Baltic amygdaloid lode about six miles south of Houghton, and the building of a new railroad from Houghton down into Ontonagon County that ran alongside the Baltic lode. Mining and railroading went together, and soon Stanton and Paine headed a web of interconnected companies that owned several new mines, mills, and a smelter—and the rail line that tied them together. By 1904 the Stanton-Paine group already controlled 30 percent of Michigan's copper production, second only to C&H's 39 percent.[4]

In 1882 John Ryan discovered the Baltic lode on property owned by the St. Mary's Canal Mineral Land Company. He leased ground from the land company to try to open a paying mine, but the attempt failed. After Ryan's land lease expired, the lode sat undisturbed until the early to mid-1890s, when a few other faltering attempts were made to mine it. By 1897 some pits exposed promising copper values, and John Stanton and other investors—who owned the nearby Atlantic mine—formed the Baltic Mining Company. Baltic purchased eight hundred acres of mineral lands along the lode and started work at the site in December 1897, under the supervision of Frederick W. Denton.[5]

At about the same time, another project that had been long been hanging fire finally received needed investment capital. In the late 1880s Charles Wright, a railroad manager, had searched for financing to build a railroad from Houghton down into Ontonagon County. This stretch of the mineral range was little developed, and a railroad, Wright thought, would encourage mining and lumbering along the route. In 1898—just as the new Baltic mine was starting up along the tentative route of the railroad—William A. Paine came aboard as an investor, and in January 1899 he helped capitalize the Copper Range Company, which was set up as a holding company. Its holdings included the new Copper Range Railroad, a wholly owned subsidiary. Another major backer of the railroad was the St. Mary's Canal Mineral Land Company, which saw many benefits to be derived from improving transportation and stimulating economic activity in the area.[6]

While starting a new railroad, William Paine, the Copper Range Company, and the St. Mary's Canal Mineral Land Company also started a new mine firm, the Champion Copper Company, which occupied twelve hundred acres of land three miles south of the Baltic mine. The Copper Range Company and the land company each owned half of Champion, and Copper Range managed its operations. Paine lured Dr. Lucius Hubbard away from his post as Michigan's state geologist and gave him the task of finding the southern extension of the Baltic lode and of putting the Champion mine on top of it. Hubbard's work was well under way by May 1899.[7]

While the Baltic and Champion mines started up on the northern and southern ends of the Baltic lode, a third mine squeezed into the middle. The Trimountain Mining Company was backed by Boston financiers led by H. F. Fay. Incorporated in 1899, Trimountain started mining after purchasing 1,120 acres from the St. Mary's Canal Mineral Land Company.[8] So by 1899–1900, three new mines sat atop the Baltic lode. The Baltic, Trimountain, and Champion were separate companies but shared the same lode and faced identical challenges and tasks. They cleared forests, built roads, removed overburden, and sank shafts. Each company hired a labor force and built a full range of mine structures plus a first wave of company housing. Each mine also made temporary arrangements for other companies to mill and smelt their product.

Baltic, Trimountain, and Champion for a brief period triplicated all their efforts. William A. Paine recognized that exploiting the Baltic lode could be made more efficient if these three operations came under the umbrella of one company so that expertise, transportation, management, and technologies could be shared where practicable. In 1901 Paine broached the idea to Stanton, who saw the wisdom of connecting his Baltic mine to Paine's Copper Range Company, its railroad, and the Champion mine. In December 1901 Paine ushered in a new Copper Range Consolidated Copper Company, and before the month ended, Copper Range Consolidated absorbed Stanton's Baltic Mining Company by swapping shares of its new stock on a one-for-one basis with Baltic stock. Similarly, investors in the original Copper Range Company received one-and-a-half shares of Consolidated stock for each share they held in the original company. To raise capital needed for future development work, Copper Range Consolidated also offered all stockholders up to 35,000 shares of its stock at forty dollars each.

Bringing Trimountain into the fold was not so easy. That company's directors felt they had options. If they delayed consolidation, perhaps they could negotiate a better deal than Baltic stockholders had received. They could remain an independent company or sell out at a higher price to some party other than Copper Range Consolidated. Its first few years made Trimountain overconfident. Its stretch of the Baltic lode was well charged with copper and sustained high production. But in 1903 the optimistic young company paid out too much in dividends at a time when its production suddenly faltered. Trimountain found itself cash short and in debt. In August 1903 Trimountain approached Copper Range about joining the consolidation. Copper Range Consolidated took in Trimountain on the same one-for-one stock swap that Baltic's shareholders had received. But consolidation, for Trimountain's stockholders, came at a price. To close the deal, they had to pay Copper Range Consolidated $850,000, the amount of Trimountain's indebtedness.[9]

Even with consolidation, Baltic, Trimountain, and Champion triplicated many efforts. Each mine needed its own shafts with rock-breaking machinery, hoists, and boilers. Each built its own stamp mill. But the companies shared some facilities: they built one machine shop and smithy, one high school, one hospital. They coordinated rail operations between the mines and their mills that stood near one another on the Lake Superior shore. And in 1903 the Stanton-Paine group incorporated the Michigan Smelting Company, whose plant would smelt the product of numerous mines. Stanton's Wolverine, Mohawk, and Atlantic mining companies held eight thousand shares of the smelter company; Paine's Copper Range Consolidated mines held the remaining twelve thousand shares.[10]

The Baltic amygdaloid lode was the last major *native* copper deposit to be discovered. It outcropped in places at the Baltic and Champion mines, but at Trimountain, glacial drift covered it with two hundred feet of overburden. The lode dipped at a steep 70 degree angle, and underground, faulted in places, it displaced to the side, making it hard to follow.[11] The copper was not uniformly disseminated but occurred in "irregular patches" separated by barren zones. Sometimes miners found the copper in the hanging side of the lode and sometimes closer to the footwall.[12] The stopes on the Baltic lode ran from sixteen to twenty-six feet wide and yielded native copper from the size of small flakes up to masses weighing several tons. Early on, the lode yielded only fourteen to seventeen pounds of copper per ton of rock. That low yield meant that, to profit, the Copper Range mines had to cull out poor rock and separate it from the stream of copper rock being hoisted. Another problem was that the hanging wall above the lode did not stand up if left unsupported for any distance. To cope with a low yield and a weak hanging, the Copper Range mines developed a new "Baltic method of mining."[13]

Baltic, Trimountain, and Champion each sank their double-tracked shafts about 1,000 to 1,500 feet apart.[14] At one-hundred-foot intervals, horizontal drifts ran out from the shafts and through the lode, preparing ground for stoping. In most Lake mines, after miners blasted up in the stopes, all broken rock rolled down to the drift and men trammed it away. The stopes were left open and the hanging largely held itself up. The Copper Range mines did not do it this way. They sorted rock in the stopes, left poor rock behind, and trammed only higher-grade copper rock to the shafts. Because they did not want to leave the hanging wall unsupported, they used poor rock to prop it up. Besides protecting men from possible rockfalls, the rock-filled stopes kept the weak hanging from dropping additional poor rock onto the mine floor and diluting the lode's copper yield.

The Baltic lode was very steeply pitched and, to exploit it, Copper Range hired wallers and pickers and devised the Baltic method of mining. This drift has been walled on both sides, and then timbermen built a heavy roof of logs and thick planks over it. Directly above this roof, miners drilled and blasted out the lode, and pickers sorted the broken rock. They passed good copper rock through chutes down to the drift to be trammed away; they left poor rock behind in the stope, supported by the drift's roof, shown here. *(Michigan Technological University Archives and Copper Country Historical Collections)*

On the Baltic lode, as miners started to raise a stope, *wallers* used poor rock to lay dry-stone walls along both sides of the drift. Copper Range favored Italians and Croatians for wallers, believing that they had a "natural liking for it." They had done similar work "in their native countries, where the building of dry-walls about farm and home is common practice." William Schacht, Copper Range's general manager, noted that "the class of work done by these men is quite remarkable."[15] The wallers laid the poor rock into walls four feet thick and eight-and-a-half feet tall. Then timbermen roofed the drift to create a platform to hold poor rock up in the stope. Men set logs into niches in the tops of the walls, running the logs from one wall across the drift to the opposite side. Laying thick planks over the logs completed the roof. When fully walled and roofed, a *drift* (gang-way) measured seven feet wide and seven-and-a-half feet tall. About every twenty-five or thirty feet along the drifts, the wallers put a circular chute or mill into their stonework on the footwall side of the drift. These mills were about five feet in inside diameter, and like the walls themselves, were laid up of poor rock, four feet thick. As stoping above the drift progressed, wallers continued to build up the chute, raising it higher, rather like a chimney. While the roof of the drift supported an ever-deeper pile of poor rock in the stope, men dropped good copper rock down the chimneylike chute; it fell to the drift to be trammed away.

The Baltic method of mining did little to change the work of miners. Using two-man drills and high explosives, they conducted regular stoping, working from one drift up toward the next. But miners shared their workspace not only with wallers, but with *pickers,* who separated poor rock from good copper rock, and with *fillers,* who spread the poor rock across the stope to produce a flat floor. The pickers worked close behind the miners, so the pickers could examine broken rock before it became too covered with dust. As they worked over a pile of rock, piece by piece, they were not expected to measure exactly how much copper each piece contained. They kept it simple. If they saw no specks of copper, they rejected the rock as waste. If they saw any copper at all, they loaded the rock into a wheelbarrow or into a car running on light-rails that could be easily moved from place to place. They pushed the rock to the nearest mill or chute and dropped it down to the drift. The pickers worked on two-man teams under the supervision of a boss who traveled from stope to stope, checking on five to eight picker teams per shift.[16] Roughly speaking, the pickers sent off to be hoisted about half the rock broken by miners and left behind the other half as waste to be spread across the stope floor by the fillers. Obviously the pickers missed some copper, but by being selective in sorting rock underground, Copper Range saved money in several ways. Because it used poor rock to support the weak hanging, it saved on timbering. It also saved by not having to pay to tram, hoist, and mill rock that would yield little or no copper.

The bottoms of the chimneylike mills stood four-and-a-half feet above the floor of the drift, so they delivered rock at the perfect height to be dropped straight into tramcars. When the Copper Range mines opened, the copper district was experimenting with mechanical and electrical means of tramming. The Baltic, Trimountain, and Champion mines made no breakthroughs in tramming technology; like nearly everybody else, they used men as beasts of burden to push loaded tramcars to the shafts for hoisting. While Quincy made significant use of electric haulage locomotives underground starting in 1901, Copper Range trammed its rock by hand until Baltic installed two electric haulage locomotives underground in 1907; Champion and Trimountain did not widely adopt the new technology until 1914.[17]

Because the Copper Range mines were new and shallow, they did not require the mammoth hoists that older, deeper mines used. Sometimes, when first sinking a shaft, the mines made do with temporary boilers or engines. When a shaft proved to be a producer, they added more permanent facilities. The mines did not always buy new or buy big. The Champion mine's first hoists could only carry 1,500 feet of hoisting rope. By 1905, at its "D" shaft, Champion had installed a Nordberg duplex hoist (with 24- by 60-inch cylinders) that carried two 3,000-foot-long hoist ropes on its 14-foot drum with conical ends. By 1910 Champion had installed essentially the same Nordberg hoist at

Four steel-frame shaft-rockhouses stand in a line at the Baltic mine in 1911. Similar vistas could be found at the Trimountain and Champion mines, also launched near the turn of the century on the Baltic lode. The steep slope shown on the left (backside) of the structure in the foreground mirrors the steep dip of the lode itself. *(Michigan Technological University Archives and Copper Country Historical Collections)*

all four of its shafts. Trimountain obtained several of its hoists used from the Arcadian mine. One of these, however, was a "practically new" 2,500-horsepower Nordberg hoist with 36- by 72-inch cylinders that could "raise a 6-ton skip 1,000 [feet], dump the rock and lower the skip to whence it started in 60 seconds." While Trimountain was buying used engines, Baltic purchased fully modern engines from Nordberg and from Allis-Chalmers.[18]

From 1898 until 1903 or so, Baltic, Trimountain, and Champion spent conservatively to outfit each of their four or five new shafts. They constructed shaft-rockhouses that covered their initial rock-handling and breaking needs, and then modified, enlarged, or replaced them within a few years if a shaft had become a good producer. In 1899 the Baltic mine erected its largest shaft-rockhouse to date, with a heavy timber frame, a skin of corrugated iron, two Blake crushers, and a 1,400-ton storage bin. Between 1901 and 1904 Baltic turned to new steel-framed shaft-rockhouses, sometimes erected by the Wisconsin Bridge and Iron Company. In 1902 Trimountain purchased three shaft-rockhouses standing at the Arcadian mine. They disassembled the structures, transported them south, and then re-erected them over the Baltic lode. By 1903 Champion had four identical shaft-rockhouses standing over its shafts, each shaft-rockhouse equipped with a steam hammer and three Farrell jaw crushers.[19]

Champion considerably upgraded its shaft-rockhouses over the mine's first decade of operation. Underground, the Copper Range mines had developed their own Baltic method of mining to cope with the peculiarities of the Baltic lode. Within its shaft-rockhouses, instead of copying what other mines did to handle, sort, and break rock, the Champion mine devised its own way of doing things.[20] The devil was in the details. All mines had access to the same machinery and tools. At Champion the trick was to arrange them in a manner best suited to its conditions and to keep rearranging them until desired results were achieved—such as reducing labor costs. Champion, over several years, succeeded in harnessing gravity feeds to greater advantage within its shaft-rockhouses and stopped dropping tons of rock onto flat floors—rock that men picked up and moved. In 1906 a rockhouse worker on the crusher floor had "about the hardest job around the mine" because he had to sort, pick up, and move great tonnages by hand. By rearranging bins, chutes, aprons and jaw crushers, and by adding air lifts for moving material around, Champion made rockhouse labor less arduous, and the company saved money at the same time. In 1906, six to eight men worked on the crusher floor at a Champion shaft-rockhouse, and it cost the company 9.7 cents per ton of rock handled; by 1910, only two men per shift handled all the copper rock hoisted out of a shaft, at a reduced cost per ton of 5.5 cents.[21]

Within a 1910-era Champion shaft-rockhouse, two skip roads emerged from the mine on a steep 70 degree angle and rose to the top of the structure. Above the shaft collar, small cranes attached to the steel framing helped men swap man-cars for rock skips—or vice versa—on the ends of the hoisting ropes, and they held unused shaft vehicles up in the air and out of the way. Because the shafts were so steep, the man-cars did not carry seats. Men rode standing up in double-decker fashion—one group stood on a steel floor that separated them from the men riding below.

Champion's pickers did all of the rock sorting underground, so the skips arriving at the surface contained all poor rock or all stamp rock. The poor rock was dumped into a small bin on a lower level of the structure. Periodically, men emptied the bin by opening gates that sent the poor rock into small tram cars on tracks. They pushed the cars outside the structure on a trestle and then dumped their contents over an inclined grizzly, whose bars were only one-and-a-half inches apart. The poor rock that passed through the grizzly was caught in a bin to be used for road construction or concrete aggregate. The oversized rock that tailed off the grizzly fell into a *raise* (a shaftlike opening) that sent it right back into the mine to fill stoped out ground and help support the hanging wall.

The hoist carried copper rock higher up the structure before dumping it into a storage bin. Champion's revamped shaft-rockhouses did not use grizzlies to sort copper rock by size: "On leaving the skip, the rock drops directly into a large bin, capable of holding about 20 or 25 skip loads. . . . In the front side of this bin are two sliding doors, one in front of each [jaw crusher], each operated by an eight-inch air lift." When the doors were raised, rock slid out of the bin onto two inclined aprons, only four feet long, whose lower ends rested on the top of two jaw crushers. A closed gate mounted at each jaw crusher stopped the flow of rock there, and men closed the door from the storage bin. A load of rock sat stationary on the short apron while men picked over it, selecting out mass copper or oversized rock. They placed these onto pans lifted up by a small air lift that ran on an overhead trolley and put them beneath a drop hammer. After being raised up by a small electric hoist, the hammer dropped onto the material below. Men dropped pieces of shattered rock through a hole in the floor to send it to the stamp-rock bin; with the help of a jib crane affixed to the outside of the building, they lowered mass copper down the side of the building to be loaded onto flatcars for shipment to a smelter. After picking out oversized rock and mass copper, rockhouse workers opened the gate at the mouth of the jaw crusher, and the rest of the load slid into that machine. Each crusher handled five hundred to one thousand tons of rock daily, breaking it to three- or four-inch size and discharging it into a stamp-rock bin. Loading chutes then delivered the rock into railcars.

The Copper Range Railroad delivered copper rock from the Baltic, Trimountain, and Champion mines to their respective stamp mills on the shoreline of Lake Superior at Redridge, Beacon Hill, and Freda. All three mills had been constructed before consolidation, so while they exhibited considerable sameness, they also exhibited interesting differences. The Baltic Mining Company built its mill first. The Stanton-Gay investors who controlled Baltic also controlled the much older Atlantic mine south of Houghton. Since the early 1870s the Atlantic had operated a stamp mill on Portage Lake, a few miles west of Houghton near Cole's Creek. By the early 1890s the Atlantic couldn't continue dumping stamp sands into navigable Portage Lake so it acquired a new mill site where Salmon Trout Creek flowed into Lake Superior. Its new mill began operation there in 1895, with its new neighborhood of worker housing standing nearby at Redridge.[22] Instead of using steam pumps to provide water for the mill, Atlantic threw a timber-crib dam across Salmon Trout Creek and piped impounded water to its mill.

When the Baltic mine began production in 1899, it sent its copper rock to the Atlantic mill to be stamped. Then it put up its own mill, right next door, which started up late in 1901. The timber dam was not tall enough to impound water for two mills, so the Atlantic and Baltic companies jointly built a new dam, located just behind the original one. They chose a dam of unusual design, only the second of its type to be built in the United States: a steel dam. The new Redridge steel dam, designed and built by the Wisconsin Bridge and Iron Company, stood on a large base of eight thousand cubic yards of concrete. Above that, everything was steel I-beams and steel plates. The dam's central section ran 464 feet long and 74 feet high. This dam put the original timber-crib dam under twenty feet of water and created a reservoir with a surface area of 150 acres. Large iron penstocks carried water off to the two mills.[23] The Baltic mill had two Nordberg compound steam stamps and two E. P. Allis stamps by mid-1902. It added two more Allis stamps by 1907, giving it a stamping capacity of four thousand tons daily. The tailings washed out into Lake Superior.[24]

The Trimountain mine, when it initiated production early in 1902, leased two stamp heads at another company's mill: the Arcadian. That arrangement ended in about a year when Trimountain's new steel-framed mill went into service at Beacon Hill. The mill, like all of those that became part of Copper Range, received stamp rock at the top end of the structure and then had a step-down design as it stretched out toward Lake Superior so that gravity carried copper, rock, and water from one step of the milling process to another. Adjacent the mill stood a boilerhouse and a pumphouse equipped with a twenty-million-gallon Nordberg pump. The Trimountain mill drew its water from Lake Superior. To avoid intake problems that might be caused by winter ice or stamp-sand deposits near the shore, Trimountain laid fourteen hundred feet of forty-inch riveted steel pipe out into the lake. The first seven hundred feet of the intake pipe ran in a trench blasted in the lake bottom. The second seven hundred feet ran on the natural lake bottom until it connected to a heavy-timbered, rock-filled intake crib anchored with heavy iron rods to the lake's sandstone bed.[25]

The machinery in the mill included four Nordberg compound steam stamps, each of which could handle a maximum of 2,500 tons of stamp rock daily. Other facilities included machine, blacksmith, and carpenter shops; an electric light plant; and a cluster of houses at the Beacon Hill townsite.

Before the Champion Copper Company had its mill up and running, in 1902 the company first leased stamp heads at the Atlantic and Baltic mills. The next year, it opened its mill at Freda, the townsite named after William A. Paine's daughter. The Wisconsin Bridge and Iron Company built the steel-and-concrete mill, which measured 178 by 215 feet and provided room for four steam stamps. A railroad trestle delivered stamp rock to the bins at the high end of the structure, and the bins chuted rock to the stamps, each of which stood on a fifteen-foot-tall concrete foundation. The Champion mill had its own way of getting water. Like the Trimountain mill, it had a triple-expansion Nordberg pump that could draw twenty million gallons of water daily from Lake Superior. But instead of laying a long pipeline out into the lake, Champion blasted a 1,020-foot-long tunnel out under the lake to bring in water.

The Champion mill initially housed four Nordberg stamps. Within a few years the mill superintendent believed the powerful steam stamps presented a problem. The copper bound up in the stamp rock was irregularly shaped, with each piece having small projections or *horns*. The stamps, as they broke the rock, knocked off the horns and discharged the copper in the form of rounded pebbles and small flakes. In hopes of keeping the copper more intact, Champion built an addition to its mill in 1905 to house two experimental gyratory crushers and roller crushers instead of stamps. The copper leaving the gyratory crushers passed onto conveyors, where workers picked out the biggest pieces by hand before the copper went to the roller crushers and then jigs and other machinery. Because the gyratory crushers left more small horns intact, the mill would have to invest less time, machinery, and money in capturing fine copper. The experiment, however, did not work. The gyratory crushers did not abrade the copper as much—but they also failed to liberate much of the small copper that left the crusher still bound up in rock. Also, this milling technique required more water, and the larger pieces of copper that flowed through the system tended to clog up certain separating machinery. So Champion replaced the experimental crushers with two compound Nordberg stamps. With a total of six stamps, the Champion mill could treat 3,500 tons of stamp rock daily.[26] Besides its stamps, pumphouse, and intake tunnel, the mill included coal-handling equipment; boilers; a 500-horsepower engine for driving mill equipment; machine, carpenter, and smithy shops; a warehouse and office; a laboratory for testing the copper content of mine rock, mineral, and tailings; a fire pump and hydrants; and a private telephone system. At the adjacent mill town of Freda, Champion built about twenty workers' houses.[27]

The Champion, Trimountain, and Baltic mills sent their mineral to the Michigan smelter on Portage Lake via the Copper Range Railroad. Instead of shipping mineral in barrels, the traditional way, they sent concentrates off in steel, bottom-dumping railcars. These special cars were emblematic of the emphasis that Copper Range and the Michigan Smelting Company put on materials handling. Just as Copper Range worked to improve the flow of materials at shaft-rockhouses, so did the Michigan smelter embody new arrangements and technologies to streamline smelting. All earlier smelters on the Keweenaw sat on level sites, where building foundations stood at the same elevation. Companies had learned the importance of verticality at shaft-rockhouses and stamp mills: take material in on a high level and then feed it down the structure from process to process. They hadn't learned to do the same thing at smelters until the Michigan smelter came along in 1903–4. Located between a hillside and the shoreline of Portage Lake—where the Atlantic mine's old stamp mill had once stood—the smelter was terraced. It rather resembled a stamp mill in the way its buildings and floors stepped down from top to bottom, facilitating the flow of materials.

George Schubert, superintendent of the smelter, wrote that they "planned the erection of a plant embodying the best of the essential and established characteristics of native copper refineries, and added new and distinct features designed to eliminate much of the manual labor."[28] Copper Range did not rely on local experts to design its smelter. The company turned to outsiders associated with the Anaconda smelter in Montana: F. I. Cairns, who had designed that smelter, and Frank Klepetco, who had superintended its operations. When the new works started up in 1904, things did not go off without a hitch. Some of the furnace designs had to be tweaked to improve their performance. But after some fine-tuning, the Michigan smelter was the largest and most efficient on Lake Superior, with a capacity of 90 million tons yearly.[29]

High-grade Lake copper sold at a premium above the electrolytic copper produced in the west because it was higher in silver content but lower in undesirable impurities. The mineral coming to the Michigan smelter had a somewhat higher arsenic content, so to maintain a premium price for its product, the smelter had to do some additional refining to eliminate the arsenic. Besides that, on the lake the Michigan smelter pioneered in

Copper Range's Michigan smelter, which poured its first copper in 1904, did not look like earlier smelters in the region, which were built on flat sites. Designed by experts from the Anaconda smelter in Montana, the Michigan smelter had a terraced, step-down design, like local stamp mills. Railcars rolling over a high trestle delivered copper mineral, plus fuels and fluxes, to the top of the smelter, and gravity and a host of materials-handling devices facilitated their flow through the process. *(Michigan Technological University Archives and Copper Country Historical Collections)*

using separate melting and refining furnaces instead of performing both operations in one furnace.

Forty-ton railcars rolling along a high trestle at the top end of the smelter delivered the mineral. Each shipment was weighed, chemically analyzed, and then stored in one of ten 150-ton steel storage bins. The facility required numerous bins and several furnaces, not just to handle a large tonnage, but to be able to take in the mineral from several mines. The smelter served not only Baltic, Trimountain, and Champion, but other mines to the north (Mohawk and Wolverine) and south (Atlantic, Winona, and Michigan). The smelter kept each mine's product separate from the others. Besides delivering mineral in bottom-dumping cars, the railroad delivered mass copper on flatcars and also transported coal, coke, limestone, and iron ore to the smelter. Because the storage bins were elevated above the rest of the site, all materials could be delivered by gravity or transfer tracks and cars to their proper destinations with relative ease.

The smelter had three melting furnaces that received the copper mineral. It took approximately two hours to charge a melting furnace, about eighteen hours to melt the charge, three hours to skim it, and a half hour to tap out its 180 tons of molten copper. Within these furnaces, crushed coal, crushed limestone, and iron ore were also part of the furnace charge: they served as fluxes to help separate the copper from the slag formed by melting rock. Furnacemen opened a door at the front of a melting furnace to skim off the molten slag floating atop the copper. That slag invariably contained some copper. Once the slag solidified, it ran through a crusher and was then sent to a separate blast furnace, which remelted the slag along with charges of mass copper received by the smelter. The blast furnace discharged an unrefined *black copper* that was returned to a melting furnace to be remelted and then refined.

After skimming off slag, furnacemen tapped the molten copper, which flowed through troughs or launders into one of two large refining furnaces, where impurities (such as arsenic, sulfur, and iron) were removed by treating the melt with a charge of soda, and by the traditional means of rabbling and poling. The refined copper was tapped into a large casting ladle moved by hydraulic controls; when the ladle tipped, it poured its molten copper into molds held by the region's first semiautomatic Walker casting machine, controlled by a man sitting in the middle of a rotating ring of ingot molds. Because different customers wanted their copper in different sizes and shapes, the casting machine could be fitted with a set of any of forty-six different molds.

Much of the smelter's product left the region in lake boats. The Copper Range Railroad transported the ingots only a mile or so east to its shipping docks, just on the outskirts of Houghton. Here, the Portage Lake shoreline was a busy place. The Copper Range Railroad's main passenger terminal dealt with human travelers. Lake boats took aboard copper that had been stored on the dock or in a large warehouse along the water. Other lake boats discharged coal at a large coal-handling facility. And near the coal docks stood the Copper Range Railroad's roundhouse, with fourteen bays for servicing locomotives and rolling stock. The railroad had put five new Baldwin locomotives into service in 1899. By 1910 it had about twenty-two engines of different makes and upward of one hundred miles of track.[30]

In their first years the Copper Range mines were nearly on an equal footing. In 1902, Baltic produced 6.3 million pounds of copper while Trimountain produced 5.7 million pounds and Champion produced 4.2. But their fortunes started to veer apart in short order. Trimountain proved the biggest disappointment. Its highest level of production came in 1905, when it sent 10.5 million pounds of copper to market. By 1910, Trimountain had the deepest shafts on the Baltic lode, with its shallowest at 1,806 feet and its deepest at 2,453 feet. But by that date, it was back to 1902's production level of 5.7 million pounds. Trimountain had already seen its best days and contraction soon set in. Through 1910 it produced 70 million pounds of copper and paid out $950,000 in dividends.[31]

Through 1910, Baltic produced a total of 132 million pounds of copper and paid out $6.5 in dividends. But Baltic, too, peaked at a young age: its production in 1910 of 17.5 million pounds of copper, pulled from four shafts 1,522 to 1,882 feet deep, proved to be its highest annual product ever. Afterward, the Champion mine eclipsed Baltic.[32] In 1910, when its four shafts were between 1,941 and 2,108 feet deep, Champion's output

Copper Range's transportation facilities occupied much of the busy Portage Lake shoreline west of Houghton. The railroad roundhouse for servicing steam locomotives occupies the foreground; beyond that, coal unloaders stand at the dock; and at another dock farther away, men load copper ingots onto lake boats. *(Michigan Technological University Archives and Copper Country Historical Collections)*

amounted to 19.2 million pounds. Through 1910, Champion lagged just a bit behind Baltic, with a total production of 131 million pounds of copper and dividends of $5.9 million. But while Trimountain's production peaked in 1905 and Baltic's in 1910, the Champion mine's production did not peak until 1916, when it produced 33.6 million pounds in a single year and paid out its all-time high of $6 million in dividends.[33]

In 1880, the native copper mines of Michigan produced fifty million pounds of copper. That figured doubled to 101 million pounds in 1890. In 1895, 1900, and 1905, Michigan's production rose considerably, from 129 to 145 to 230 million pounds annually, before falling a bit to 221 million pounds in 1910.[34] The rise in Michigan's total production was partly due to the new Baltic, Trimountain, and Champion mines coming on line, and partly due to production from other new or reorganized mining companies up and down the mineral range. The Wolverine mine, north of Calumet, had never been an important producer in the 1880s. Reorganized in 1890, it made about five million pounds of copper annually by 1901 and ten million pounds by 1908. Nearby, the Mohawk Mining Company, organized in 1898, made 6 to 11.2 million pounds of copper annually between 1903 and 1910. Just south of Houghton, a new Isle Royale Copper

Company, incorporated in 1899, took over several failed or failing properties and produced a high of 7.6 pounds of copper by 1910. Down in Ontonagon County, three new firms that organized in 1898–99—the Adventure Consolidated, the Mass Consolidated, and the Michigan Copper Mining Company—assumed control of numerous properties worked earlier in the century. These three new firms never became long-term successes, but early in the twentieth century each contributed about two million or more pounds of copper to Michigan's production.[35]

While new mines started up, others closed down. Perhaps the greatest loss was the Atlantic mine, which from the mid-1880s through 1905 had produced three to five million pounds of copper annually. Then, because of cave-ins, the mine closed suddenly in 1906.[36] Overall, there was no dramatic increase in the number of active mines in the district. Fifteen mines operated in 1890. That number fell to twelve in 1895 and recovered to fourteen in 1900. In 1905 and 1910, eighteen and then nineteen companies recorded production. Although the mines did not become much more numerous, they did become bigger. About fifteen of the mines operating in 1885 produced less than a half-million pounds of copper per year. That same year, only four companies produced more than three million pounds. Between 1890 and 1910 the number of mines producing less than a half-million pounds per year fell to three or fewer, while a dozen companies produced at least three million pounds per year, including eight that produced more than seven million pounds annually.[37] Employment figures mirrored the production growth. Houghton County's total mine, mill, and smelter employment, which was something less than 5,000 in 1880, reached 7,310 in 1890 and 13,971 in 1900, and peaked at 17,974 in 1909.[38]

During this era the district saw some lean years in terms of profits, but also some very rich ones. The mines' fortunes rose and fell with the price of copper. In the early 1890s the entire American economy suffered poor years and so did the copper market. Prices declined from 15.8 cents per pound in 1890 to 9.6 cents in 1894, which was the lowest price ever paid for Michigan copper. Prices rebounded strongly, however, near the turn of the century, hitting 16.6 to 17.6 cents per pound in 1899–1901. Prices then declined to 15.7 cents in 1905 and just 13 cents in 1910. Dividends tended to rise and fall along with copper prices. The district as a whole paid out about $3.3 or $3.4 million in dividends annually in the early 1890s. With the price spike at the turn of the century, the mines enjoyed their most profitable years to date, hitting over $12 million in dividends in a single year, with Quincy and C&H accounting for most of this. With the drop of copper prices to just 13 cents per pound by 1910, dividends fell to $6.9 million.[39]

Despite the surge in terms of production, employment, and dividends, the mines in some important ways fell behind. They did not expand as rapidly or as much as the overall American copper industry. Michigan had once absolutely dominated that industry but had lost its mantle to Butte, Montana, "the richest hill on earth," in the late 1880s. Michigan produced 84 percent of the nation's new copper in 1880, when no other state produced more than 5 percent and Montana produced only 2 percent. But by 1890 Butte accounted for 42 percent of America's production, and Lake Superior dropped to 38 percent. That percentage continued to fall with each passing decade: by 1900, 24 percent; by 1910, only 20 percent. In 1910, Michigan was only the third largest production state, trailing Montana and Arizona. Other states accounting for at least 5 percent of the nation's production by 1910 included Utah and Nevada, with California just missing that mark.[40] While many indicators of industrial well-being had risen after 1890, other signs, especially the closer one got to 1910, pointed toward troubling times ahead.

Importantly, shortly after the turn of the century, C&H, Quincy, and Copper Range all operated under college-educated superintendents. Practically trained men—who had worked their way up the ranks to become managers—had had their day; now the day belonged to academically trained engineers. At C&H, that engineer was James Mac-Naughton, hired in 1901 by the company patriarch, Alexander Agassiz. MacNaughton had graduated in civil engineering from the University of Michigan. In 1905 Charles Lawton, with a degree from the Mechanical Department of the Michigan Agricultural College (now Michigan State University), took the reins at Quincy. That same year, Frederick W. Denton took charge of Copper Range's mines. Denton had graduated from the Columbia School of Mines and had taught after 1890 at the mining school in Houghton.[41] The new leaders were modern practitioners of mining. They believed in cost accounting, in efficiency, and in the essential need to increase worker productivity. When they looked underground at their mines, they believed that, to remain competitive, new methods had to be substituted for some traditional practices and technologies. One of their mutual goals was to trim labor costs by cutting the number of miners employed. Their desire to eliminate miners drove them to champion the one-man drilling machine, which in 1913 would stir up a hornet's nest of labor unrest.

13

❖ ❖ ❖

PATERNALISM REVISITED

The Baltic, Trimountain, and Champion Mines

By the time the Copper Range mines opened, the Lake Superior copper industry had a half-century-long history of constructing houses, providing doctors and hospitals, supporting churches and schools, routing roads, and picking up the garbage. The Baltic, Trimountain, and Champion mines did not break with tradition. Like their predecessors, they engaged in a wide range of paternalistic practices.

Companies in the 1840s and 1850s pushed back the wilderness to plant small settlements on the mining frontier. The Copper Range mines faced the same task. Although parts of the Keweenaw were heavily industrialized and populated by 1900, the Baltic lode ran beneath a heavy forest. Reportedly, only one log house stood within three miles of the Baltic mine when it began working in 1898.[1] In 1899–1900 the earliest workers at the Champion mine blazed 21,260 feet of new roads, cleared timber, scrubbed out underbrush, dug exploratory pits and trenches, and began removing the overburden from atop the lode.[2] To shelter these men, the mines erected rude *camps,* or temporary bunkhouses that slept about fifty laborers. A bit later, they put up many men in boardinghouses, which were better constructed and furnished. Then, as full production got under way, each company built from 150 to 250 family dwellings.

Julia Hubbard, the daughter of Lucius Hubbard, who had helped track the Baltic lode and then managed the Champion mine, described the earliest bunkhouses at that mine:

> The camp . . . was one enormous room with double bunks around the sides where the men slept. Down the middle ran a long table with benches where

the miners ate, and after supper, played dominoes and checkers or struggled with pencil smudged letters to the folks at home. At the right of the entrance were two huge wood stoves and a storeroom filled with barrels of apples, potatoes and flour, and sacks of beans and whatnot.[3]

Boardinghouses followed on the heels of the rudimentary camps. These wood-frame structures, two stories tall, housed up to seventy men. With sleeping rooms upstairs and common rooms on the first floor, the boardinghouse, as described at the Trimountain mine, was "fully equipped with furniture and utensils."[4] By the time the Copper Range mines erected one or two boardinghouses (sometimes later converted into hotels), they had also laid out street grids and started building family houses.

At Baltic, the company laid out a small grid of avenues and streets, about six running north-south and two running east-west. One avenue honored company officer John Stanton by borrowing his name: Stanton Avenue. The others were simply numbered. In 1898 Baltic rushed into construction two large dwellings for single men, plus fifteen small log houses and five frame houses for families. Presumably the company intended the log structures for "ordinary" married workers and the frame structures for bosses or managers. Construction slowed over the next few years while the company prepared itself for higher production. Baltic abandoned log construction and built eight frame houses in 1899, only four in 1900, and twelve in 1901 (when it also built six boardinghouses). The location's peak building boom was in 1902–3, when Baltic put up fifty and then thirty new houses. Ten additional units went up in each of the next two years as the company's building boom ended. Between 1898 and 1905, Baltic erected 138 houses and 9 boardinghouses or rooming houses.[5]

In 1899 the Trimountain Mining Company ran 10,200 feet of graded roads and cut swaths through the forest to create 12,000 feet of ungraded "winter" roads. To open up lots for housing construction the company put in some nine avenues running northerly and another four streets running perpendicularly. In 1899 Trimountain erected three houses for mine officials, and a mining camp for fifty men, which the company soon replaced with a larger, more permanent boardinghouse. In 1900 Trimountain built fifty more houses, and in 1901 it put up another twenty-four. By 1903 the Copper Range Consolidated Company's annual report stated that 154 company-built dwellings stood at Trimountain, each sitting inside a company-built fence.[6] Like many earlier companies, to save itself construction and maintenance costs, Trimountain encouraged employees to build their own houses on leased company ground. According to a map of the location, by 1908 eighty-five privately owned dwellings stood on company land. They were not sprinkled throughout the location but stood in clusters along select blocks. One cluster of privately owned dwellings had its own name: Italy. This neighborhood, no doubt initially occupied by Italian immigrants, stood off by itself on the far side of the mine proper and across the Copper Range Railroad tracks.[7]

South of Trimountain stood Painesdale and the Champion mine. This mine location grew to become the largest of the Copper Range properties. The Baltic lode, the Copper Range Railroad, and the line of shafts at the mine sliced across the southeastern corner of the site, with schools, churches, and the bulk of the housing lying uphill of the mine, to the north and west. Champion enforced a sense of regularity and order on Painesdale by laying a rectilinear street grid over the northern half of the site, fitting the grid in where it could over rolling, uneven terrain. As the company blazed streets to set off lots, instead of merely numbering them, it named them after people (Hubbard, Hulbert, Goodell) or after other mines (usually smallish, unsuccessful ones, such as Evergreen, Toltec, or Laurium). The northernmost end of Painesdale was four or five blocks wide. Because of variations in the terrain, the grid as it moved south pinched down to a width of only two blocks before expanding again to three or four. The southern half of Painesdale fell off down a steep hill toward the mine and lacked a readily discernible, orderly plan. Little pieces of a street grid could be found here and there, but several curving roads meandered down the hill, where the housing was less concentrated.[8]

Champion built a considerable number of dwellings over a short period, starting with a mine camp in 1899. By the end of 1900, 20 five-room houses, one general manager's house, and three boardinghouses accommodated workers of differing class and marital status. Two of the boardinghouses for single men had a kitchen, dining hall, and sleeping rooms; the third contained no kitchen or dining hall but offered sleeping rooms, a washroom, plus a doctor's office and hospital room.[9] Horace Stevens's *Copper Handbook* reported that eighty dwellings stood at Painesdale in 1901. Thirty additional dwellings went up in 1902 and another thirty in 1903. By the end of 1903, about 154 company dwellings occupied Painesdale, and that number climbed to 250 by 1911.[10] Champion, too, encouraged the building of employee-owned houses on its land. By 1912, fifty-five to sixty privately owned houses occupied lots from the northernmost reaches of the mine location, Adams Street, down to the southernmost neighborhood of Seeberville.[11]

The Baltic, Trimountain, and Champion locations had the look and feel of earlier ones. The houses, mostly small and a story-and-a-half tall, were constructed of wood, with masonry (mostly poor rock) foundation walls. Although larger, two-story boardinghouses stood at the locations, family housing predominated. The architecture of these houses reinforced notions of orderliness and regularity, but not rigid sameness. Runs of identical houses often occupied a given street. Meanwhile, a second street's houses were all alike, too, but different from the first street's houses. Inside, ordinary company houses had some amenities such as plastered walls but lacked running water, bathrooms and toilets, sewers, central heating, and gas or electrical service. Outside, houses along a given street occupied lots of identical size and shared a common setback from the street. Wood fences surrounded most lots, if not all. None of the streets were paved; none had sidewalks.

The Copper Range mine locations did differ in some details from earlier ones. Instead

An early resident wrote that, in laying out Paines-dale, the Champion mine preserved trees to create a more attractive landscape. Historical photos, including this 1938 aerial view, indicate that few trees stood in neighborhoods laid out for ordinary workers. The southern half of Paines-dale, where upper-level and lower-level managers lived, retained more forest. *(Michigan Department of Natural Resources)*

of ruthlessly clearing the surrounding forest, the Champion mine preserved many tall maples at the location.[12] At other mines, large trees hardly existed. The Copper Range locations had no fifty-year-old dwellings standing a few blocks away from five-year-old company houses. The housing all looked new because it was. It was so new, in fact, and so hurriedly put up, that for a few years many houses were founded only on cedar posts sunk into the ground. Only later did Copper Range dig out basements and put masonry foundations under them.[13] Copper Range built more double houses (today called *duplexes*) than did most mines. Because they shared one wall, double houses expedited construction and saved on costs. Instead of siding all houses with painted clapboards, Copper Range cedar shingled many dwellings, especially houses taking the saltbox form. The Copper Range locations, in the first years, also included more boardinghouses than established mines; the companies needed these because no commercial village stood nearby and ready to help house the large number of single men being hired.

Older mine locations displayed a hierarchy of houses that had evolved over decades as one wave of construction followed another. The Copper Range mines exhibited a similar diversity, but their houses did not become different through some long evolutionary process. The hundreds of houses standing at the Copper Range mines had been constructed in ten years or less. They differed because the mining companies believed in difference. They had a hierarchy of workers—who needed to be put up in different classes of houses. Housing at Painesdale indicated an employee's rank in many ways. Did his house have a masonry foundation, or did it still stand on cedar posts? How big was his house lot? Was it 40 by 60 feet, 50 by 100, 100 by 200, or 250 by 200?[14] Did a man's company house have a porch? Did it have closets in the bedrooms? How about a bathroom, central heating, or electricity? Did it have low or high ceilings? How many rooms or outbuildings did it have?

In 1902, Champion built ten "temporary" houses in its new neighborhood of Seeberville, by its "E" shaft, as far from the eventual center of Painesdale as a worker's house could get. When one looks at one of these houses in particular, one sees that the company founded the small, four-room, one-and-a-half story, gable-end dwelling on cedar posts—and it stayed on cedar posts. The company never went back to put a basement or masonry foundation under it. The house was only sixteen feet wide and, as originally built, probably about twenty-four feet long, with just a kitchen and one communal room downstairs. At some point (maybe when running water was added to the house) Champion moved the kitchen to a small addition off the back, covered by a shed roof. The original four rooms in the house were either plastered or planked and wallpapered and had wood floors. The kitchen had printed linoleum or oilcloth on the floor and exposed wooden planks on the walls and ceiling.

Outside, a ragtag fence of wire, horizontal boards, and vertical sticks surrounded the Seeberville house. Champion did not side the dwelling with clapboards or shingles, but with rough, unpainted horizontal boards that did not overlap. At the kitchen addition a few furring strips and nails held up the tar paper that covered the kitchen's exterior to limit cold, wintertime drafts. Besides the kitchen, a bedroom (fifteen by eleven feet) and dining room occupied the first floor, and two more fifteen- by eleven-foot bedrooms took up all the second, where a steeply pitched roof cut into the ceilings and left little headroom. A lone window under each gable end provided all the natural light and ventilation that the cramped bedrooms received. In 1913 seventeen people lived in this house: a Croatian miner, Joseph Putrich; his wife, Antonia; their four children, all under the age of four; a hired girl; and ten male boarders, all Croatians. Putrich's house, and others like it in Seeberville, constituted Painesdale's poorest class of company housing.[15]

Houses of various sizes and types clustered in other Painesdale neighborhoods.[16] Champion built four-room front-gable houses that were 16 feet wide and 26 feet long. They had a dining room/living room and kitchen on the first floor and two bedrooms upstairs. Champion built another style of four-room house: narrow saltboxes that were only 16 feet 4 inches wide and 26 feet 3 inches long. It also put two saltboxes side by side to make a double house, with each unit having four rooms. The double saltboxes were 30

Some of the Champion mine's poorest workers' houses occupied the Seeberville neighborhood near Painesdale. This house stood on cedar posts, and its skirt of vertical boards was intended to keep cold winter winds from whistling under the house's main floor. Rough, unpainted planks—not clapboards or shingles—sided the exterior walls. Occupied by a Croatian family and many Croatian boarders, the house became a murder scene during the strike of 1913–14. *(Michigan Technological University Archives and Copper Country Historical Collections)*

feet wide and 32 feet deep. Sometimes they made the each side of a double saltbox form into a six-room house by adding an additional wall to break up the front, downstairs room into two rooms: a parlor and separate dining room.

Champion erected many houses where the front door, instead of being located under the double-pitched gable, was located in a long wall of the house, under a straight line of eaves.[17] One common *side gable* house, 25 by 21 feet, had six rooms: sitting room, dining room, and kitchen, with a pantry downstairs and three bedrooms above. These side-gable houses were a cut above the four-room houses. Besides offering more rooms, they had entrance landings at the tops of the outside steps leading up to the front and rear doors. They also had outside cellar doors that led to full basements, and more windows to let in natural light (six on the first floor and four on the second).

Champion built seven-room houses of the front-gable type; these were a full two stories tall and could be expanded to eight rooms if occupants wanted an additional bedroom in the attic. They had an entrance landing at the front, where turned posts supported a small storm-shed roof over the door. When a workers' family entered the front door, off to the left stood the parlor, and behind the parlor stood the dining room. Straight ahead a hallway ran back to the kitchen; to the right, a flight of stairs ran up to the second floor, where four bedrooms and stairs to the attic were located. This house form had select amenities that smaller houses lacked, such as bedroom closets and a full second story with higher, flat ceilings. Besides having a pantry off the kitchen, the seven-room gable-end house also had an attached woodshed off the back, which could be used for storing fuel or as a mudroom.

The earliest Painesdale houses all lacked running water. Champion started piping water into some houses as early as 1902–3, but only into kitchens and perhaps basements. Thanks to upgrades in company pumps, water towers, and distribution mains, by about 1907 all company houses had running water, but workers' houses remained minimally plumbed and without bathrooms.[18] For many years, residents used privies at the back of house lots; they washed up or bathed in the kitchen, like settlers a half-a-century earlier, or availed themselves of public baths. One important gulf that separated a worker's house from a manager's: one had no bathroom; the other one did.

Champion erected its earliest "better" houses for managers at "E" location, so called because it was close to "E" shaft, at the foot of Hubbard Street. One of general manager Lucius Hubbard's three daughters, Julia, remembered that "E" location looked like it had been carved out of virgin forest.[19] A ball diamond occupied a nearby field, but parts of the forest survived. Julia recalled that her mother hung a hammock "between two enormous trees in a lovely wooded spot not far from the house." In 1899–1900 the mine erected twenty houses at "E" location, which were painted gray and enclosed with fences. The houses were described as five-room frame dwellings, 20 by 30 feet, but ten of them carried a 10- by 12-foot addition, and one, occupied by Hubbard and his family, had a 16- by 25-foot addition.[20] Julia Hubbard reminisced about living in this house for a year or two, before her family moved to a grander one, custom designed and built for her father.

Julia remembered that when you entered the front door, a stairway on the right led up to the bedrooms, and the parlor was on the left. In winter the parlor held a nickel-plated stove; a register in the ceiling let heat from the stove pass up and into her parents' bedroom. In summer months the stove was moved out and a piano moved in. Behind the parlor was a combination dining room and family room. This room had the only decent light in the house, provided by an oil lamp that hung over the dining-room table. Behind the dining room was a "nice roomy kitchen" where a live-in domestic cooked for the family. From the back door of the kitchen, "a boardwalk led to a privy."

If the houses at "E" location initially lacked indoor plumbing and bathrooms, these deficiencies were soon corrected after Champion built water towers and ran water lines. The original five-room, 20- by 30-foot house in "E" location evolved into a seven-room, 20- by 46-foot house—and one of the rooms was a full bath, with a sink, tub, and toilet, on the second floor.[21] Other amenities set off these seven-room houses from others at Painesdale. These gable-end structures had front porches. On the first floor, they had a living room, a dining room, and a kitchen with pantry. Two chimneys allowed for a stove in the kitchen and another in the living room. Upstairs, three bedrooms shared space with the bath. Importantly, the house had two sets of stairs to the second floor. The rear stairs led up to the back bedroom—and doors could be closed to isolate that bedroom from the others and from the bath. In other words, this house was set up to allow for servant's quarters.

The men in charge—like general manager Hubbard and mining captain John Broan—only lived in "E" location until Champion built larger, finer houses for them. Champion sited the top managers' houses uphill from "E" location, situating them along Hubbard Avenue and Algomah Streets. These houses included up to twelve rooms. In 1903 the Hubbards moved into a Copper Country mansion designed by architect Alexander C. Eschweiler. The Boston-born Eschweiler, son of a mining engineer, had spent his youth on the Keweenaw amid the mines and ran his own architectural firm in Milwaukee.[22] Eschweiler designed a number of important structures for the Copper Range and other Houghton County clients. The dwelling he drew up for the Hubbards, which stood at 31 Hubbard Avenue, was the biggest, most elaborate house in Painesdale.

Instead of standing on a poor-rock foundation, the Hubbard house, standing on a huge lot, was founded on red Jacobsville sandstone. A porch with a central doorway and pediment dominated the symmetrical façade of the two-and-a-half story, side-gabled house, which had two-story-tall bays flanking the front door, each capped by a steep gable at the roofline. Constructed in a manner influenced by the Craftsman style, the Hubbard house was clapboarded on the first floor and shingled above. Besides sheer size, the house embodied many architectural indicators of Hubbard's status, including brackets under the eaves and corbeled brick chimneys. Also, a two-story carriage house and stable accompanied the main building.

Inside, the Hubbard house was 48 feet 10 inches by 53 feet in plan. The basement contained a boiler room and three coal bins (for boiler, pea, and kitchen coal), plus laundry and drying rooms, a storeroom, and vegetable cellar. On the first floor, a main hall with quarter-sawn oak flooring and a coat closet was flanked on one side by a 24- by 16-foot living room with a fireplace, and on the other side by an 18- by 16-foot dining room with a built-in china closet. At the rear of the first floor, a side veranda occupied one corner and a kitchen, icebox, and pantry occupied the other. In between, off the rear hall, stood a library, a lavatory, and two sets of stairs, one for family and one for servants. Five bedrooms and two full baths occupied the second floor. A commodious master bedroom had one of the bathrooms all to itself, accessed by a private door, and a spacious 8-foot by 3-foot 6-inch closet.[23]

Copper Range grouped six other managers' houses along Algomah Street; each had individual façade treatments and plans.[24] The house at no. 2 Algomah Street was not huge—one-and-a-half stories with only seven rooms—but it carried many amenities

The Hubbard house, named for the first mine superintendent who lived there, stood in a Painesdale neighborhood lined with manager's houses. Designed by Alexander C. Eschweiler, a Milwaukee architect with Copper Country ties, this 1903 dwelling was clearly Copper Range's grandest company house. *(Michigan Technological University Archives and Copper Country Historical Collections)*

lacking in an ordinary worker's house. It had an enclosed porch and a vestibule and closet inside the front door. The first-floor living room had a fireplace and sliding pocket doors, so the room could be opened to the hallway and dining room or shut off from those spaces. Swinging doors led from dining room to panty to kitchen. The back door at the kitchen had a storm entrance to protect the interior from cold drafts during winter. Upstairs, all three bedrooms had large walk-in closets. The house had a full bath on the second floor, as well as a toilet in the basement. It had central heating, full plumbing, and electrical service.

The larger, two-and-a-half story "Official's house" at no. 3 Algomah had four bedrooms and similar features: porch, fireplace, pocket doors, pantry, and full bath—plus additional rooms in the attic. No. 6 Algomah, a nine-room house designated as the "Physician's residence," had a full porch, Jacobsville sandstone foundation, fireplaces in the living and dining rooms, pocket doors, a full bath on the second floor, and sinks in three of the five bedrooms. The attic also included two rooms for servants' quarters. The top Copper Range managers not only ran corporate paternalism at the mine, they benefited by living in fine houses under that paternalism.

After they improved their earliest houses by adding running water and putting basements and masonry foundations under them, the Copper Range mines stopped short of equipping them as fully modern houses. Ordinary workers' houses did not get central heating, with a furnace and radiators; did not get bathrooms; and did not get electricity. The mines had electric light plants by 1901, but very few houses were wired over the next decade.[25] Copper Range sent power to mine shops, to a few street lamps, and to managers' houses. Yet it did not want to get into the electric utility business. As late as 1913, few people in Baltic, Trimountain, or Painesdale had electricity, and they paid twelve cents per kilowatt-hour for it (which was an average charge for the time). F. W. Denton, general manager of Copper Range, wrote, "We are not in any manner catering to the public demand for current, and as soon as any other source of supply is available we shall cease to supply the few customers that we now have."[26]

Copper Range did not strive to be in the vanguard of the modernization of worker housing. It did not try to top the Calumet and Hecla company's housing, but to match it. Instead of installing furnaces, bathrooms, and electrical wiring, it opted for more traditional means of servicing employees in company houses: it didn't charge extra for water; it sold fuel to householders at competitive rates (four dollars per cord of wood and four-and-a-half dollars per ton of coal); it picked up the garbage as needed and scavenged neighborhoods of trash; and it maintained houses at no charge to their occupants.[27] This maintenance work, however, sometimes lagged. In 1915 Copper Range planned to round up a crew of schoolboys in the summer to paint company houses that hadn't been painted in ten years. The new paint would protect the houses while making the neighborhoods look better.[28]

Copper Range charged modest rents that helped keep a worker's cost of living in check. Rents at the Baltic, Trimountain, and Champion mines averaged eighty to eighty-five cents per room per month—lower than the typical dollar per month per room that most mines charged.[29] Copper Range rented houses to married workers; it never built enough houses to shelter everyone; in allocating houses, it paid particular attention to a man's occupation and ethnicity; and it allowed householders to take in boarders to earn additional income. (In 1910, a total of 247 householders at the Champion mine took in 500 boarders.)[30] Also, like other mining companies, Copper Range had a wide range of houses suited to its hierarchy of workers. By 1913, 18 of its 607 houses at Baltic, Trimountain, and Champion had only two or three rooms, 101 houses had four rooms, 182 had five, 112 had six, 93 had seven, and 89 had eight. Only 12 houses, reserved for top bosses or managers, had nine to twelve rooms.[31]

In the first years of the twentieth century, Copper Range dealt with many ethnic groups. In 1905, for instance, the Champion mine had 1,027 men on its payroll, including 366 Finns, 225 Englishmen (mostly Cornish), 198 Austrians, 92 Italians, 50 Poles, 22 Irishmen, 18 Swedes, 11 Germans, and 10 French Canadians, 10 Americans, and fewer than 5 each of Norwegians, Croatians, Russians, Prussians, and Greeks. Of those men, 450 were married and vied for only 214 company houses. By 1912, when Champion employed 1,160 men and rented out 252 houses, the ethnic/nationality mix had changed. Now the mine had 297 Austrians, 285 Finns, 200 Englishmen, 103 Americans (mostly the American-born sons of immigrant parents), 83 Lithuanians, 79 Italians, 78 Poles, 10 Germans, 10 French Canadians, 9 Swedes, and fewer than 5 each of Irishmen, Russians, and Syrians.[32] In allocating houses the Champion mine could not precisely match a worker's job, status, and ethnicity with a particular size or style of house, nor could it segregate workers into permanent ethnic enclaves around Painesdale. Nevertheless, mine managers had a sense of whom their better workers, or "best families" were, and offered those employees better houses.

The "best families" at the Champion mine included Americans, Englishmen, Cornishmen, Irishmen, Swedes, Scots, and Germans, but not the Finns, Austrians, Italians, Armenians, or Lithuanians.[33] "Best families" included mining captains and assistants, miners, and some timbermen, but few to zero trammers, wallers, or pickers. Among surface workers, "best families" came from the ranks of managers, clerks, physicians, mining engineers and assistants, and skilled tradesmen, such as machinists, blacksmiths, and electricians. Champion generally did not rent company houses to unskilled surface workers, such as rockhouse men.

When an ethnic group moved up into better jobs, it also moved into company housing. When Finns first arrived, mining companies relegated them to unskilled jobs, especially tramming. But as Finns put in their time and became numerous, they started taking jobs as miners. In 1905, many Finns at Champion still worked at unskilled jobs: 112 trammers, 86 copper pickers, and 13 timbermen. Of those, only three received the keys to a company house. But of 111 Finns classified as miners, 52 lived in a company house. By 1912 another ethnic group had started moving up: the Austrians. Champion rented houses to only three Austrians in 1905, who were skilled surface workers. No Austrians worked as miners, and the 86 Austrian trammers, 59 copper pickers, and 21 wallers were locked out of company housing. By 1912, when 42 Austrians held miners' jobs and 26 were timbermen, a total of 39 occupied company houses.[34]

Champion moved Finns and Austrians into lesser houses at the mine—but at least they had one. The same could not be said for newly arrived Lithuanians. In 1912 F. W. Denton made his Lithuanian workers happy. They had "humbly" petitioned him for a day off on 20 July, "it being our National Holiday." Denton told them that "I take pleasure in granting your request." That same year, because they worked below the rank of miner, none of the 83 Lithuanians at the mine earned a company house. Four years later, it would be the Armenians. In lieu of providing houses, Champion mine management figured these immigrants might "need something in the nature of a camp located out of sight and yet near enough the mine to keep them from being afraid of being molested."[35]

Baltic, Trimountain, and Champion adhered to local tradition by providing medical services to workers and families. Each mine had its own physician and dispensary, and their related stamp-mill settlements (Redridge, Freda, and Beacon Hill) had an assigned physician, too. Unlike companies that made participation in a medical program mandatory only for underground workers, Copper Range made all workers enroll. And instead of charging single men fifty cents per month for medical coverage, or half of what married men paid, the company required all workers to pay in a dollar per month.[36] For that rate, a worker and his family had access to doctor visits, medicine, and hospitalization (the latter carried an additional rooming charge).

The mines and mills shared a single hospital sited in Trimountain in a large house formerly occupied by the Trimountain mine agent. F. W. Denton's two-and-a-half story house stood empty after he moved to Painesdale, so in 1905 Copper Range began converting the structure into a hospital. The Trimountain hospital differed from earlier mine hospitals in important ways. Medical standards, practices, and equipment had advanced a great deal, and William Paine wanted Copper Range's hospital to reflect those advancements and be "first class in every particular." While the hospital looked like a late Victorian house on the outside—with its full-length porch, bay windows, trap-door monitors, and decorative shingling on the gable ends—on the inside it had a modern design and new equipment.[37]

Copper Range, a proud company run by wealthy men, did not want to hide its good works or live in the shadow of the Calumet and Hecla mine. On the weekend that Copper Range dedicated its hospital in April 1906, Dr. Victor C. Vaughn, dean at the University of Michigan, delivered the opening address, and all Copper Country residents were invited to tour the facility. The *Daily Mining Gazette* reporter who walked through the structure with its chief physician, Dr. W. K. West, reported, "No hospital in the county has a more complete apparatus for their surgical or medical work than the one at Trimountain." That "apparatus" included an X-ray machine; an operating room with extra electric lights and white enameled walls; a scrub room next to the operating room, equipped with foot-pedal controls for the water faucets, to keep physicians' hands cleaner; an elevator and dumbwaiter running from the basement to the third floor; an internal telephone system connecting all rooms; and a call-bell network that rang at a central nurse's station and indicated the source of the call.

The thirty-room hospital, flanked by a nurses' residence and an icehouse, sat on a piece of high ground between blocks of company housing and the mine proper. Its basement contained a kitchen, pantry, storerooms for drugs and other sundries, and a morgue, which could be entered directly through an outside door. The hospital did not need a furnace and boiler in the basement because steam piped from the mine heated it. The first floor held waiting, receiving, and consulting rooms; the pharmacy; an accident ward for treating small wounds; the staff dining room; and the chief physician's office and library. This floor preserved the original oak wall paneling of the agent's house. Operating and scrub rooms occupied part of the second floor, which also contained wards, private rooms, and bathing facilities for men. Doctors kept offices there, as did the chief nurse.

Copper Range reserved the third floor for the treatment of women. That floor, too, had wards, plus private rooms and baths for patients willing to pay for them. Early company hospitals primarily treated traumatic injuries suffered by men on the job; they treated very few women. The more modern Trimountain hospital differed in two important ways. First, it was not so restricted to treating on-the-job injuries but cared for patients suffering a wide range of medical problems—people who, in earlier decades, would have been tended to by company physicians in their homes. Secondly, the Trimountain hospital admitted more women and regularly treated the mothers, wives, or daughters of company employees.[38]

As a supplement to its medical/hospital program, Copper Range administered the "Club," or the mutual aid-and-benefit society its employees had started. The company collected dues, paid out benefits, and invested any surplus money on behalf of the members. Copper Range did not contribute to the fund; only the Calumet and Hecla mine paid into its aid-and-benefit society. At Copper Range, all men at the mines (but not the mills) paid in 50 cents per month. In return, the Club paid out benefits in the event of an accidental injury or death. Unlike most benefit societies, this one did not offer aid in cases of sickness or disease, but only in trauma cases. The disabled worker, starting six days after his injury, received a dollar per day for up to six months. A committee of managers and workers decided on the benefit to be paid to those suffering a permanent disability, such as the loss of an eye, an arm, or a leg. If a man was ambulatory but could not do his old job, Copper Range often found him a lighter job. If a man was killed on the job, his family received $600.[39]

Like mining companies before them, Baltic, Trimountain, and Champion got the schools they wanted, and those schools were good, impressive edifices for their time, rivaling any in the Copper Country. By 1903 a two-and-a-half-story grade school stood at each of the three mines; in 1903 a library opened across the street from the Champion mine's grade school in Painesdale; and in 1910 a fine high school opened close to Painesdale's grade school and library, creating a large educational campus in the midst of a mine town. In their architecture, massing, and materials, the library and high school made it obvious that major wealth was associated with Copper Range's small settlements.

The Baltic Mining Company built its school in 1900 and sold it five years later to Adams Township for a bit over $5,000.[40] It had a central pavilion and entrance, flanked by symmetrical wings and topped with a bell cupola. The school's wood-frame superstructure was handsomely done, with large windows and narrow clapboard siding on the first and second floors (painted in contrasting colors), and cedar singles on the gabled pavilion, the attic's monitors, and the large hipped roof. The school stood nicely balanced and attractively textured. When opened, it enrolled 75 students but could accommodate up to 360 pupils. In 1903 a nearly identical school went up in Painesdale. In 1902 Trimountain built its 400-pupil school, similar in size and materials to the Painesdale and Baltic schools, but lacking their symmetry and architectural elements, such as the bell cupola.[41]

In 1903, thanks to William A. Paine, who donated $40,000 to the project, Copper Range opened a library near the Painesdale grade school.[42] Christened the Sarah Sargent Paine Memorial Library, after the benefactor's mother, and designed by Milwaukee architect Eschweiler, the library was handsomely done up in the Jacobean Revival style, using locally quarried red sandstone.[43] The Paine Memorial Library (supplemented by small branch libraries at other Copper Range locations) rivaled the Calumet and Hecla library built in 1898. The library (85 feet 10 inches by 50 feet) had reading rooms for children and adults. Its collection boasted six thousand books in English and other languages, such as Finnish and Croatian, plus it subscribed to numerous papers and magazines. Mining company officials sat on the library's board, and the mines contributed funds for staffing and maintaining the library and purchasing reading materials.[44]

The library also served as a community center. It had a game room where men could smoke and play cards. (Later, in the 1920s, the library added a billiard room.) An auditorium with a raised stage occupied the top floor and hosted parties for local clubs, speeches, dances, and musical performances.[45] In the basement, in an era when society was setting higher standards of personal hygiene, Copper Range families availed themselves of public baths at little or no charge. They bathed there with greater ease and probably better results, than achieved in company houses that had running water, but no tubs or showers.

Children at Baltic, Trimountain, and Painesdale, if they continued their education past grade school, first attended high school at the Atlantic Mine. The Adams Township school board, controlled by mining-company officials, voted in 1907 to build a new high school in Painesdale.[46] As the *Daily Mining Gazette* noted in 1912, the biggest taxpayer in the district, the Copper Range Consolidated Mining Company, wanted "the children of its employes [sic] to have every educational advantage." For one dollar, Copper Range gave the school a fifty-year lease on a building site next to the Paine Memorial Library. Erected and equipped a cost of $125,000, the six-year high school (grades 7 though 12) opened for the 1909–10 academic year. The *Daily Mining Gazette* said it was "undoubtedly the finest mining camp school in the United States."[47]

Eschweiler, the library's architect, designed the high school, too. Like the library, its walls were laid up handsomely of red sandstone. The school, 170 feet long and 60 feet wide, had a tall basement with two full stories above. Done in the Neo-Tudor style, it had a central pavilion and entrance, projecting wings on the ends, and a long battlement carried from end to end.[48] Besides the usual classrooms, the high school incorporated an auditorium on the second floor, chemical and physical laboratories, manual training facilities for boys, and a "domestic science" laboratory for girls. The school also included a gymnasium, with separate baths for girls and boys.

Students from Painesdale and Trimountain walked to the high school. Children from Baltic, the village of South Range, Atlantic Mine, and the mill towns of Freda, Beacon Hill, and Redridge road the rails to school. There was no escaping students'

One of many imposing buildings on the Keweenaw that were erected of locally quarried red sandstone, Painesdale High School opened for the 1909–10 school year. It enrolled students in grades seven through twelve who lived in over a half-dozen mine and mill towns controlled by Copper Range, plus the village of South Range. *(Michigan Technological University Archives and Copper Country Historical Collections)*

working-class roots, when steam boilers at the Champion mine heated their school, and a Copper Range Railroad train took them from home to school and back.[49] The curriculum included music and Latin, yet the twelve teachers did not offer their three hundred students any "faddish" courses of study, or any choice, for that matter. The proscribed curriculum, which included no electives, was

> based on the idea that the pupils of the Painesdale high school are the children of wage earners and probably will be wage earners themselves. Few of them can look forward to college, so while they are in the high school they will get that education that will best fit them for tackling life immediately after leaving the high school.

The boys took manual training to prepare themselves for industrial work, while girls learned skills—cooking, sewing, and laundry—that in the short run might make them employable and in the long run would make them better wives and mothers. The school was an institution of the Progressive Era, to be sure, as it provided for the education, health, and hygiene of its students. Boys and girls did "regular Swedish exercises for the strengthening of their bodies." And "every Monday, under proper supervision, every pupil in the Painesdale high school takes a shower bath."[50]

Baltic, Trimountain, and Champion catered to the religious needs of workers. The Stanton Methodist Episcopal Church and the Evangelical Lutheran Church went up in

For decades, students living at the Trimountain and Champion mines walked to Painesdale High School. Others came from more distant places—the Baltic and Atlantic mine locations, South Range village, and the stamp-mill locations of Freda, Redridge, and Beacon Hill, out on the Lake Superior shore. Students from these communities commuted on school trains run by the Copper Range Railroad. *(Michigan Technological University Archives and Copper Country Historical Collections)*

Baltic, whereas congregations in Trimountain worshipped at the St. James Methodist and Finnish Lutheran churches. In Painesdale, William Paine served as benefactor once again, contributing to the $10,000 cost of the Albert Paine Methodist Church, named in honor of his father and located near the schools and library. Most Cornish and American families worshipped there. Finns in Painesdale built an Apostolic Lutheran Church, and Croatians worshipped at their Sacred Heart Catholic Church.[51] All the churches were substantial constructions, standing tall on high foundation walls of sandstone or poor rock. These walls permitted for a more impressive, elevated church entrance at the tops of stairs and helped create a high-ceilinged full basement for meetings and socials. Of the churches along the Baltic lode, the Methodists built the grandest, largest ones. Handsome as they were, these wood-frame, clapboarded and shingled Methodist mine churches were *first generation* structures built by young, smallish congregations. They paled in comparison to contemporaneous churches built by older, larger congregations in commercial villages such as Houghton, Hancock, Calumet, and Lake Linden. There, early in the twentieth century, some congregations erected their second- or third-generation churches, which they often built of red sandstone, finished out with very fine stained glass and ornate interiors.

The Baltic, Trimountain, and Painesdale locations had what the companies wanted them to have and lacked everything else. They had houses, streets, schools, churches, and libraries. They had fire halls and post offices. They had water towers and Copper Range Railroad depots. They had some social clubs and fraternal organizations, such as Finnish Temperance Halls in Baltic and Painesdale, and an Independent Order of Oddfellows Lodge in Painesdale, which also served as an "opera house" and provided meeting rooms for other organizations. They had company-operated bowling alleys, skating rinks, and ballfields. Trimountain had a bandstand for summer concerts.[52] They all had establishments that lodged visitors and perhaps workers: Siller's Hotel in Painesdale, the Trimountain Hotel and Boarding House, and the Baltic Hotel, with twelve to fifteen sleeping rooms and indoor toilets.[53] Still, only a few commercial operations dotted the mining landscape. Few and far between, they never came close to constituting a downtown.

All the Copper Range mines had a general store. By 1900, older mining companies on the Keweenaw had left the store business; they generally deemed the company store unnecessary because of the growth of neighboring villages that offered multiple general stores and specialty shops. The Copper Range mines, because they opened before the village of South Range existed, all felt obliged to open stores. Distance, too, was a factor in opening stores, especially at Painesdale, which was over seven miles from Houghton. Having local stores at the Range locations made it more convenient for people to live there. Some of these outlets remained company stores, while the mines turned others over to independent merchants. Copper Range liked to remain in control of a few stores so they could compete with independent merchants, particularly those in South Range, if they started charging too much for goods.

At its location, the Baltic Mining Company ran a wood-frame general store for a while before turning it over to the South Range Mercantile Company, which was a partnership formed in 1901 by three local investors: H. Stuart Goodell, John B. Dee, and James R. Dee.[54] Goodell, the largest investor in the mercantile company, relocated from Houghton to Painesdale to better oversee store operations along the Baltic lode. He was the son of the agent of the St. Mary's Mineral Land Company, which had supported the development of the Copper Range Railroad and sold mineral lands to the Baltic, Trimountain, and Champion mines. Goodell also married Charlotte Hubbard, a daughter of Lucius Hubbard, the geologist and first general manager of the Champion mine—so he was well connected to the mining companies.[55] The Dees were well-known Houghton businessmen: John was secretary and treasurer of the Peninsula Electric Light and Power Company; James, in addition to managing the power company, superintended a coal and coke company and the local offices of the Michigan Telephone Company and Western Union Telegraph. He also helped establish the Douglass House in Houghton, the village's best hotel, and promoted local sports, especially hockey.

Trimountain opened its store in 1899 and operated it for years. The Trimountain company store sold all kinds of meats, vegetables, and other foods for human consumption, plus animal feeds, kitchen and tableware, men's boots, shoes and clothing, and other sundries. The store at the Champion mine carried the same wide range of goods and, like the Baltic store, was operated by the South Range Mercantile Company. All three stores at the mines shared space with the local post office, so residents could make one trip to pick up their mail and buy food or personal items.

Aside from Siller's Hotel and the general store, the Painesdale mine location included only a few other small businesses: a "sausage house," barbershop, and several candy and small grocery stores. Several of these little businesses probably operated out of first-floor storefronts in structures built by employees on company land, where the owner lived above the store and perhaps took in boarders. The land leases no doubt enabled Copper Range to step in and close any activity that violated company standards. F. W. Denton sent a stern warning to five small store operators in Painesdale: stop selling cigarettes to minors, or I will cancel your leases. Denton told Frank Santoni, caught selling beer out of his house, to stop such sales and stop allowing visits by disorderly men, or he would lose his lease, too.[56]

The Quincy, Pewabic, and Franklin mines had Hancock to provide workers with a wide range of living choices, goods, and services. The Huron and Isle Royal mines had Houghton; Calumet and Hecla, Tamarack, and Osceola had Red Jacket village. Workers at Baltic, Trimountain, and Painesdale could trade at South Range, a new village developed by a Houghton businessman, Reginald Pryor, a civil and mining engineer with financial interests in mines, lumber, and banking.[57] In 1902—sensing new business opportunities arising from mining start-ups along the Baltic lode—Pryor and three associates incorporated a new Wheal Kate Mining Company. They borrowed the name from a mid-nineteenth century mine that had failed and acquired 240 acres of the old Wheal Kate mine, which was next door to Baltic.

The stated purpose of the new Wheal Kate Mining Company was not only to mine, mill, and smelt copper, but "to acquire, hold and dispose of all such real estate as may be necessary or convenient."[58] In fact, as noted by Stevens's *Copper Handbook* in 1903, Pryor's company was not a mining company at all, but "mainly a townsite company."[59] It soon operated as the South Range Townsite Company, and on 17 May 1903 the company paid for an advertising supplement in the *Daily Mining Gazette* that touted the new community's promise. By that date, new roads in and around the village had been cut, and a street grid, four blocks wide and fourteen blocks long, had been laid out. A map of the new village showed that nearly 190 lots had been sold, and 23 structures had been built or were under construction. A Copper Range Railroad depot sat near the middle of the projected downtown. A large double storefront of brick and a brick bank neared completion, and several general merchandise stores and two taverns with hotel accommodations for thirty to fifty men had opened.

South Range claimed advantages for businesses and families. Businesses were advised to locate in South Range because it was the commercial settlement nearest the Copper Range mines and their big payrolls. Families should purchase property and build or buy

In 1946 a Copper Range steam locomotive stopped between South Range's Kaleva Temple (left) and its Copper Range rail station (right). The Knights of Kaleva, a Finnish immigrant fraternal organization, started as a lodge in Hancock in ca. 1899 and moved to South Range in 1904. Members of the secret society were more likely to be merchants than staunch unionists. To help finance their Temple, the Knights at different times rented part of it out to serve as a dentist office, theater, restaurant, clothing store, hardware store, and post office. *(E. A. Batchelder, Musser Collection, Lake States Railway Historical Association)*

houses in South Range because home sites measured fifty by one hundred feet, making them "little farms" big enough for chickens, vegetables, and fruit trees. Village boosters noted that South Range was close to three shallow, cool mines that were more comfortable to work in. They were said to be safer, too, than older mines with their numerous rockfalls. Boosters pointed out that in South Range a man's "home was his own." He could work in whichever Copper Range mine he wanted, and if he switched employers, he could change jobs without changing his residence. And if he quit or was fired, the company could not require him to sell his house to another employee or move it off company property.

South Range never grew to be as large as promoters hoped. Still, it served as a new place to locate for recent immigrants wanting to start small businesses, and those businesses offered shopping and entertainment options to Copper Range employees and families. By 1906 South Range claimed 1,177 residents, officially incorporated as a village, and held its own elections.[60] By the end of that year the village had three organized bands and one church, the Finnish Apostolic Lutheran. It had the South Range Bank, founded by Pryor and presided over by L. L. Hubbard, the former general manager of the Champion mine. It had a Gas, Light, and Fuel Company, a Western Union Telegraph office, building-supply company, and a hardware store. Among its several dealers in general merchandise were a Laborers' Commercial Co-operative Store and a Croatian Co-op. South Range had a Cerrutti Brothers meat market, a restaurant, a billiards hall, and an assortment of saloons run by Italians, Finns, and Croatians. It had the Sterling Spring Bottling Works and a branch of the Bosch Brewery. A jeweler, a tailor, a photographer, and a Jewish dealer in dry goods called South Range home, as did a physician, a dentist, and a lawyer.[61]

The *Engineering and Mining Journal* of 28 December 1912 carried an article by Claude T. Rice titled "Labor Conditions at Copper Range." Rice described many of the paternalistic practices at the mines. He noted their recreational facilities, schools and library, aid fund, medical programs, public baths, and housing. All this paternalism, Rice noted, meant that Copper Range could pay lower wages because the company house and doctor represented an important, additional compensation for work. The paternalism attracted married men and a "better class" of men, whose wages went "to the merchant and the butcher instead of to the saloonkeeper and the gambler." "Provision for families," Rice wrote, "brings families, and families . . . beget stable conditions as far as labor is concerned."[62] Because of the paternalistic treatment they received, "the miners stay contentedly in these Lake Superior mining camps." But Rice failed to notice that many men worked at jobs below the rank of miner and thus received less favoritism from Copper Range. He failed to see that single men outnumbered married men at these mines despite all their emphasis on families, houses, churches, and schools. He failed to appreciate the trouble brewing in the South Range mines over new technologies. As a consequence, in December of 1912 he could not predict that, on 14 August 1913, company-hired thugs would send a hail of bullets into a Champion mine house in Seeberville, killing two men and seriously injuring two others.

14

❖ ❖ ❖

HOLDING ON

Corporate Power in an Age of
Social Change, 1890–1912

During the pioneer era, ambitious mining companies arrived on the Keweenaw in search of copper and wealth. Just as they felt their way along, incrementally, learning about the geology of the region and how to mine, mill, and smelt their copper, so did they feel their way along in the social sphere. The companies out of necessity became community builders and paternal employers. They never wanted, however, to do everything for everybody, and as local society grew more mature, as commercial settlements and local governments rooted in, the mining companies were content (and perhaps even relieved) to divest themselves of some social responsibilities. Most got out of the business of provisioning settlers by farming or running their own stores. They turned over to government such tasks as keeping the peace and road building. They let the county Superintendents of the Poor assume primary responsibility for local indigents. And because they did not want to house all their own workers, let alone everybody else in the region, they turned over to house builders and boardinghouse operators the tasks of sheltering many of the men, women, and children who migrated to Lake Superior.

Still, the mining companies remained keenly interested in social developments at the mine locations proper and in nearby communities. They had a large stake in the orderly development of the society that attended their industry. The companies saw themselves as the primary institutions of life and work on the Keweenaw. They were willing to let "life go on," as long as it proceeded in a manner that did not challenge their interests. But when things didn't go their way, the companies sought to wield influence, if not absolute control, over that society. Sometimes they merely sought to curtail an individual's behavior, for example, nip a problem with alcohol in the bud; sometimes they chased away a prostitute; in the Civil War era, with street crime and lawlessness on the rise, they established and drilled a short-lived militia group to protect order.

In the quarter century after 1890, the companies often felt that change threatened their hegemony. They had lost, in the late 1880s, their leadership role in the U.S. copper industry, and their share of that industry declined decade by decade even while their copper production rose dramatically. The mining companies at the same time sensed an erosion of their leadership role in Keweenaw society. For decades, much had gone right. Industry and society had progressed in tandem, matured together, and the Keweenaw had earned for itself the reputation of having the most harmonious labor-management relations in the American metals industry.[1] But by the late 1880s, there was the growing suspicion that some things were going wrong.

One early sign of trouble came at Calumet and Hecla (C&H), a mine that was a giant tinderbox. A fire shut C&H down in 1887 and then, the next year, another catastrophic fire occurred. The fires struck just when organizers for the Knights of Labor union were active in the region. C&H and other local mines were cognizant that social discord, unionism, and industrial violence were on the rise in America. In 1886, the year of the Great Upheaval, 600,000 workers were involved in nearly 1,500 strikes and 140 lockouts. Some of the worst violence occurred not that far from Lake Superior. In Chicago the Haymarket Riot involved molders at McCormick Harvester; near Milwaukee, immigrant ironworkers and the state militia clashed.[2] Against this background, C&H grew very suspicious of the origins of its fires. It thought that among its ranks there might be an arsonist who wanted to harm the company, not work for it. The company posted a $10,000 reward to snare the perpetrator. The reward went unclaimed; C&H discovered no arsonist. The tendered reward, though, symbolized a new era of suspicion and mistrust.[3]

The mining companies found it easier to dominate the Keweenaw when its population remained small. Only about 1,000 resided on the entire Keweenaw in 1850; about 20,000 lived there by the early 1870s. When the mining companies boomed in the late nineteenth and early twentieth centuries, so did the population. By 1890 the population of Houghton County reached 38,000. That figure climbed steeply to 66,000 in 1900 and to 88,000 in 1910.[4] For the companies, local society became more difficult to herd, not only because of the large population increase, but because of the makeup of that population. From the last decades of the nineteenth century on into the twentieth, the immigrants who fueled the population growth distinctly differed from earlier ones.

To a considerable degree, the immigrants who arrived on Lake Superior were the same ones coming to the United States as a whole. When immigrants from England (including Cornwall), Scotland, Ireland, and western Europe (especially Germany) were the most frequent arrivals to the United States, they were also frequent arrivals at the mines. But by the late nineteenth century, the immigrant flow from these regions had

In 1890 a cooperative store opened at Tamarack, and at least four other co-ops opened along the mineral range before 1910. Ethnic workers, especially Finns, started co-ops to become less dependent on company stores and regular merchants. They gained more control over what they bought and how much they paid. The Mass Co-operative in Ontonagon County is show here in 1919. *(Michigan Technological University Archives and Copper Country Historical Collections)*

slowed. In their place came southern Europeans, especially Italians, and eastern Europeans, such as the Croatians or Slovenians. These eastern Europeans, from their end of the Austro-Hungarian Empire, were often lumped together as "Austrians" by Americans lacking the ability or the desire to differentiate among them. The Finns of northern Europe comprised another group that arrived in large numbers on Lake Superior.[5]

These new immigrants, unlike the old, were more likely to come from nonmining traditions, more likely to come from rural areas or small villages, and more likely to be farmhands. They were less likely to be acclimated to industrial work and less likely to be skilled workers. As the mining companies expanded in the 1890s and beyond and sought greater efficiency and higher productivity, the men coming for jobs at the mines were not well suited for them. The men found themselves toiling in an underground that was most strange and threatening. Many objected not only to the arduousness of the work, but to its hazards. This was especially true of some new ethnic groups (especially the Finns) who tended to support social and political movements that the companies found an anathema, such as unionism and socialism.[6] They seemed more prone to challenge working and living conditions.

On the industrial side of things, the turn-of-the-century mining companies had no problem embracing modernity. The Michigan College of Mines had opened its doors in Houghton in 1886. When the companies had first hired engineers from the likes of M.C.M., they did not use them as vitally important managers or experts. Instead, they often used them as glorified surveyors who did the mines' underground mapping. But by early in the twentieth century, the companies hired engineering graduates for the modern knowledge they could bring to bear on operations. With the help of new engineers, the mines pushed new cost-accounting measures and embraced notions of efficiency, productivity, and scientific management.[7] The companies did not want to be running nineteenth-century mines in a twentieth-century world. But on the social side of things, the same companies saw modernization as a mounting challenge, as a source of tension and pressure, as a set of social changes that might cost them money rather than make them money.

Pedestrians walking the streets of the copper district's major villages saw modernization in the architecture and buildings. Before 1890 some downtown structures had been constructed of poor rock or other masonry, but most had been framed in wood and stood only two stories tall. Often a false front was added to on the gable end to try to make a structure appear grander than it really was. In the two decades after 1890, the most important villages made themselves over, especially Houghton, Hancock, and Calumet. Out near Portage Entry, or Jacobsville, local quarries opened up that produced a handsome red sandstone. As the towns modernized and reconstructed themselves, Portage Entry sandstone, along with brick, became a building material of choice.[8] Shoppers, worshippers, tradesmen, and miners encountered impressive masonry structures along their local streets.[9]

In a period of economic growth, banks, hotels, shops, and restaurants went up. At Calumet in 1900 the grand, publicly owned Calumet Theatre, done up in brick in the Renaissance Revival style, went up on one end of town, while the tall, Gothic Revival, red sandstone St. Anne's Church, supported by its French Canadian congregation, went up on the other end of town. Together, they served like bookends for the modern Red Jacket village, and in between them stood the Vertin Brothers department store, representing the modern, expanded version of the general store of the nineteenth century. In Houghton and Hancock, as well as in smaller towns like Lake Linden, similar transformations occurred.

Local streetscapes changed in many ways. Sidewalks were added, lifting citizens' shoes out of the dust, dirt, or mud. More streets came to be paved with brick, asphalt, or concrete by early in the twentieth century. By 1910, nascent auto dealerships occupied storefronts for the first time, and automobiles rambled down the streets, often alarming animals and pedestrians and eliciting the wrath of constables. Starting in 1900, streetcars, too, took up their part of the street and eventually connected the major villages in the center of Houghton County.[10]

In the village of Red Jacket, the lower floors of the three-story-tall sandstone structure contained a liquor store and wine cellar. The smaller, less pretentious store was home to Hand You & Co.—a dealer in "Chinese and Japanese Fancy Goods." The small Chinese population in Calumet early in the twentieth century also ran laundries and faced discrimination, as evidenced by one local newspaper headline: "Local Chinks Velly Busy." *(Michigan Technological University Archives and Copper Country Historical Collections)*

In 1900 this municipally funded town hall (left side) and theatre (right side) opened in the village of Red Jacket. Many of the nation's most famous entertainers, shows, bands, and orchestras traveled north to Lake Superior to play in the magnificent Calumet Theatre done up in the Renaissance Revival style. *(Michigan Technological University Archives and Copper Country Historical Collections)*

Important harbingers of modern life were not found at street level, but below or above it. Beneath the streets, especially after the 1890s, ran water lines, sewer lines, and sometimes gas lines. Above the streets, atop poles, ran telephone and power lines. Improvements in communications and power were hallmarks of modernization. In the villages near the mines, these improvements often got off to a small start in the late 1870s (for telephones) and 1880s (for electrical power). But not until the late 1890s and early 1900s did telephones and power lines become commonplace and serve ordinary homes, as opposed to public or commercial spaces.[11]

As part of modernization, people wanted and expected more. With the expansion of commerce and advertising, they became more active consumers. They were in a mood to buy things rather than make things or make do without things, and being a bigger and better consumer required a larger income. Members of the working class hoped to live in better housing. Many in the Copper Country still crowded into modest dwellings where boarders often took up much of the space, intruded into family life, and made privacy a luxury. But those same boarders provided the family with income, and that income helped fuel the dream of someday acquiring more possessions and moving into a bigger and better house.[12] Many families had fewer children now—the average number of children dropped from about seven in 1800 to about four by 1900.[13] Fewer children meant that parents had to spread their income over fewer dependents. They might be able to spend more money on each child, or they might spend it, someday, to purchase their own home, or even an automobile.

With the advent of modern utilities, the attainment of greater comfort and convenience became a cultural goal of the early twentieth century. The house was to be transformed. Electricity provided cleaner light that was safer and needed far less tending to compared with kerosene lamps. Gas, too, was cleaner and required less work than did wood fires used to heat the stove and the house. Inside plumbing and sewer hookups meant far less work, particularly for women, who no longer had to haul water both in and out of the house in buckets. Toilets were much desired because they eliminated the chamber pot indoors and the nasty privy outdoors—and the walk out to the privy, too. Water lines, sewers, and toilets improved public sanitation and health and led to higher standards of personal hygiene and cleanliness.[14] People expected to be healthier and to benefit from more professional doctors, often armed with a specialist's knowledge and more medical science than earlier practitioners.

To residents, all these changes fell under the heading of progress. To the mining companies, these same changes caused headaches and problems. In 1910, companies rented out three thousand houses; the vast majority of them had been built in an era of no utilities and few amenities. Employees now wanted utilities retrofitted into their company houses. The men, women, and children living at the mines expected their lives to be improved. As consequence, modernization burdened the mining companies, the largest landlords on the Keweenaw.

The mining companies were not alone in facing the pressures of modernization or in coping with the challenges posed by rising standards of life. By the end of the nineteenth century the nation as whole addressed quality-of-life issues. Growth and economic and social change had been beneficial to some, especially to the four thousand or so millionaires who already resided in the United States. But to others, including the great wave of immigrants who had recently arrived, the United States was not living up to its promises. In the wake of rapid industrialization and urbanization, problems of inequality, political corruption, squalor, poverty, urban crime, unsanitary living conditions, pestilence and disease, and business excesses had become legion. Turning toward their problems rather than away from them, many Americans believed that wrongs needed to be made right, that the downtrodden deserved a better life. Others acted out of fear—fear that socialism and rebellion might rear up unless the needs of the indigent and of working-class men, women, and children were better served. Out of these concerns arose the social and political reform movement known as Progressivism, which played out between the last years of the nineteenth century and America's entrance into the First World War.[15] In the Progressive Era, politicians, scientists, technologists, and other professionals and experts applied their knowledge to identify and correct social problems. Central to Progressivism was the belief that life could and should be made better for all.

In many places, Progressivism started at the city level in response to local problems. Then it moved to the state level and finally on to the national level. Progressivism entailed new social programs, institutions, ordinances, and laws. In Michigan, Progressivism led to legislation that restricted child labor, shortened the workday, and established workers' compensation. On the national level, it led in 1910 to the creation of the U.S. Bureau of Mines and in 1913 to the creation of the cabinet-level Department of Labor and the U.S. Commission on Industrial Relations. As America moved into the Progressive Era, the mining companies on Lake Superior sensed their autonomy was eroding. Unfettered big business had had its day. They would have to adapt to new political and social realities. The copper-mining companies most keenly felt reformist zeal—and an overturning of their "old time rules"—in the realm of mine safety.[16] The pressure to change came from local courts, attorneys, and juries; from the Michigan state house; and from the U.S. Congress. This pressure surfaced as early as 1887 and wrought a near revolution, smashing the old-time rules by 1912.

Serious accidents and fatal injuries had been a macabre measure of economic growth. The more men working underground, the more men who died. For decades, for every thousand men working underground for a year in the Lake mines, five would die. In the 1850s, at least twelve men died underground, and fifty-four died in the 1860s. That death toll nearly doubled to 106 underground fatalities in the 1870s, and it nearly doubled again in the 1880s, when the mines suffered 195 losses of life. In the 1890s, as the industry rapidly expanded because of the rise of the electrical industry, the death rate escalated and claimed 284 men. Early into the new century, as the industry climbed toward peak production and employment, it also set new fatality records. Between 1900 and 1909, 511 men died in the copper industry. From 1905 through 1911, the mines killed an average of sixty-one men per year, or more than one per week. At the end of this unfortunate run, one out of every ten men killed in the entire U.S. metal-mining industry died on the Keweenaw Peninsula.[17]

Charles Lawton served as agent at the Quincy mine through these deadliest years. Under what Lawton called the "old time rules," work-related deaths had not led to condemnation of the mining industry. Society in the nineteenth century did not pay particular attention to men dying at work, because death was commonplace. Death claimed many young women during childbirth (that was their "occupational" hazard), and year after year communicable diseases swept away young children. In the early 1880s, children under the age of five accounted for 75 percent of all recorded deaths in Houghton County.[18]

Under the old-time rules, mining was seen as inherently hazardous. Workers took precautions, yet accidents happened and men died. People believed in accidents and not so much in finding fault or laying blame. When a man died, he was at the wrong place at the wrong time—when a powder charge went off prematurely or a piece of the hanging came down. If anyone could be blamed, usually it was the victim himself. He did something careless or stupid and paid the ultimate price. A key tenet of the old-time rules was that familiarity bred contempt. Men, after working long in a mine, stopped worrying about hazards that had not yet hurt them. When they treated a hazard with contempt and ignored it—that's when it killed them.[19]

As personal injury law, or the law of torts, evolved in the nineteenth century, courts tended to codify the old-time rules. Three legal doctrines made it difficult for accident victims or their survivors to sue an employing company and win. These were the doctrines of assumption of risk, of contributory negligence, and the fellow-servant rule. The doctrine of *assumption of risk* was that a worker, when accepting a job, accepted all the hazards associated with that job. If you assumed the occupation of miner, you assumed the risks of falling rock or blasting accidents. Under *contributory negligence,* if a victim could be shown to have been negligent on the job, even in the slightest way, then he, and nobody else, was responsible for the accident. And under the *fellow-servant rule,* if the unsafe behavior of one worker, even a boss, caused the death of another worker— then the fellow servant, not the employer, was to blame. Under such doctrines, plaintiffs found it difficult to beat a mining company in court, so few even tried.[20]

This all started to change in the late 1880s. One chipping away of the old-time rules occurred in Michigan in 1887, when the state legislature debated passing a bill with provisions for limiting child labor, shortening the workday to eight hours, and creating a system of county mine inspectors. Legislators scrubbed the child-labor and eight-hour-day sections, but they passed legislation creating mine inspectors, who were to be appointed by county boards of supervisors. The law required a mine inspector to examine each mine at least once annually, and he had the power to "condemn all such places where he shall find that the employees are in danger from any cause, whether resulting from careless mining or defective machinery or appliance of any nature." The inspector was to report each year on "the number of mine accidents occurring during the preceding year causing either death or injury to persons." Also, he was to judge whether each fatal accident occurred "through the fault or negligence of employers" or "through the fault or negligence of employees."[21]

By no means did the early mine inspectors on the Keweenaw overturn the old-time rules regarding mine accidents, fault, and liability. For starters, the county supervisors who appointed the inspectors were often mine managers or friends and allies of same. The inspectors, such as Josiah Hall in Houghton County, were experienced mining men who had spent their careers believing in the old-time rules. Hall, who had been a mine captain and a mine superintendent, almost always found that fatal accidents were unavoidable or caused by careless victims and not by their employers.[22]

The mine inspectors for years never wielded as much power as they could have, or should have, at the copper mines. They were not proactive, but reactive. They did not seek out safety; instead, they investigated deaths. They totally ignored the provision of the state law that required them to investigate less-than-fatal injuries, so their reports never painted a full picture of the personal costs men paid for working in this industry. But their reports, despite all their failings, did provide for the first time an accurate count of how many men were killed, who they were, how old they were, where they came from, how they died, and what they were doing. Until those reports were made, many underground deaths had gone unnoticed. Local newspapers paid scant attention to victims, and half of them in the nineteenth century were not recorded in local county death records.[23]

Society accepted fatal accidents in part because most of them killed just one, or maybe two, men at a time. A lone miner died under a fall of rock, was blasted, or was run over by a skip. Maybe he just tripped and fell several hundred feet. Keweenaw society became more critical of the industry's safety record, and less accepting of the old-time rules, after the region's worst fatal accident in 1895, when thirty men and boys perished of smoke inhalation in the Osceola mine. Most mines, including Osceola, did little timbering, so workers did not consider fire a prominent hazard. That belief led to the Osceola catastrophe. A shaft caught fire, and while some fought it, others, not fearful of the fire spreading, sat down to eat lunch. But the fire did spread along the shaft, and by the time the workers tried to escape through a neighboring shaft, it was too late. Smoke rose up the shaft on fire and drifted across the upper levels of stoped-out ground and down the adjacent shaft that the men thought was a safe exit. Instead, they died there.[24]

The Osceola fire claimed three times as many victims as the second most tragic mine accident on the Keweenaw, and it focused local attention on mine safety. Nationally, around the turn of the century, some significant accidents occurred at coal mines that moved worker safety onto the political agenda. In Utah, 201 died in a single incident;

in Pennsylvania, 239; and in West Virginia, 361. Such accidents, occurring as they did in the Progressive Era, prompted calls for reform and the creation in 1910 of the U.S. Bureau of Mines.[25] This new bureau collected data, employed scientists and engineers to improve mine safety, and trained mine workers in safe practices. The Bureau of Mines did not regulate the mining industry, but mine managers on Lake Superior recognized that this new federal presence put their operations under more scrutiny. At about the same time, in 1911 Michigan altered its system of county mine inspectors. No longer were they appointed by county boards of supervisors, which were dominated by mining interests. Instead, each county selected its mine inspector by popular election. This change made the mine inspector answerable to the general public. And instead of being required to visit a mine just once per year, by law an inspector checked on each mine every sixty days.[26]

Turn-of-the-century courts also challenged the old-time rules of mine safety and the law of torts. More Americans believed that rapid industrialization had been unfair, that too many workers had given up their lives, or their health and wholeness, for too little recompense. This change in social values encouraged more injured or maimed employees, or the heirs of killed workers, to take companies to court—and more often, as juries became more liberal, they won.

The Keweenaw reflected this national trend after it became home to attorney Patrick O'Brien, a champion of the injured or killed miner. O'Brien was a local boy who had gone through the Calumet schools and then attended law school in Indiana. While he was there, his father died in the C&H mine. When O'Brien graduated, he started specializing in personal injury lawsuits, first in the iron-mining region around Superior, Wisconsin. In 1899 he returned to Copper Country, took up residence in Laurium, and started suing mining companies. His business did well enough that he added a partner, Edward LeGendre, in 1910.[27]

For decades the law of torts one-sidedly favored employers. In the Progressive Era, this changed. Companies not only found themselves going to court more often—they also lost more often and sometimes paid out large sums. On the issue of liability and compensation, the pendulum was clearing swinging away from the mining companies and over toward the workers. The mine companies and their attorneys tried to hold the line against the growing number of lawsuits. As more cases came up, they adopted a harder stance. They stopped offering charity to accident victims because some interpreted charity as an admission of responsibility or guilt. They stopped initiating out-of-court settlements, too, because they started to believe that every time a worker got a dime out of them, it encouraged another worker to threaten suit. They fought every case they thought they had a chance to win and went toe-to-toe with plaintiffs' attorneys, especially O'Brien and LeGendre. But over time, all the lawsuits and losses wore the companies down. By 1910 they believed that every injury or death would lead to a bitter trial or difficult negotiation, and all the litigation, turmoil, legal fees, and monetary damages encouraged the companies to find a new way out: workers' compensation.[28]

Michigan enacted workers' compensation in 1912. The program gave something to workers and companies. If injured, workers were now entitled to payments without having to pray for charity or go to court. The law buffered workers and their families from financial calamity in the wake of a fatal or serious accident. For the first time, the companies were responsible for providing monetary awards to the dependents of fatal accident victims (a death benefit, for example, of $2,700) and awards to the injured and maimed (such as $4,000 spread over five hundred weeks to a permanently disabled miner). The mining companies, in turn, avoided endless lawsuits and gained a cap on the amount an employee or his heirs could receive for a given injury or death.[29]

Only five years before, the mining companies would have bitterly fought any legislation that made them pay for accidents. But by 1912 they more than acquiesced in workers' compensation; they actively supported it. When the stark fact hit home that they were now responsible for workers' compensation coverage or insurance, the companies finally abandoned one last vestige of their old-time rules—their belief that accidents were unavoidable. Once they had to pay for accidents, they finally believed in accident prevention. Starting in 1912–13, for the first time they rushed into print booklets such as "Rules and Regulations for the Protection of Employees." This safety literature listed things to do and not do at work, and it opined that workers and employers needed to work cooperatively to improve safety conditions. Companies suddenly espoused a deep concern for workers' welfare and set up Safety First programs up and down the mineral range. Companies initiated first-aid training, placed emergency medical supplies around the workplace, tapped the U.S. Bureau of Mines for safety training, and added "safety engineers" to their payroll.[30]

As the mining companies saw their hegemony in the district being nibbled away by new peoples, new ideas, new laws, and the Progressive Era, they continued to practice paternalism to try to keep the peace with many of their workers. They still saw paternalism as a way of instilling loyalty among the skilled workers they most wanted to keep. C&H even expanded its paternal programs. In the last three years of the nineteenth century, when C&H enjoyed its most profitable years to date, the company suspended workers' payments into the company's aid fund and medical program. For a time the company carried all the costs of these programs, saving its working-class families $54 of annual fees (representing about three-fourths of a month's pay).[31] The company in 1904 also instituted its first retirement plan. But true to their usual mode of operation, C&H managers kept this plan firmly in their grasp; they ran the pension plan in secret so they could control the demand for this benefit. Men over age sixty who had worked for C&H for at least twenty years were eligible for consideration for a pension but were not automatically entitled to one. The company did not publicize this program. General manager James MacNaughton decided how much money the company would spend on pensions, who would get one, and who would not.[32]

Besides owning over 3,000 houses, the mining companies owned the land under

1,750 houses that workers had erected on company property.[33] Because they politically controlled their locations and ran them as they saw fit, the era of modernization inevitably applied pressure upon the companies. Workers expected to enjoy a higher standard of living than prior generations. They asked their employers for new amenities and services, to be available across the location to renters and home owners alike. The mining companies had to address these requests but refused to rush headlong into giving workers everything they might have wanted. When the mines built new houses, in many ways they were decidedly bigger and better than those built from the 1840s through the 1880s. But in their newest houses, and certainly in their oldest ones, the companies dragged their feet when it came to outfitting, or retrofitting, them with modern utilities such as running water, toilets and sanitary sewers, central heat, electric light, or gas.

The companies offered stopgap measures that made some modern amenities available to workers and their families, even if not at their houses. Between the mid-1890s and 1920 the three largest companies on the range—C&H, Copper Range, and Quincy—all built libraries for workers. Curiously, they attached to each library a bathhouse, where workers and their families, for a nominal charge, could bathe.[34] The bathhouses took some pressure off the companies, which were trying to avoid connecting all houses to water and sewer lines.

When companies made utility upgrades, they sometimes did so in an odd and seemingly inefficient manner—yet one that saved them money. C&H at the turn of the century owned over seven hundred houses. The company was not about to upgrade them all at once at company expense. If the occupant of a company house wanted a phone, fine. But the worker had to pay all installation charges, and when the worker left, the phone line remained. C&H for a while started to wire its houses for electrical service but backed off and shifted much of the economic burden onto the occupants. An employee who wanted electricity had to contract with a local electrician for the work. An employee who left the company house could take his light fixtures with him, but the wiring had to stay. In the case of water and sewer lines and inside plumbing, tubs, and toilets, C&H made these things available over time—but not to everyone everywhere. When it ran new water and sewer lines down a street of company houses, it did not automatically connect them with each house and install a host of sinks, tubs, and toilets. Instead, it often made people ask individually for such amenities. That way, C&H clearly remained the boss; you got your first indoor toilet, typically set in a basement corner, if you asked for one and the company decided it was time for you to get one.[35]

C&H did better than most companies in modernizing its housing stock. By 1913 its 764 frame company houses all had inside water faucets and 325 had sewer connections. Across the district as a whole, only about half of the company houses had running water by 1913, and far fewer than that had toilets and sewers. Of the 607 company houses at the Copper Range mines, all had running water, but none had sewer connections. The

In 1898, C&H erected a handsome library for workers and their families that origi-nally included a bathhouse in the basement. Here, members of the working class could use their time in wholesome ways—education and personal hygiene. By 1900 the Library contained 25,000 volumes; by 1906, men took 24,675 baths here annu-ally, and women took 7,951. A circus parade passes by the library in 1916. *(Michigan Technological University Archives and Copper Country Historical Collections)*

179 company houses at the Mohawk mine lacked both amenities. At Quincy the managers' houses had water and electricity by 1913, but 440 worker houses still had neither.[36]

An important test of wills occurred when another symbol of modernization arrived at the turn of the century: electric streetcars. Eventually the streetcar line would run from Houghton, through Hancock, up to Laurium and Red Jacket village, and beyond to Mohawk and over to Lake Linden. C&H keenly felt that the proposed streetcar line challenged its hegemony in its own backyard. Shoppers, churchgoers, merchants, and seekers of entertainment wanted streetcars. But C&H's president, Alexander Agassiz, did not want streetcar lines running in the vicinity of his mine. For one thing, he did not want streetcars to interfere with the free passage of C&H rock and coal cars on its own railroad. And he did not want streetcars to make it too easy for people to congre-gate in large groups—groups of men who might engage in union or other anticompany activities. If they wanted to attend a meeting, make them walk. Agassiz mustered his company's troops to try to fend off the streetcars. He pitted local society's desire for one modern thing against another. C&H pumped and sold water to both Red Jacket village and to Laurium. C&H hinted to both villages that the company might cut off their water supply over this issue. Laurium also wanted to run a new sewer line that crossed under C&H property. Agassiz told his mine agent, S. B. Whiting, to tell Laurium that "they can have a sewer or R.R. [railroad] but not both."[37]

Agassiz lost the contest; the streetcar line was built. This political tempest, albeit short-lived, was significant. For Agassiz, the incident represented an erosion of mining-company power, an erosion of local society's willingness to place their own interests below the interests of the mining company. From 1900 to his death in 1910, Agassiz and C&H took measures to prop up their dominance in the region and to show more corporate resolve and fight in the face of discontented employees and unwanted social changes. Other companies followed C&H's lead.

The companies had escaped much of the labor violence that had beset other mining districts facing union battles in the late nineteenth and early twentieth centuries, and traditionally they had seen unions as more bothersome than threatening. The companies generally chose not to overreact publicly to unionization attempts, including those undertaken in 1903–4 and again in 1908–10 when the Western Federation of Miners set up local unions in the district.[38] The WFM had been a major foe of mining interests elsewhere, and over time its leadership had grown more radical, more socialist. Rather than trying to stamp out unionism and socialism in a speedy, direct, and confrontational manner, the mining companies headed off these threats in more covert ways.

The mining companies engaged in a range of interrelated activities meant to minimize the threat posed by radical ideas or persons. The Quincy mine, and probably several other companies, hired company spies whose secret services they acquired through detective agencies. Spies worked alongside men, both underground and on the surface; they frequented the same bars and hangouts. Then they told tales about them in letters sent to the detective company. The detective company typed the letters up and mailed them off to the mine agent, who read them as a glimpse into the working-class culture that he suspected harbored ill-will and bad politics.[39]

At C&H, if they suspected a man of unionism, they did not want to make him a cause célèbre, so they did not emphatically march him out of the mine. The company sought some other reason, rather than unionism, for firing him and then did it. C&H headed off unionism in other ways. It would not hire men coming to Lake Superior from other mining regions known to be on strike; it definitely did not want to mistakenly hire any men deported from union battlegrounds like Colorado or Idaho. C&H was also loath to hire men, such as foundry workers, who came from industrial centers such as Milwaukee, where labor-management turmoil ran deep.[40]

In contesting unionism and socialism, the mining companies—especially C&H—thought they could thwart unwanted ideas and actions by eliminating select people

Mining companies believed that Finns were their most fractious employees—the ones who complained about working conditions and embraced radical ideas like socialism. Many conservative Finns wanted to get along in Copper Country society, but other Finns were outspoken critics of capitalism and strong proponents of unionism. Starting in 1904, in Hancock they published a Finnish-language paper, Työmies (The Worker) that espoused working-class causes. *(Finnish American Heritage Center, Finlandia University)*

thought most likely to preach or to succumb to such ideas. They practiced ethnic discrimination. By early in the twentieth century, mine managers all across the Upper Great Lakes—at copper and iron mines both—increasingly blamed Finns for labor problems. C&H began purging many Finns from its payroll, replacing them with southern and especially eastern Europeans deemed more tractable. By 1913, Finns comprised the largest foreign-born ethnic group in Houghton Country. But if you looked at the 2,200 underground workers at C&H, Finns (both foreign and native born) were only the fourth largest ethnic group represented there. Their employment trailed that of the Austrians, Italians, and Cornish.[41]

The mines practiced ethnic discrimination in many ways: in hiring and firing, in the level of jobs awarded, and in the number and type of company houses rented. At C&H, superintendent MacNaughton believed that discrimination was a smart way of doing business. To avoid having to hire the "riff-raff" off the streets of America's larger cities, he knew he needed thousands of Europeans. But as he informed the new U.S. Department of Labor and the Immigration Service, he wanted some Europeans and didn't want others, for reasons ranging from physical stature to politics to work experience. He most preferred Swedes, Germans, northern Italians, and the eastern Europeans, who on Lake Superior were called "Austrians." He did not want "fruit-peddlers and the like," persons of small stature and soft hands, characteristics he associated with Greeks and other Mediterraneans.[42] MacNaughton was emphatic to federal officials: "We do not want Finlanders." Besides discouraging Finnish migration to the region, limiting the job opportunities for Finns, and firing many, the mining companies also sought to control this group by covertly having a financial interest in a conservative Finnish newspaper. Leery of most Finnish papers, which they saw as radical, the mine managers wanted to assure that the region's Finnish population had access to a paper more favorable to companies, capitalism, and assimilation. So they secretly underwrote the publication of such a paper: *Päivälehti* (Daily Paper).[43]

Overall, in the two decades after 1890, as the mining companies expanded dramatically in terms of production and employment, they understood that not all was well. They no longer dominated the copper market. They worked deep mines with high production costs. They operated in a social setting where the interests of the mining companies, and the interests of society as a whole, were more at odds than before. The mining companies found it more difficult to work their will on Lake Superior. At their locations they found it more difficult, and sometimes impossible, to keep modernization at bay. Much vigilance seemed needed to keep union men and socialists off payrolls and local streets. Companies found they could not dodge Progressive Era reforms. They had to deal with the Bureau of Mines and had been pushed into things like workers' compensation and a Safety First movement.

By 1900 to 1912 the companies lived in a world of new conditions—one they felt less comfortable in, and one that caused them discontent. Soon, another big challenge to their authority would arise in 1913—the region's greatest labor strike. The strike became a test of wills that the mining companies decided they would not, and could not, lose. By prosecuting it with vigor and determination, they would once again establish themselves as dominant within the region.

15

❖ ❖ ❖

SHOW THEM WHO'S BOSS

The Strike of 1913–1914

For over half a century, the Lake Superior copper mines had few major battles with workers. That is not to say that harmony always ruled. Labor shortages, labor surpluses, falling copper prices, rising copper prices, industrial contraction, and industrial expansion—these things often tested the peace because they tended to separate workers and managers. What was good for one usually worked against the other. Consequently, strikes and attempts to unionize had occasionally disrupted operations at the mines.

The International Workingmen's Association organized in the region in the early 1870s, and at the same time, the Portage Lake mines and Calumet and Hecla (C&H) weathered strikes in 1872 and 1874. During the turbulent 1880s, a decade of many major labor revolts elsewhere, the Knights of Labor spawned a labor movement at the mines. The region was hit by a trammer revolt in 1893, and Quincy had another one at the turn of the century, when it tried to cut trammers' wages after introducing electric haulage locomotives underground. In 1903–4 the Western Federation of Miners initiated its first unionization attempts in the Copper Country. Shortly thereafter, walkouts plagued Quincy in 1904 and 1905, and a three-week strike erupted there in 1906. That same year, Finnish trammers struck the Michigan mine in Ontonagon County, hoping to achieve higher wages and force the hiring of some Finnish trammer bosses. Instead, they ran into sheriff's deputies, a riot ensued, and two Finnish trammers ended up dead.[1]

Without doubt, labor problems were on the rise after 1900, but still the region had not witnessed any sustained labor rebellion on a broad front. The Finnish trammer strike of 1906 resulted in two deaths, but that conflict was short-lived and occurred at a small producer, miles away from the core mining district. Not until mid-1913 did the Copper Country experience its first epic labor battle. It is always called the "strike of 1913–14," but in truth it was more than a strike. It was a major social upheaval.[2]

The mining companies felt themselves under pressure on several fronts. Local society, state and federal legislatures, and courts seemed bent on overturning all the old-time rules the companies had thrived under. Economically, they were the deepest mines in the country, working the lowest-grade copper deposits. Many lodes were pinching down, and their copper values were falling off. The companies felt vulnerable. They needed to become more competitive by reducing their production costs. They needed efficiency and higher productivity. They needed scientific management, cost accounting, and new technologies.

Underground workers felt just as vulnerable on their side of the equation. They worked for large companies whose employees numbered in the thousands. Everything they saw and heard told the newer arrivals from Europe—the Finns, the southern and eastern Europeans—that their bosses at best merely tolerated them because they needed them so badly as the mines expanded after 1890. The men knew they were often scorned, denigrated, and held in low esteem by their bosses because the men weren't skilled enough, weren't steeped in mining traditions, weren't fully assimilated into the industrial culture they had migrated to.

Many men chafed at the unequal distribution of paternal benefits doled out by the companies. They groused about having lower pay and longer hours than miners in Butte, Montana. They believed the modern gospel of efficiency, for them, spelt only trouble because it meant work speedups. They felt the underground, warmer now, maybe in the 80s, was an uncomfortable—and unsafe—place to work. At Quincy, "man-made earthquakes" or air blasts happened all the time, making the men nervous and fearful. Across the industry, like clockwork, about a man a week died on the job.

The breech between labor and management grew wider in the early twentieth century. Still, an uneasy peace might have continued for years except for one thing: the one-man drill. The companies thought they had to have it, even if they had to force it into place. Workers thought they had to resist this machine at all costs. The one-man drill brought conflict to a head because the math was simple. If it could do the work of a two-man machine, then the mines would need half as many miners. Besides promising to devastate the ranks of miners, the one-man drill promised to devastate future miners— the trammers who hoped one day to become miners. Miners and trammers had long lived as two classes of workmen who had little truck with one another. Miners were the superior ones, and the companies treated them as such. But now, miners felt more akin to trammers than ever before because they felt mistreated by their companies. This new technology was aimed at them. The threat of this new machine forged a bond between two groups that now shared strongly felt grievances.[3]

The one-man drill wasn't the sole cause of the strike of 1913–14, but it was the strike's trigger. The Western Federation of Miners (WFM) pulled the trigger and fired the first shot. The mining companies fired right back with a total commitment to win this confrontation. Once the union started the strike, the companies became bent on finishing it. This was their chance to reassert their control in the region and show everyone who was boss. The strike began on 23 July 1913 and lasted until 12 April 1914. In between, all hell broke lose on the copper range.

Until early in 1913 the WFM had only modest success in establishing local unions at the mines. WFM membership lagged. But it became apparent during the winter of 1912 that companies were preparing to replace two-man drills with one man machines. As was usually the case on Lake Superior, labor unrest slept during cold months, only to emerge during warm months. Early in the spring and summer of 1913, the one-man drill galvanized the men, and membership in the WFM mushroomed from a thousand to about seven thousand. Virtually all these members worked underground. Surface workers at the shops, mills, smelters, and railroads stayed out of the fray. Into the fray went Finnish, Croatian, Slovenian, and Italian trammers, not unexpectedly, since they were long deemed the fractious ones. But into the fray, too, went the miners, including Cornishmen not accustomed at all to such rebellion.[4]

The WFM had a tradition of radical and staunch unionism. That tradition got them into this fight, kept them in this fight, and made them pay too great a price for inevitably losing this fight. The WFM, headquartered in Colorado, had won and lost many battles in the West; it wanted to expand into the East and organize in Michigan. But signing new members into nascent locals and calling for a strike there against entrenched mining firms—those were two entirely different things. The executive board of the WFM, led by Charles Moyer, knew that the union had less than $23,000 in the bank and that the likes of C&H, Copper Range, and Quincy would be hard to beat in any showdown.[5] The executive directors of the WFM did not initiate the strike of 1913–14, nor did they want it at that time.[6] This strike came about because the WFM locals on the copper range got out in front of the union's national leadership. The Keweenaw men wanted a strike in the summer to thwart the introduction of the new drills. The locals held a membership referendum early in July—should we ask to meet with the companies to press our grievances? Should we strike if the mines refuse to meet with us or concede our grievances? Backed by a majority of "yes" votes to both questions, the local WFM leaders sent letters, not approved by the national WFM, to the mining companies. The letters stated grievances—and gave warning. Should the companies "follow the example given by some of the most stupid and unfair mine owners in the past" and refuse to meet with WFM representatives, then they would lose the opportunity "to have the matters settled peacefully" and could expect a major strike.[7]

The mining companies did not respond to their letters. A response would have been tantamount to recognizing the existence of the WFM, and they were not about to do

Alexander Agassiz once fought to keep streetcars out of Red Jacket village and Calumet Township, in part because they would make it too easy for men to travel to
anticompany meetings. On 8 June 1913, a large crowd of union men attended a
meeting at the Calumet Theatre called by the Western Federation of Miners, a union
whose ranks were swelling up and down the mineral range. In their midst ran one of
the dreaded streetcars that Agassiz had tried, unsuccessfully, to keep out. *(Michigan
Technological University Archives and Copper Country Historical Collections)*

that. The fateful course was set. The locals had thrown down the gauntlet, the companies
ignored the locals, and the WFM's executive directors were loath to rein in the locals or
stall the promised strike. This strident union couldn't retract threats without losing face,
so the men took to the streets.

The WFM locals had specific goals in mind. First and foremost, they needed to force
the companies to recognize the WFM as the collective bargaining agent for workers. No
longer would the companies be able to set wages and hours and benefits on their own;
they would have to negotiate these things and enter into labor agreements. Beyond that,
the WFM's most important aims were a minimum wage of three dollars for all underground workers, an eight-hour workday (down from the current nine hours), and two
men on all drills.[8]

The WFM knew the mining companies would not willingly concede anything, so
their principal tactic was to force negotiations by shutting down the mines and inflicting
economic pain upon their employers. At the negotiating table, the companies would have
to bow to union demands in order to get their mines back up and running. While bat-
tling the companies, the WFM hoped to muster popular support. The members wanted
the public to see them as the downtrodden whose demands were warranted and just.

The companies did not go off in separate ways to battle the WFM. C&H managers and attorneys called the shots. "Big Jim" MacNaughton led the way, and other companies fell in behind. The companies' goal was to lose nothing important over the strike and to regain greater control over the region. They had recently seen their hegemony challenged in many ways, and for a few years they had been treading gingerly on the issue of adopting one-man drills. They would kick the WFM out and bring in the one-man drill. While waging the battle, they would present themselves as beleaguered companies, as good corporate citizens with a long history of benevolent paternalism. They were being hounded by outside labor agitators, socialists, and anarchists, who were ruining a good thing for all the hardworking men in the district. As the strike wore on, it became apparent that the companies intended not just to win, but to crush this flirtation with unionism. Early in the strike, Big Jim MacNaughton had written to his company's president, Quincy A. Shaw Jr., that the mines were "all of one opinion, namely, that the Union must be killed at all costs." Late in the strike, MacNaughton pressed for a dramatic kill: "If we want to be insured against a repetition of this thing within the next 15 or 20 years we have got to rub it into them now that we have them down."[9]

Over many months, the companies never gave the union any hope of a settlement. They refused to mediate the situation and never sat down with the WFM. Pleas to negotiate a strike settlement came from Michigan's governor, the federal labor bureaucracy, and the U.S. Congress, but all came to naught. The mining companies adopted the demeanor of imperial arrogance. They were using this strike to take back control of their region. Governor Woodbridge Ferris gave up hope of brokering some sort of negotiation: "When James MacNaughton says that he will let grass grow in the streets before he will ever treat with the Western Federation of Miners or its representatives, I believe what he says." A U.S. Congressman, during an investigation held on the Keweenaw early in 1914, asked the MacNaughton about letting some high authority, some disinterested power, arbitrate the dispute. Would he not listen to the governor or even the president? In response, MacNaughton reached into his pocket and brandished its contents at the Congressman: "This is my pocketbook. I won't arbitrate with you as to whose pocketbook this is. It is mine. Now, it would be foolish to arbitrate that question. I have decided it in my own mind."[10]

While forever refusing to negotiate, the companies did acquiesce in a short shutdown the union forced upon them. The companies knew they could weather a loss of income far better than their employees. They knew they were wealthy, while the WFM was poor. WFM headquarters had less than $23,000 to its name. On the opposite side, C&H had sixteen million pounds of copper ingot to sell and an additional twelve to thirteen million pounds of mineral that, when smelted, would yield another eight million pounds of copper.[11] The company had $1 million in cash and another $1.27 million due in bills receivable. So the mining companies welcomed a shutdown because it would lay some economic hurt on the strikers—while it would aggravate loyal men suddenly shut out from their work, thus increasing their antiunion sentiment. After accepting a brief shutdown, then the mines would reopen on a limited scale, using imported men or *scabs* brought into the district on guarded, armed trains. Loyal workers would join the scabs in restarting production. As the strike dragged on, more disenchanted strikers would abandon the union and return to work. Finally, the union would quit. At that time, the companies would take back rank-and-file strikers, those who had learned their lesson about organized labor the hard way and swore off allegiance to or membership in the WFM. But they would never take back strike leaders.

Over the run of the strike, the opponents contested high moral ground. Who was the victim in this struggle and who was the villain? Who occupied high moral ground and who just lost it? The WFM lost high moral ground right at the start. The union felt it had to shut down the mines—and there was no polite way to accomplish that. On the afternoon of 23 July and on the morning of 24 July, the strike started up at the C&H mine with street skirmishes and violence.[12] Strikers used fisticuffs, brickbats, and clubs and hurled insults to chase away the men coming off one shift. They battered some heads and bodies and got what they wanted—the violence caused Big Jim MacNaughton to close the mine. But the companies got what they wanted, too. The rebellious men roaming the streets had resorted to violence to work their will. It looked like the companies were the victims; the strikers, the villains. At C&H, MacNaughton could not have scripted a better opening scene in terms of public relations. He wrote to company president, Quincy Shaw:

> If we had planned the whole affair beforehand we could not have played into our own hands any better than the strikers did. The mob violence practiced by them put us in the best possible position. Outside of the ranks of the strikers themselves there is absolutely no sympathy for them anywhere.[13]

Late at night on 23 July, concerned over the safety of men and mine property, Mac-Naughton (who happened to head the Houghton County Board of Supervisors), met with Houghton County Sheriff Jim Cruse, a man MacNaughton believed would handle this strike in whatever manner the companies asked of him. Having just won high moral ground, MacNaughton wanted to avoid, for the time being, arming hundreds of quickly deputized men at C&H with guns; let them have billy clubs, nothing more.[14] Cruse and MacNaughton decided to appeal to the Democratic governor, Woodbridge Ferris, to send the Michigan National Guard up north. Cruse sent a telegram to Ferris at 2:00 a.m. on 24 July, and later that day Ferris detailed virtually the entire National Guard, some 2,765 men, up to the mines.[15] Soon the mines looked like military encampments. Men bivouacked in tents pitched around C&H mine and atop Quincy hill and elsewhere. Foot soldiers stood watch and drilled; mounted cavalry rode the streets.

With the coming of the National Guard, an uneasy calm held around the mines over the next few weeks while the companies acquiesced in their temporary shutdowns.

At 2:00 a.m. on 24 July 1913, a day after the strike started, Houghton county sheriff James Cruse telegraphed Governor Woodbridge Ferris, asking him to send troops to protect the peace and property at the copper mines. Governor Ferris responded by ordering the Michigan National Guard up to the Copper Country. Guardsmen, including those shown in their tent encampment in the shadow of C&H's Red Jacket shaft, bivouacked right at the mines. *(Keweenaw National Historical Park Archives)*

Some men left the district. Some, after briefly flirting with the WFM, started to defect, while others held firm. Hostile confrontations were infrequent, thanks to the presence of the National Guard, but Governor Ferris couldn't keep the guard up there for long. Besides the cost of it, he worried about Guardsmen being seen not as peacekeepers, but as strikebreakers. The Guard started to withdraw half its forces in mid-August, three weeks into the strike. The mining companies, in anticipation of the withdrawal, deputized six hundred local men loyal to their cause. They also hired fifty men—the companies called them guards, the WFM called them thugs or goons—from the Waddell-Mahon Corporation in New York City.[16] The deputies and Waddell men would protect the companies when they reopened.

Putting loyal men, WFM defectors, and imported men underground, and starting to hoist rock again, was more an attempt to break the union than an attempt to begin serious production. Companies started reopening their mines just when the National Guardsmen began leaving. This poorly timed initiative raised tensions and excited confrontations between strikers and the men guarding the mines, the poorly trained deputies and the hired muscle, the Waddell men, out of New York. At the Champion mine, one confrontation led to a hail of bullets and two deaths.

On 14 August, after drinking beer in South Range for some time, strikers John Kalan and John Stimac walked several miles back toward the company house where they boarded in Seeberville, one of the Champion's neighborhoods. Nearing home, they took a familiar path across mine property, but that property, near major shafts, was deemed off-limits to strikers and was patrolled by a company watchman.[17] The watchman tried to head off Kalan and Stimac, they exchanged some verbal insults, and then the strikers continued on their way. A little later, a Waddell man, Thomas Raleigh, talked to the watchman and learned of the strikers' belligerence and trespass. Raleigh gathered up several more men and marched his armed squad over to Seeberville, intent on gathering up Kalan and Stimac and taking them back to the mine. When the men refused to go, a fight broke out. The fight led to gunshots, all coming from the company men who fired round after round into the crowded boardinghouse. Bullets struck four men, all Croatians: Steve Putrich, Alois Tijan, Stanko Stepic, and John Stimac. Tijan died instantly, and Putrich died the next day.[18]

With these two killings, high moral ground shifted to the WFM and its men, who now claimed their first martyrs. The companies had decried strike leaders as outside troublemakers, but their outside thugs had perpetrated gun violence. About five thousand mourners showed up at the funeral for Tijan and Putrich, and their deaths reenergized the strike, strengthened the union's resolve, and intensified the conflict. Strikers were encouraged by the WFM's promise to pay strike benefits of three dollars per week for a single man, up to seven dollars weekly for a married man with five or more children, and a maximum, in case of emergency, of up to nine dollars per week. To raise this money, the union levied assessments of up to two dollars per month on all its members outside Michigan, plus it borrowed money from other unions and established a defense fund and accepted contributions. With martyrs and a promise of soon-to-be-received benefits, the strikers gained confidence and resolved to stay the course.[19]

By mid to late August, the mining companies had gained confidence, too. Thanks to early union defectors, part of the strike movement had already collapsed, and the companies had reopened their mines, albeit on a modest scale. Mine managers thought the union was doomed, and that the strike would collapse when summer gave way to fall and winter loomed on the horizon. Surely the men would capitulate by then. In the meantime the companies would increase the pressure upon strikers by importing more replacement workers, which would allow them to augment underground operations.

It turned out that the union and the companies were both wrong in how they perceived the future course of the strike. But the union was more wrong than the companies. In effect, this strike was over in mid-August as soon as C&H, Copper Range, and Quincy resumed mining. There was nothing the union could do after that to force the unwavering companies to the bargaining table. They weren't going to do that, period. Instead, they continued to take back disaffected union men, brought in more replacements, and slowly built up their production levels. Basically, with the deaths of the Seeberville men, the visit of union president Charles Moyer to the Copper Country shortly thereafter, and the providing of strike benefits, the union raised the stakes of the strike and prolonged it. But since

During the strike, both sides had their martyrs. On 14 August 1913, men guarding the Champion mine in Painesdale fired into a Croatian boardinghouse in Seeberville, killing strikers Alois Tijan and Steve Putrich. A few days later a well-orchestrated funeral parade moved along the main commercial street in Red Jacket. The Western Federation of Miners made sure that the killings did not go unnoticed. *(Michigan Technological University Archives and Copper Country Historical Collections)*

the mines had already reopened, the strike was lost. Failing to recognize that, and reinforcing the strike rather than winding it down, cost the union and its members dearly.

For their part, the mine managers were right in recognizing the tactical importance of reopening the mines, but they were wrong in underestimating the amount of fight left in the WFM membership, whose holdouts were predominantly Finns, Hungarians, and Croatians. The companies expected the strike to collapse much sooner than it did. The dragging on of the conflict, however, cost the union far more than it cost the employers.

In September the labor struggle intensified. The WFM did more parading, more picketing, and more taunting of the men going underground. Striking men were not the only ones taking to the streets. Their women were there, too, mothers, wives, sisters, and even daughters. They hurled insults, and sometimes they hurled bucketsful of human excrement, drawn up from privies, at the scabs.[20] Women and men were arrested in considerable numbers at times. The National Guardsmen still in place struggled to keep the peace but sometimes exacerbated the conflict. "Big Annie" Clemenc, a local, became a labor hero when a Guardsman raked her wrist with a bayonet while she marched draped in an American flag. Up at Kearsarge, north of Calumet, Guardsmen fired their weapons when confronted by two hundred marchers protesting the use of scabs. One bullet resulted in a serious head wound to a fourteen-year-old girl, Margaret Fazekas. To many, the Guardsmen, too, had lost high moral ground and too closely resembled the deputies and thugs hired by the mining companies.[21]

The mining companies continued to import workers in September and October. They housed them in hastily built bunkhouses near the shafts and protected them with deputies. By October the National Guard force had fallen off to only a couple hundred men, but Sheriff Cruse had deputized twelve hundred men. C&H alone had four hundred "Deputies and Hotel Men" to protect life and property. High atop its 150-foot-tall No. 2 shaft-rockhouse, the Quincy mine erected an observation station so guards could watch over the entire mine site.[22]

Besides using hundreds of deputies and hired thugs, the companies sought to maintain control of the explosive situation through the courts. The seventeen mining companies being struck had a covert agreement to have their legal battles during the strike handled by C&H's principal attorneys, the firm of Rees, Robinson & Petermann. When the strikers increased their street actions, which coincided with the companies' importation of more men, the attorneys sought an injunction to prohibit picketing and the harassment of men going to work. This request for an injunction went to Judge Patrick O'Brien—who as a practicing attorney had often sued and beaten the companies in personal injury lawsuits.[23] O'Brien's politics clearly put him in the camp of the workers, but as a judge he wanted to reinforce respect for law and order. He wanted to honor the strikers' rights to assembly and to free speech but also wanted to avoid dangerous confrontations in the streets. As a consequence, his judicial decisions tilted one way and then the other, winning him enemies and critics on both sides.[24] Initially, on 20 September, he granted the companies their injunction, but then he lifted it nine days later. Rees, Robinson & Petermann requested a second injunction, and this time O'Brien denied it. The companies' attorneys then appealed to the Michigan Supreme Court, and on 8 October the high court reinstated the injunction.

Shortly thereafter, picketing and taunting led to sporadic street violence, and O'Brien ordered the injunction enforced. Local officials happily obliged, and in two instances they arrested 141 strikers at Allouez and another 68 at Mohawk.[25] When those arrested appeared before Judge O'Brien, he tilted back toward the men. Instead of throwing the book at them, he encouraged them to obey the law and opined that circumstances surrounding the strike had goaded them into violating it. In short, if they had been treated better, they would have behaved better. Instead of immediately prosecuting the men, O'Brien released them on their own recognizance. Companies and their allies now faulted O'Brien as coddling agitators, and many citizens were growing tired of the turmoil around the mines, which O'Brien seemed reluctant to quash.

By the end of October, the WFM's leaders knew they were whipped. They had raised money to sustain the battle; they had intensified it in the weeks and months after the Seeberville killings; they had given locals the gate to prosecute the strike with vigor in the streets. Now they quietly looked for a face-saving way of ending it. Union officials turned to Judge O'Brien to see whether he might help them broker a deal to end the strike. They would quit, they said, and give up their demands, if the companies would just agree to take men back without requiring them to withdraw from the union. O'Brien met with Jim MacNaughton to discuss the deal. MacNaughton, knowing that victory was certain, was in no hurry to end the strike in a conciliatory manner. He nixed the deal.[26]

The companies at nearly the same time gained an ally in an organization called the Citizens' Alliance. Covertly, the mining companies and their attorneys helped form this organization and then helped fund its activities and its publication, called the *Truth*, which sought to counter the WFM's publication, the *Miners' Bulletin*. The Citizens' Alliance wore the mask of a grassroots movement of Copper Country residents who simply wanted to restore peace and tranquility to the region. But the Citizens' Alliance had no doubt as to who had destroyed peace and tranquility in the first place: the union, whose "poisonous propaganda of destructive socialism, violence, intimidation, and disregard of law and order" had menaced the Copper Country long enough: "The Western Federation of Miners must go."[27]

The Citizens Alliance claimed 5,236 members who condemned the union and wanted it gone. To show solidarity, all members were supposed to wear a Citizens' Alliance button or pin when out in public. On 27 November the Citizens Alliance turned up the heat on strikers by publishing its first issue of *Truth*, which was distributed free of charge throughout the district. A day later the mining companies did their part to undercut the union. They voluntarily announced that, beginning 1 December, they would introduce an eight-hour workday and new grievance procedures for men who felt they had been wronged.[28] The mining companies, not long before, had fought against an eight-hour workday but knew it was going to be forced on them sooner or later as a part of Progressive Era reforms. So they chose to surrender it in the midst of the strike. They surrendered it, so they said, not because of the pressure applied by the WFM, but because the companies felt it was the right and just thing to do. The companies protected their claim to high moral ground by benevolently presenting the men with a shorter workday and new ways to address wrongs. The companies hoped their moves would encourage remaining strikers to bolt the union and return to work before hard winter set in. Surely, the union would now collapse. But it didn't happen. Instead, the worst month of the strike lay just ahead: December of 1913.

On 5 December, Judge O'Brien held the final hearing for 139 persons arrested in late October for harassing company workers at the Allouez streetcar station. Their actions had violated the court's injunction, and O'Brien found them all guilty. Then, he sought justice for the strikers in his own way: he suspended all their sentences. Going beyond that, O'Brien opined that the men were not criminals but "were engaged in a heroic struggle for the mere right to retain their membership in a labor organization."[29] The mining companies had incited these men by doing everything possible to "increase their bitterness and hostility."

Judge O'Brien awarded high moral ground to the strikers. The Citizens' Alliance responded quickly and strongly criticized O'Brien for virtually inviting the strikers to follow a violent and lawless path. The next day, on 7 December, strikers perpetrated an act of violence that set off a firestorm of ill-will and sent the district reeling.

Arthur and Harry Jane, brothers in their twenties, had left the copper district early in the strike to avoid the region's troubles. They chose to return and came back to the Champion mine on 6 December. They boarded with a fellow Cornishman, Thomas Dally. At 2:00 a.m. on 7 December, bullets ripped into houses at Champion, including Dally's. The Jane brothers, both hit, died immediately, and Thomas Dally died of a head wound the next day. In a neighboring house, a rifle shot wounded a thirteen-year-old girl. The men doing the shooting, trying to scare off scabs and returning workers, were two rank-and-file union men, Hjalmer Jallonen and John Juuntunen, plus John Huhta, once the secretary of the South Range WFM local, and Nick Verbanac, a paid WFM organizer.[30]

The shooters' identities were not known at the time, but the Citizens' Alliance responded to the murders with alacrity and decisiveness. Their fears had been realized. O'Brien's liberal treatment of the WFM had virtually sanctioned lawlessness. As a consequence, good Cornish workers lay dead at the Champion mine while the WFM's "murder inciting mercenaries" ran free. Antiunion sentiment boiled over, the fire fanned by the English-language press and by the Alliance. On 8 December, the *Daily Mining Gazette* in Houghton shouted, "FOREIGN AGITATORS MUST BE DRIVEN FROM THE DISTRICT AT ONCE." Fearing anarchy, the Citizens' Alliance nearly endorsed vigilantism. The Alliance quickly organized large rallies in Calumet and Houghton on 10 December. They added special trains to bring in the crowds. The companies not only paid for the trains; they also let their employees off work so they could attend the rallies and hear good citizens condemn the WFM's "poisonous slime" and its "reign of terror."[31]

In the weeks that followed, the last of the die-hard union men and their families endured more ostracism and condemnation. Sheriff's deputies and Alliance members harassed them; many WFM houses and enclaves were raided; some serious but nonfatal shootings occurred.[32] Still, the union locals did not collapse. Even Big Jim MacNaughton at mighty C&H begrudgingly tipped his hat to the tenacity and steadfastness of the strikers. The workers had hung tough longer then he thought they could or would. On 17 December he wrote Quincy Shaw that "the leaders are holding them together very well and but few of them are returning to work."[33] The next day, on 18 December, managers of the various mines tried another gambit to convince men to quit their union; they announced that all men who did not return to work immediately would lose their jobs permanently; there would be no taking them back. During this very troubled Christmas season, local merchants appealed to MacNaughton to stay this plan; a week before Christmas was no time to enforce such a harsh ultimatum. Make the deadline for returning to work after Christmas, they asked. Make it after the first of the year. MacNaughton agreed.[34] He, too, was tired and worn down by the long strike. He was "terribly blue and depressed at times," and this delay promised a bit of respite over the holiday.[35] But it was not to be.

On Christmas Eve in 1913, strikers and their families, mostly Finns, plus many Croatians and Slovenians and some Italians, enjoyed a Christmas party on the second floor of Italian Hall in Calumet. It was meant as a moment of celebration in a time of trouble. Adults and children gathered early in the afternoon, and by 2 o'clock, some 175 adults and 500 children partied upstairs in the hall. They heard some speeches, sang some carols. Children got to visit Santa Claus, and each received and enjoyed a rare treat, however modest it was. The party had started to break up, and many had left, thankfully. Because at about 4:30, all hell broke loose. People—adults and children—stampeded for the stairs and raced down to the exit doors. A few in the lead pushed the doors open and escaped. But, behind them, people tripped and fell. They were buried and crushed under the next to come down, who were buried by still others coming down. As a result of the panic, seventy-three people, packed in tightly and incapable of breathing, died in the stairwell late that afternoon, including about fifty Finns and twenty Croatians and Slovenians. About sixty of the dead were children from two to sixteen years of age.[36] Italian Hall claimed far more lives than any mine accident or any other catastrophe ever on the Keweenaw.

The exact sequence of events at Italian Hall, from the start of the panic through the screams and cries and death moans, cannot be reconstructed with certainty. In the midst of such panic, who was the calm and collected one, the careful observer, who witnessed the event from the very start, who survived the event, and who then recalled it accurately and completely? In the hypercharged political and contentious era of the strike, who in the Copper Country could tell the story of Italian Hall honestly without having it colored by months of inflammatory incidents or by an allegiance sworn to the WFM or the Citizens' Alliance? Truth was hard to come by in December of 1913. For some, like the editors of the important Finnish newspaper *Työmies* (The Worker), published in Hancock, what happened at Italian Hall was murder, pure and simple.[37] For others, it was a terrible prank, gone horribly awry. For still others, it was an inexplicable tragedy. Some thought the sheer loss of humanity would serve to unite the two sides of the strike; common grief would draw them together. Instead, the opposite happened. The deaths and accusations and intemperate language drove the sides ever further apart.

Usually, the Italian Hall disaster is laid at the feet of a cry of "Fire!" that triggered the mass hysteria. There was no fire, but surely something triggered the stampede.[38] Many stories of Italian Hall were recounted in homes and bars, in local English and Finnish newspapers, in other papers across the United States, and in a three-day-long coroner's inquest where seventy people testified. Much later, the story was told by folksinger Woody Guthrie in his song "1913 Massacre." Among the stories recounted locally in the hours and days immediately following the disaster, some said a man came upstairs

Seventy-three people, mostly children, died in the Italian Hall disaster at Red Jacket on Christmas Eve, 1913. After attending funerals and joining in a lengthy funeral procession out to Lake View Cemetery, a crowd of mourners stood along one of two mass graves where many of the victims were laid to rest. Twenty-eight Protestants lay side by side in one grave; twenty-five Catholics in another. *(Michigan Technological University Archives and Copper Country Historical Collections)*

wearing a Citizens' Alliance button on his coat and yelled "fire!" to set off an alarm and ruin the party. Others said they saw the man, heard the cry, but saw no button. Another story had it that deputies or Citizens' Alliance men placed obstacles near the exit doors, hindering safe escape. Various conspiracy theories immediately surfaced, which in a twisted way intermingled with everyone's grief and shock. Facts were hard to come by in the aftermath of the disaster; people couldn't even agree on just how many had died. But accusations abounded. The president of the WFM, Charles Moyer, happened to be at the mines this Christmastime. He infuriated many procompany local residents when, like *Työmies,* he almost immediately pinned the blame for the Italian Hall disaster on a Citizens' Alliance member's cry of fire.

In the name of charity, Copper Country residents instinctively and immediately set up a relief fund earmarked for families of the dead. Many contributions came from mine officials, merchants, and supporters of the companies. By noon, the day after Christmas, $25,000 had been collected. Then the president of the WFM offered up a huge and hurtful snub. On behalf of the strikers and their families, living and dead, Moyer rejected any aid or charity. He made it clear that "no aid will be accepted from any of these citizens who a short time ago denounced these people as undesirable citizens."[39] The managerial class had always refused to recognize the WFM; now the WFM refused the right of that class to share in the grief and tragedy of Italian Hall.

The denial of their grief, the refusal of their charity, and the charges leveled at the Citizens' Alliance terribly angered many antiunion residents and called forth a harsh and decisive reaction. On the evening of 26 December, a half-dozen men went to Charles Moyer's room at the Scott Hotel in Hancock. In the scuffle and manhandling that ensued, Moyer suffered a gunshot wound to the back. The men then hauled him outside and delivered him to another forty men who dragged him over the bridge to Houghton and tossed him on a train bound for Chicago.[40]

The Italian Hall tragedy caused the U.S. Congress in February to hold an investigation on the Keweenaw into conditions in the copper mines.[41] But outsiders, even Congressmen, had little effect on the course of things after the tragedy and Moyer's violent deportation. The companies weren't budging an inch in their determination. And remarkably, many members of the WFM locals weren't budging either. They wanted to continue the fight into the spring of 1914. But Moyer and other WFM leaders knew that the strike was not only hopeless—it had severely hurt the union. The WFM had already spent $800,000 on the strike, including almost $400,000 of mandatory assessments paid in by WFM members elsewhere. The union found itself basically bankrupt and with an angry membership. Still, Moyer would not simply tell the locals in Michigan to quit; he couldn't allow himself to do that. Instead, the union's executive board decided to cut benefit payments to the strikers, a move they announced to the Michigan locals. In response, the 2,500 or so continuing strikers voted to call it off. The strike, which for all intents and purposes was lost by the WFM in mid-August 1913, when the mines reopened, finally ended on Easter Sunday, 1914.[42]

Although justified in many ways, the strike of 1913–14 was hardly a fair fight. The WFM did not stand a chance against the entrenched, well-financed mining companies, which acted with unwavering resolve to hand the WFM a decisive defeat. And in the end, the companies got what they wanted and the workers got what they feared: the one-man drill became the new standard, half the miners lost their jobs, production did not suffer, and productivity increased.[43]

Still, the region and the mining companies paid a steep price to see this battle of wills conclude all too slowly. They suffered nearly nine months of social upheaval where tensions ran high and sometimes lives were lost. The incident left bitter divisions in society, created wounds that would not heal. The district lost its reputation as a mining region with surprisingly good labor-management relations. The companies won their favored drilling machine but lost many good men who beat a path out of the district rather than weather the strike. And most surely, the replacement workers or scabs they brought into the district were not as good or as experienced as the men who left, or the men the companies wouldn't take back because they had been too actively involved with the WFM to be "forgiven." Just after the Italian Hall tragedy, in January 1914, down in Highland Park, Michigan, Henry Ford announced his five-dollar-per-day wage and bonus plan. A man could toil up on Lake Superior in a hostile social environment and in a dangerous work environment for about three dollars a day—or he could board a train to Detroit, work in the most modern auto plant, and make more money.[44] A more or less permanent exodus of workers had started. From the strike on, rather than staying in the mines, many workers gave them up and headed to the auto plants of Detroit or Flint to make Fords or Buicks or, a bit later, Chryslers.

For the companies, especially Jim MacNaughton and C&H, which had called all the important shots in the strike, the labor rebellion was seen as their big chance to show who was boss. For two decades their power and control had been nibbled at from all sides, it seemed. Sometimes they had to swallow their pride; sometimes they had to change their course to adapt to the new world in which they worked. The strike gave the companies an opportunity to reassert themselves in a dramatic and forceful way, to define for themselves just how they were going to act in order to unleash a harsh blow against unionism.

The companies asserted themselves and won the day, but in truth, they weren't as strong and all-controlling as they had felt themselves to be in the nineteenth century. All the force and muscle they brought to bear against the WFM could not protect them from changes to come, some of which briefly floated the mines to their highest levels of operation, but most of which tossed them about. Continuing change offered the mines a host of challenges and soon put them into a long period of decline, which all the resolve and arrogance in the world could not head off.

16

❖ ❖ ❖

MAKING THE HARD TURN

From Growth to Decline

The strike of 1913–14 was over. The mines and their attendant communities had just lived through their most tumultuous period ever. But there would be no return to normalcy for this mining district. The mines and their men did not resume old ways, quickly mend fences, and pick up where they left off. The companies won the strike handily but still couldn't control their fate. After the strike came an unsettling period of six or seven years—one marked by both highs and lows. At the end of that period—by the early 1920s—the Lake Superior native-copper industry was in permanent economic decline. After seventy-five years of growth, the mines entered a period of nearly fifty years of winding down and, ultimately, of closing down.

The mining companies had little time to celebrate their victory over the Western Federation of Miners before international events dealt them a blow. The poststrike mines had hardly hit their full productive stride when the First World War broke out in Europe. The war immediately interrupted trade and commerce, which negatively affected the copper market and copper prices. In 1914 the price of copper fell 2 cents from the previous year's level and stood at only 13.6 cents per pound. This drop forced cutbacks on the Lake mines, and some even suspended operations. Then, as the war continued on for several years, world markets adjusted to it, transportation adjusted to it, and copper, used in munitions, proved a valuable commodity in high demand.

Copper prices climbed to 17.3 cents per pound in 1915 and then shot up to 28.2 and 29.2 cents in 1916 and 1917. Recognizing the economic bonanza being offered to them, the Michigan mines pushed production levels to all-time highs. They had pro-

duced 166 million pounds of copper in 1914. They produced 258, then 267, and then
256 million pounds of copper annually in 1915, 1916, and 1917. These years, in terms
of annual production, represented the absolute peak for the mines. These years also represented their peak in terms of profitability. The mines had paid out a mere $1.6 million in
dividends in 1914 (their lowest total since 1887), but dividends soared to $18.7 million
in 1916 and then to the all-time high of $23.9 million in 1917.[1]

The mining companies knew these glory years were a brief wartime bubble that
wouldn't last. Before war's end, prices retreated to the 18-cent-per-pound range. Then,
between 1920 and 1921, the price dropped precipitously from 17.5 to only 12.5 cents
per pound. During the war, copper had been overproduced and stockpiled. After the war,
holders of copper dumped it on the market, glutting it. At only 12.5 cents per pound,
the Michigan mines couldn't make money. Many closed or severely restricted operations
in the early 1920s, and in 1921, for the very first time since 1849, not a single company
paid out as much as a dollar in dividends.[2] Throughout most of the 1920s, the price
never climbed back above 14.6 cents per pound, and the Michigan mines were in trouble
over the long run. They sought to recover themselves, and in many ways over the coming decades, they did. But situations beyond their control, such as the Great Depression
of the 1930s, knocked them back down, and no short-lived bonanza, like the tremendous price-and-profit spike of the First World War, ever lifted them way up again. After
1920–21 the mines were never as large, profitable, or important as they once had been.
Instead of managing growth, they now managed decline. They sought to maximize their
final returns on their investments in mine, mill, and smelter plants before closing them
down.

This industrial decline declared itself in myriad ways. Michigan had produced about
80 percent of the nation's copper in 1880, 45 percent in 1890, 25 percent in 1900, and
20 percent in 1910. This drop-off continued. In 1916 the Lake mines hit their all-time
high production of nearly 267 million pounds, but that represented only 13.3 percent of
the nation's new copper. During the late 1920s the mines accounted for about 10 percent
of national production, and by the Second World War era, only about 5.[3]

Production fell off similarly, from the 1916 high of 257 million pounds to 161 million in 1920 and only 92 million in 1921, which represented the copper district's smallest annual production since 1889. Production recovered somewhat through the 1920s
to reach 186 million pounds in 1929, but then came the Great Depression. In 1933 the
mines sent to market only forty-seven million pounds, their lowest production in fifty-
four years. Again, the mines fought back against the Depression's economic hardships
and recovered somewhat, but the production level they reached by the late 1930s and
into the Second World War era was only about half the level of 1929—only about ninety
million pounds—and that figure fell off sharply to only sixty-one and then forty-three
million pounds in 1945 and 1946.[4]

Dividend payments presented a similar picture—the long-term trend was decline,

During the Depression, Keweenaw County used Civil Works Administration and Works Progress Administration funds to build roads, parks, and other amenities to encourage tourism, which was seen as a future source of revenue. Near Copper Harbor, the new nine-mile-long Brockway Mountain Drive climbed 726 feet above Lake Superior and offered motorists a beautiful overview of woods and water. The rustic Keweenaw Mountain Lodge, along with its log cabins and nine-hole golf course, offered tourists a place to play, eat, and sleep. *(Michigan Technological University Archives and Copper Country Historical Collections)*

with just a few good years in the mix and some very bad ones. After the disastrous year of 1921, when no dividends were paid, the industry managed to attain dividend levels of $4 to $5 million per year in 1925 to 1927. They paid out $7 million in 1928 and $11 million in 1929, when copper brought a high price for the decade of 18 cents per pound. Then came the Great Depression and crashing prices: only 8 cents in 1931, 5.6 cents in 1932, and 7 cents in 1933. Some dividend lows were just $112,000 in 1931, $392,000 in 1934, and no dividends, again, in 1935. By the late 1930s and through the Second World War era, a "good" year had come to be redefined as one with dividend payments in the range of $2 million to $2.5 million.[5]

The companies had wanted to shrink their employment ranks between the 1890s and 1913. They aimed new technologies such as electric haulage locomotives and one-man drills at the goal of enhancing productivity and hoped the new technologies would enable them to run a leaner workforce. It had worked. Total copper industry employment stood at about 18,000 in 1910 and at 15,000 by 1915. Thereafter, national and international economics, and their faltering fortunes, did the companies' labor trimming for them. Mines closed; others shrank. The industry employed about 12,000 in 1920, 9,000 in 1925, 7,500 in 1930, 4,500 in 1935, 3,500 in 1940, and only 3,000 in 1945. Yes, the

Since the 1840s, Keweenaw mines had opened and closed, and communities had grown only to decline or even disappear. In tough times, whenever it was hard to take care of the living, taking care of the dead was often neglected. Overgrown cemeteries like this one dotted the landscape. Relatives sometimes provided inexpensive perpetual care at a grave site by covering it with a thick layer of stamp sand; nothing would grow on that. *(Michigan Technological University Archives and Copper Country Historical Collections)*

industry lived on for decades after entering into permanent decline in the early 1920s, but on a vastly reduced scale. The mines were only shadows of their former selves.[6]

The fortunes of industry and society had been inextricably linked since the 1840s. Where and when the industry had grown, mine locations and nearby villages had grown. Where and when the industry declined or closed, a pall fell on neighboring settlements. The Copper Country had created ghost towns all along. While the industry boomed here, it went bust there. Towns such as Copper Harbor, Eagle Harbor, Eagle River, and Ontonagon had withered, or at least lost all growth potential, when nearby mines in the hinterland struggled or completely failed. Mine locations, all up and down the range, had done the same. While a Cliff or Minesota mine flourished, so did its location. Houses went up and schools and churches. A thousand or more people moved in. When the host mine closed, a thousand or more people left, and the schools, churches, and houses went vacant, and over time, went down.

Meanwhile, other places such as Houghton, Hancock, Calumet, and Laurium flourished because their adjacent mines continued in operation for decades and multiplied many times over in terms of their production and employment. But after 1920, even these communities suffered. No village, no part of the economy, no social institution went unaffected by the onset of the decline in the mining industry. People out of work, or tired of dangerous, low-paying jobs, got up and left, taking their families with them. The children of mine workers—even if their parents stayed in the Copper Country— left the district after finishing school. They moved to places offering better job prospects and a better future. The population loss was precipitous, especially between 1910 and 1950, and did not begin to stabilize (at much lower levels, however) until the 1960s and beyond. Houghton County, the center of the mining industry, had also been the Copper Country's center of population. Houghton County had a population of 88,000 in 1910. It lost 16,000 residents by 1920, and another 19,000 by 1930, when its population stood at about 53,000. The county's population dipped still further to 48,000 in 1940 and only 40,000 in 1950. In the 1960s and 1970s, it stabilized at about 35,000 residents, meaning that it lost 60 percent of its population over fifty years.

The smaller population units within Houghton County evidenced the same decline. In the vicinity of the Quincy mine, Hancock's population dropped from 9,000 in 1910, to 7,500 in 1920, to 5,800 in 1930. Above Hancock, on Quincy Hill, where many mine workers lived in two adjacent townships, Quincy Township's population declined from 1,500 in 1910 to 734 in 1930, while Franklin Township declined from 5,700 residents in 1910 to 2,600 in 1930. In 1910 the total population of Hancock plus Quincy and Franklin Townships was 16,200; by 1950 it fell to only 7,300. Large population losses also were suffered on the margins of the Calumet and Hecla (C&H) mine. The village of Calumet (or Red Jacket) declined from 4,200 in 1910, to 2,400 in 1920, and to 1,560 in 1930. Over those same decade breaks, the village of Laurium declined from 8,540 to 6,700 to 4,900, while Calumet Township (home to most company-owned housing) fell from 32,850 to 22,370 to 16,050.[7]

The social decline harnessed to industrial decline could be seen in the dirt lanes or streets at most locations. Sometimes company housing along those lanes stood all boarded up. Sometimes the houses were gone. Entire neighborhoods were gone, and only cellar holes remained. In 1910 the equalized assessed valuation of all personal and real property in Houghton County was $93.4 million, by the dark year of 1921 it had plummeted to $64 million, by the mid-1930s it fell to only $17 to $19 million, and by 1946 it stood at only $14 million.[8] Institutions, too, eroded as the population declined. At the peak of the mines' operations around 1910, when they employed workers from dozens of ethnic groups or nationalities, the Copper Country boasted some eighteen daily or weekly newspapers (including religious ones), printed in the English, Finnish, Swedish, Italian, and Slavic languages. Thirteen papers survived until 1931, ten until 1941, and nine until 1951. But finally the bottom dropped out, and all the foreign-language papers dropped out, leaving only three English-language papers by 1960 and only two by 1970—the *Daily Mining Gazette* of Houghton and the *Native Copper Times* of Lake Linden.[9]

Architects, contractors, and businesses erected much new construction in Houghton, Hancock, and Calumet between 1890 and 1910. Then the boom ended, and
almost all the architects moved away. In Houghton, shown here about 1940, the
automobiles and advertising signs changed—but the buildings themselves changed
very little, decade after decade. *(Michigan Technological University Archives and Copper Country Historical Collections)*

Along the streets of Calumet and Hancock, on the margins of some of the best and
most important mines in the district, traffic to the shops and stores slowed after 1920
as the economy contracted. Great growth had been the hallmark of 1890–1910. Towns
had rebuilt themselves of stone and brick, constructing new shops and stores and banks
and hotels. Communities erected numerous new churches and schools, as well. But after
1920, new construction proved novel, proved rare. As the population declined, closings,
rather than openings, became the norm.

While many left the area to find better futures elsewhere, those who remained in
the Copper Country's communities fought to keep what they had, such as their local
newspapers in their own language. Although all but two of eighteen papers died out by
1970, the sixteen that died did not end quickly with the onset of industrial, economic,
and social decline, even though the decline was clearly irreversible. Many of them fought
on, lingered on for years, even if they never thrived. Much of local society was like that
after 1920, into the Great Depression, through the Second World War years and beyond.
Local society tried to conserve and keep; it did not suffer losses easily. So for decades the
population held on to and supported myriad churches, even though their congregations
had shrunk, and myriad schools, even though their enrollments had steeply declined.
In some ways, local society did not reach an equilibrium—where its reduced number of
institutions finally matched its downsized population—until 1960 and after.[10]

While society coped with broad based, general decline, so did the surviving mining companies. They, too, did not capitulate, did not quit easily. They owned thousands of houses, many deep shafts extending between one and nearly two miles underground, tremendously expensive surface plants at the mines, numerous railroads, massive stamp mills, and several large smelters. All of these things still had economic value, still had life left in them, as long as the companies managed to stay afloat. The companies coped all the time with the economic problems of mining at great depths, falling yields, high production costs, stiff competition from other copper districts, and an eroding base of skilled laborers. They coped with whatever calamities the national or world economy threw their way, especially the steep recession after the First World War, followed a decade later by the Great Depression. And each individual company coped with its own particular circumstances. At C&H, for instance, that meant dealing with the end of mining on the Calumet Conglomerate lode, while at Quincy that meant dealing with the structural instability of its hanging wall and its confounded air blasts.

To extend their working lives as long as possible, the mines operating after 1920 tended to follow some general strategies. This was no time to lavish money on physical plants or new technologies, but the companies spent money judiciously if they thought it would help them make money and extend their lives. They did not just play out the old; they also added some new.

Underground, they sought productivity increases and costs savings in the blasting of rock by experimenting with new types of one-man drilling machines and new explosives and detonators. Companies also sought a greater level of mechanization in the mucking and tramming of rock. In mines whose lodes were of lesser pitch—where broken rock did not roll freely down to the drift—men had climbed up into the stopes to shovel or push down the rock. To perform this work, companies turned to stope scrapers, which were attached to cables and moved by small compressed-air–powered engines called *tuggers*. The cable, by winding or unwinding on a drum, pulled the scraper to the top of the stope and then dragged it down the footwall, where it collected the broken rock in its way. Similarly, in drifts at the bottoms of stopes, mines used scrapers to drag rock up a ramp and into tramcars. In some places, companies turned to power shovels, which resembled small steam shovels, to fill tramcars. Electric haulage locomotives became more widely used to deliver tramcars to the shafts. The newer haulage locomotives, instead of being of the trolley type, were often battery powered.[11]

In hard times, companies tended to work their best and forget the rest. At one time, during their boom years, to maximize production they took out ground having marginal copper content. No more. The companies needed to be more selective and mine rock that would earn them money, not lose them money, after blasting, tramming, hoisting, milling, and smelting it. That often meant trimming operations underground, leaving more poor-grade rock behind, and even closing shafts that threatened profitability instead of contributing to it.[12]

In hard times, companies tended to work their men harder and to supervise them more carefully. They put more bosses underground. They set higher work quotas—a man had to drill a greater footage of shot holes per shift or load more tramcars per shift. Sometimes they flirted with bonuses. Men who exceeded the work expected of them received additional pay for the extra holes drilled or the extra rock trammed.

On the surface, like in the underground, companies trimmed operations. Because of their diminished production levels, they no longer required much of the equipment erected during their turn-of-the-century boom. Operating too many shafts or too many steam stamps or too many mills or too many rail lines made no economic sense, so companies winnowed down. They mothballed hoists, boilers, and other facilities, and when convinced that machinery had no future use, they sold or scrapped it. They mothballed neighborhoods, too, emptying some out while concentrating workers in other parts of the location.

Sometimes the companies found it wisest to shut down altogether, whether for a short period or an extended time. Some, indeed, quit for good, but companies like C&H, Quincy, and Copper Range, when they found it too expensive and unprofitable to sustain production, retreated from it, but not permanently. They didn't close for good, auction off all their equipment, sell off their lands, or lay off every last worker. But they did freeze operations or put them into a kind of suspended animation. Maybe they kept only a skeletal crew to do maintenance and repair, to do the work necessary to preserve equipment so it could be used again. Maybe they kept the pumps running. Maybe, so as not to lose all their workers, they split up whatever few jobs they had, particularly in the worst of times (the early 1920s and the early to mid-1930s), and gave them to the married men they most wanted to keep.[13] Maybe they opened up new ground, not for mining—but for farming or gardening—and encouraged their idled men to be as self-sufficient as possible in hard times. In the depths of the Depression, in mid-1933 the ranks of the unemployed in Houghton Country swelled to 8,800 men, in part because, as a survival strategy, the Quincy mine had closed altogether, as had the Isle Royale mine, and Copper Range and C&H had closed many of their operations, too.[14]

Not many new mines opened during this era, but it was not for the lack of trying. C&H pinned hopes for its future on finding new mineral lands. Besides acquiring large tracts of property, it staffed a geology department. By 1919 and going into the 1920s, C&H had five or six geologists on staff, working under the guidance of a Harvard consultant, geologist L. C. Graton. The geologists not only conducted field surveys, they also spearheaded a thorough research project into the history of the district—where companies had been, what lodes they had found, how much copper they had taken out, and why they closed.[15] C&H's work formed the basis of an important publication, the U.S. Geological Survey's Professional Paper 144, *The Copper Deposits of Michigan,* published in 1929.

Unfortunately for the company, C&H's geological work did not result in the company's launching of any mines of great size or profitability. C&H's chief rival, the Copper Range Consolidated Mining Company, also hunted for new ground—and Copper Range found some, a huge deposit not of traditional native copper, but of copper sulfide, located in Ontonagon County about seventy-five miles south of the heart of the traditional native-copper range. Copper Range began production at its new White Pine mine in the early 1950s, and White Pine stayed in operation until the mid-1990s.[16] Ironically, C&H had once owned this property but had given up on it before Copper Range acquired it.

In the general absence of freshly discovered native-copper deposits to work, a few companies, with C&H in the lead, benefited by opening new "mines" at the lake bottoms just outside their stamp mills.[17] Instead of looking to extract copper from new mines or from old mines six thousand to nine thousand feet deep, they looked for copper one hundred feet beneath the water surface. They reclaimed the copper once lost in milling: they sucked up the underwater stamp sands, pumped them to a new reclamation plant, reground them, and then captured the copper by a combination of gravity and chemical processes. The copper from reclaimed sands cost considerably less to produce than the copper from deep mines, so reclamation was a welcome technology in hard economic times.

Throughout most of their history, the mines had rarely gone beyond the mining, milling, and smelting of copper. They made it; they sold it. They hadn't entered into the production of much else, either on or off the Keweenaw Peninsula. But desperate times called forth some innovative measures. The companies started to diversify, to enter into new ventures. If they stopped making new copper, they might smelt copper scrap. If they decided some land would never be needed for mineral or wood, they might sell it off as vacation property to city people or to local residents who wanted camps. If a closed shaft filled with water, start a water company and sell it to local communities. Open or acquire a company that used copper to manufacture tubing, pipes, or ferrules. Reuse part of your closed plant at the mine as a production facility for chemicals.[18]

The companies had sensed looming economic problems shortly after the turn of the century. They knew their fortunes were in decline by 1910 and absorbed the 1913–14 strike in part because they felt they had to: they had to install the new drills; they had to reassert their authority and control in the face of competitive challenges. But as soon as the strike ended, the firms, all of them, rode a roller coaster. They went down at the start of the First World War; they went up to the very peak in the middle of that war; they swooped downward during the recession following the war; they climbed back up throughout the twenties and then dipped lower than ever during the Great Depression. Any time they climbed up after 1920, the highs were brief and the peaks were low. During the worst of times, faltering companies got off the ride, leaving in production, by the Second World War era, only the Isle Royale and Quincy mines, and select operations of C&H and of Copper Range.

17

❖ ❖ ❖

THE QUINCY MINE

From Struggle to Shutdown

The Quincy mine came close to not surviving the 1920s. On top of the general economic woes suffered by all the Lake mines, Quincy had its bursting rock pillars underground, its *air blasts,* sometimes called "puffs," "old rousers," "radiator rockers," "shake-ups" and "man-made earth quakes." Pillars started to shatter or shear off in 1906 when the mine was five thousand feet deep, and the problem continued for well over twenty years. By 1920, over four hundred air blasts had cost Quincy some $4.5 million, plus they frightened workers, made them "timid," and taxed the problem-solving abilities of mine managers, captains, and engineers. They also strained relations between the mine agent, Charles Lawton, and company officers, especially William Parsons Todd, who served as vice president from 1912 until 1924, when he became president.[1]

Air blasts occasionally awakened Lawton at night when they shook his bed in the agent's house on Quincy Hill. The air blasts frightened Lawton, especially when he thought of "the number of men that are underground and the great possibilities for disaster." Air blasts threatened the life of the mine agent himself when he had to go underground to inspect damage, "when all the timbers are cracking and snapping and the hanging wall [is] going off in reports like cannons, and especially when you really do no know just how quick you may or may not 'Go to Glory.'"[2] As an engineer, Lawton wanted to fix his mine's structural problem. W. Parsons Todd, on the other hand, never believed that an affordable solution would be found. Lawton, if he could, wanted to prevent the next air blast. Todd was more willing to roll the dice. He hoped the next air blast would be the last. Someday all the pillars that were ready to burst would burst; all the hanging that was ready to fall would fall. Then the mine would be in equilibrium again and secure.[3]

While Lawton and Todd debated the merits of various courses of action, such as filling abandoned stopes with sand to help support the hanging (something that Lawton argued for and Todd against), the company did go ahead with some other fixes to its problem. To protect shafts as they went deeper, Quincy started leaving fifty-foot-wide rock pillars alongside them in 1906. The company increased them to 200 feet by 1913 and to 225 feet by 1920. Along shafts in the mine's midregion, where they passed through stoped-out ground and were unprotected by pillars, Quincy built heavy-timber cribwork, filled with poor rock, to guard against crushing.[4]

After the air blasts started, laborers began laying poor rock into ten-foot-wide rib walls at the bottoms of stopes. By 1924 these rib walls, increased to a width of twenty feet, ran along the stope tops, too. To protect the hanging wall in the stopes, Quincy left regularly spaced rock pillars, similar to those it left alongside its shafts. As stoping miners advanced away from the shaft, at regular distances they jumped past rock—no matter how rich in copper it was—in order to leave a support for the hanging. While this measure provided more security in the stopes, it also left much copper unmined. So Quincy in the late 1910s adopted an advance-and-then-retreat mining system. As the miners worked out along the drift, they left regular rock pillars. Once they finished stoping to the end of the drift, they pared down the size of the pillars, "robbing" them of copper, while retreating back toward the shaft.[5]

During its period of greatest growth, 1890 through about 1905, Quincy vastly expanded its underground works and sank new shafts. Shortly thereafter, it closed some shafts—first the old No. 4, and then, during the strike year of 1913, the No. 7 shaft on the southern end of the mine and short-lived No. 9 on the northern end.[6] Electric haulage made it possible to close No. 4, whose rock could be trammed to other shafts. No. 7 proved disappointing in terms of its production, plus underground workers disliked that shaft because of its poor ventilation and frequent rockfalls. Most importantly, though, No. 7, at a depth of 6,497 feet (all attained during the short period of 1900–11) bottomed out at the edge of Quincy's property. So, after stoping out all its levels, Quincy closed No. 7. On the opposite end of the mine, the No. 9 shaft never really proved itself as a producer, so Quincy closed it, too.[7] The mine went into the 1920s with just three hoisting shafts—Nos. 2, 6, and 8—and, in the hard economic times of 1922, it mothballed No. 8.

Quincy trimmed its operating costs by closing shafts, but the closures left the company more vulnerable to air blasts. If the hanging collapsed over a shaft, the fall could severely limit Quincy's hoisting capacity and inflict a substantial loss of production. In 1922, air blasts crushed the No. 6 shaft for a run of twenty levels, and it took three and a half months to recover it. An air blast crushed No. 6 again along ten levels in 1924, which put the shaft out of commission for six weeks.[8] Quincy in the 1920s could ill-afford such production losses or the heavy costs of returning No. 6 to service. But the worst was yet to come. In July 1927 the No. 2 shaft caught fire at the fifty-third level. Quincy's men couldn't extinguish it, nor could a more experienced firefighting crew from Calumet and Hecla (C&H) that went down with breathing apparatus into the smoke-filled No. 2 shaft. To smother the fire, Lawton ordered that steel fire doors at all mine openings be shut and sealed. Mining ceased. The U.S. Bureau of Mines' fire experts arrived on the scene from Duluth and backed Lawton's move to seal off and shut down the mine. After some false starts—attempts to reopen the mine before the fire was totally out—the fire was finally smothered, and on 15 August, men went underground to begin recovering 2,500 feet of shaft, from the fifty-third level up to the eighteenth.

Late in October, while a timber crew with 150 years of underground experience still worked to repair the fire damage, a second calamity hit: a series of air blasts. The hanging in the shaft collapsed, killing seven men beneath it. It was the worst fatal accident in the history of the mine. A series of less destructive air blasts rumbled on, chasing away rescue workers and impeding the full recovery of No. 2, which was closed for all of 1928.[9] At a time when other Lake mines became profitable again and got back on their feet, Quincy struggled desperately. It had to put No. 8 back in service while No. 2 remained down—and, more importantly, the deadly air blast finally forced the company to switch over fully to the retreating system of mining: go all the way to the ends of long drifts, start stoping there, and work back toward the shaft.[10] Stoping and copper-rock production virtually ceased while Quincy's miners did all the development work of shaft sinking and drifting needed to ready new levels for the retreating system. In 1928, Quincy's production plummeted to just 1.2 million pounds of copper. The last time its production had been that low, the calendar had read 1859.[11]

From the strike through the 1930s, compared to the changes Quincy made in its mining methods to cope with increasing depth and air blasts, all other underground changes were of lesser import, and none reversed the company's declining fortunes. It did introduce a variety of new machines, some of which mechanized tasks that until then had always been done by hand. It terms of breaking rock, Quincy adopted a new generation of one-man drills in the 1920s that were made by Ingersoll Rand or Chicago Pneumatic.[12] Beginning in the late 1910s, the company finally mechanized the task of cutting up mass copper. It is not clear whether Quincy's first "copper cutters" were pneumatic chisels or air-operated twist drills, which became the region's copper-cutting tool of choice by the 1920s. The operator drilled a series of closely spaced holes across the mass and then charged each hole with a high explosive and blasted the mass apart.[13] To replace human labor in mucking up rock, in 1918 Quincy started using air-operated tuggers to move heavy scrapers along the mine floor in stopes and drifts. The dip of its Pewabic lode was less steep at depth, and broken rock no longer rolled freely down the stopes to the drifts for tramming. Instead of adding more men up in the stopes to push down the rock, Quincy used a tugger, which first pulled a heavy scraper up to the top of a stope and then dragged it back down, mucking out the rock in its path. At the bottom of stopes, Quincy experimented with small power shovels to pick up the rock and dump it into cars. When those failed, Quincy turned to tuggers and scrapers to pull rock up ramps and into tram-cars. In 1930, for the first time Quincy also used conveyor belts to fill its tramcars, which were now pulled by electric haulage locomotives powered by batteries instead of overhead trolley wires.[14]

Aside from its haulage locomotives, Quincy still made small use of electricity under-ground. Electric lights were scarce; men through the 1930s lighted their way with the carbide lamps first introduced in 1912–14. Quincy did turn to electric power to aid in unwatering and ventilating the mine. In 1916 the company began abandoning its bailing skips and started using electric pumps to unwater its levels. Also, by the mid-1920s Quincy supplemented natural ventilation by employing about a dozen electrically powered fans underground to send air through flexible tubing into some of the hottest, stuffiest parts of the mine.[15]

On the surface, high profits during the First World War years prompted some new construction. In 1917, Quincy built a two-story brick-and-sandstone clubhouse for workers. Quincy lagged behind C&H and Copper Range in providing such a facility, but in a war year when it paid out its largest annual dividend ever (nearly $2 million), Quincy thought it a good time to build.[16] The first floor, split into men's and women's departments, had bathing facilities with tubs, showers, and toilets. The plumbing in the clubhouse substituted for the indoor plumbing still not found in Quincy's com-pany houses. The second floor held a reading room, lecture hall, and by 1918, a library. Quincy also erected its last wave of company houses during the wartime boom. It built story-and-a-half saltbox houses at its mill town, Mason. At the mine, it erected its largest worker houses ever, both single and double houses, built to plans purchased from Sears, Roebuck & Company. The single Sears houses, a full two stories tall, had three rooms on the first floor and four on the second. Quincy surrounded each house lot with a nice picket fence, and to promote American patriotism during the war, it planted a flagpole in front of each house. Both sets of new houses were seen by the company as necessary upgrades in its housing stock; Quincy needed them because the company was having trouble keeping skilled, married workers on its payroll.[17]

In 1917 the E. P. Allis hoist at No. 2 shaft—then 7,289 feet deep along the incline—was near the end of its rope. Throughout its history, Quincy had shown a proclivity for squeezing every last foot of depth out of old hoists before replacing them. Not this time. After receiving a proposal from the Nordberg Manufacturing Company late in 1916, Quincy ordered a new No. 2 hoist because it promised to raise a higher tonnage of rock per lift faster and at less cost. Nordberg was supposed to deliver the hoist in fourteen months, but restrictions on the production of heavy machinery during the First World War delayed receipt of the engine until late 1919. While waiting for the hoist, Quincy erected its new hoisthouse, which in itself was a distinctive bit of industrial architecture in the Copper Country. The fireproof building was constructed almost entirely of rein-forced concrete, with red-brick veneer and arched windows on the walls, and a green tile roof. The hoisthouse cut a distinct contrast to earlier ones on the site that had been built

Between the late 1850s and the 1930s, a series of steam hoists at Quincy's No. 2 shaft lifted copper rock from as deep as 9,200 feet, measured along the inclined shaft. Hoist engineer Frank Nancarrow stands on the operator's platform of the last engine to lift from No. 2—the Nordberg Manufacturing Company hoist that went into service in 1920 and out of service in 1931. *(Historic American Engineering Record Collection, Library of Congress)*

of wood, then poor rock, and then Jacobsville sandstone. The building included large doorways and a heavy overhead crane to facilitate the erection of the massive hoist, which took nearly a year to assemble. Erecting crews started on the hoist in December 1919, and it lifted its first rock in November 1920.[18]

Quincy's No. 2 Nordberg hoist was the region's last hurrah for massive steam-powered equipment. The hoist's grooved, cylindro-conical drum, thirty feet in diameter at the center, could carry ten thousand feet of one-and-five-eighths-inch wire rope in one layer on the cylindrical portion and one conical end. This cross-compound hoist was actually four engines in one. It had two high-pressure cylinders on one side of the drum (mounted on an inverted-V frame so that each engine served as a leg for the hoist to stand on). These cylinders had a bore and stroke of thirty-two by sixty-six inches and ran on 160 pounds per square inch of steam. Two low-pressure cylinders on the opposite side of the hoist had a bore and stroke of sixty by sixty-six inches. The fuel-efficient engine reduced hoisting costs in two key ways. First, it was a compound engine that used its steam twice. Steam at high pressure drove the smaller pistons on one side of the hoist. Then, when that steam was exhausted at a lesser pressure, instead entering the atmosphere it was piped across to the second side of the hoist, where it drove the larger pistons. The low-pressure side of the engine also ran with a condenser, meaning that steam on a power stroke was pushing against a vacuum on the opposite side of the piston, so it was capable of more work.

Quincy's boom days were over before this ca. 1925 photo was taken. Shot from No. 6 and looking southward, this view shows the ca. 1900 complex built around the blacksmith and machine shops; the No. 2 shaft-rockhouse erected in 1908 (with the observation platform added to the top during the 1913–14 strike); and the No. 2 hoisthouse built for the 1920 Nordberg hoist. *(Chuck Pomazal/Michigan Technological University Archives and Copper Country Historical Collections)*

The double-acting pistons provided eight power strokes per revolution of the drum. The drum, turning thirty-four revolutions per minute, hoisted skips enlarged to ten-ton capacity at 3,200 feet per minute, or thirty-six miles per hour. Altogether, the engine and its house represented a grand investment of $370,000—made during the profitable war years. But Quincy never fully reaped the rewards of that investment because air blasts, shaft collapses, and economic hard times limited the hoist's use and cut short its working life.

Quincy also invested in its mill and smelter. The company kept up with technological change in terms of acquiring new equipment for the fine grinding and washing of copper rock. It installed Wilfley tables to capture a higher percentage of fine copper. These worked well, and buoyed by First World War copper prices, Quincy planned to expand both mills in order to house more Wilfleys. These additions were set to go online in 1920, but the copper market collapsed. Instead of expanding, Quincy contracted. To save on operating costs and to match its milling capacity to a reduced mine output, Quincy closed its No. 2 mill in January 1920.[19] It mothballed the equipment and waited for a time when the mill would be needed again. That time never came.

In 1923 it invested in a new electrical plant at the mill. The company used more electricity now and believed it could produce it at less cost than what a local utility was charging. At Torch Lake it built a powerhouse for a General Electric 2,000-kilowatt steam turbine. The turbine did not require its own boiler but was powered by the exhaust steam coming from the Allis stamps.[20] By the late 1920s, Quincy made an additional

During the Depression, the Works Progress Administration created the Federal Arts Project (FAP). Gerhard Bakker, a German-born artist who worked out of Milwaukee, joined the FAP. He specialized in wood and linoleum cut prints and was interested in industrial subjects. In 1936 Bakker visited Quincy Hill and produced this print of the No. 6 many-gabled shaft-rockhouse, which had been sitting idle since 1931. *(Fletcher Studio/Waupun Historical Society)*

improvement to its sole operating mill on Torch Lake. For the first time it used something other than gravity separation to capture fine copper. In 1927 A. W. Fahrenwald of the U.S. Bureau of Mines published work on a new oil-flotation process for recovering copper. This process agitated fine copper and rock particles in a tank filled with water and an oily frothing agent. The agitation created bubbles, which rose to the top of the tank. These oily bubbles had an affinity for fine copper, captured it, and carried it to the top of the tank. Quincy quickly adopted oil flotation, and the process allowed the company to capture an additional three pounds of mineral concentrate per ton of rock stamped.[21]

In 1920 Quincy made substantial improvements at its smelter—again, just in time for a major downturn. Quincy fell in line with other modernized smelters in the district; it replaced its four or five small reverberatory furnaces with two larger ones. The first furnace melted the charge and sent its product to the second refining furnace. A reduced force of furnacemen could perform all needed charging, rabbling, poling, and tapping, and all the molten copper flowed from a single source into molds carried on a semiautomatic casting machine.[22]

Quincy's officers and managers no doubt second-guessed many of the decisions they made from the strike years through the 1920s. Wartime profits encouraged them to

While shut down between 1931 and 1937, the Quincy Mining Company did not maintain these houses in its Lower Pewabic neighborhood. They deteriorated as they sat empty. Today, only cellar holes and poor-rock foundations mark where these houses once stood. *(Farm Security Administration Collection, Library of Congress)*

spend in ways that afterward looked suspect. But the company did not have the luxury of forecasting the future. It did what it thought was best at the time, and Quincy continued to believe it needed to invest in new technologies to sustain itself. The company did not manage decline by retreating from its problems; it invested in its industrial and social infrastructure to try to secure a longer, better future. But in many ways, the plan simply didn't work, especially once the Great Depression arrived.

On 22 September 1931 the Quincy Mining Company closed its works on the Pewabic lode, which it had mined virtually without stop since the late 1850s. It ceased milling and smelting, too, because the Depression had driven copper prices down too only 7.5 cents per pound, while Quincy's production costs were 11 cents per pound. In 1931 the mine's expenses were about twice the amount of its sales, which stood at only $585,000. The mine couldn't continue until copper prices recovered substantially.

The Quincy mine sat idle from September 1931 until June 1937. Over those years the company reorganized itself to garner needed capital to keep the idled mine in decent condition and, later, to prepare it for reopening. Quincy lost some lands on the Keweenaw for failure to pay taxes; some of its stock was forfeited for nonpayment of assessments called in to help caretake the mine. While struggling to survive, the company still tended to the needs of jobless families at the mine location. It plowed fields so residents of Quincy Hill could plant crops. Quincy let many former employees stay in company houses rent free, and others, never on the company payroll, could occupy a Quincy house, too. [23]

When copper prices improved in 1936, Quincy started making plans to start up again. It called in an assessment on each share of stock to finance the purchase of supplies, the unwatering of the mine, and the repair of underground and surface works at the mine and mill—but not at the smelter. The smelter remained idle, and Quincy arranged for C&H to smelt its mineral. The "Resumption Expenses" to put mine and mill back into service amounted to over $313,000.[24] This investment seemed merited because copper had climbed back to fourteen cents per pound. Still, copper prices did not go high enough in the next few years to cover the mine's costs. The company called in another assessment on the stock to help it get by. When the Second World War arrived, Quincy benefited from government price controls and intervention in the market.

At the start of the war, the Lake mines produced only 5 percent of the nation's new copper. Although the mines were marginal producers, the government supported them so that during the conflict they could sustain or exceed their prewar production levels. The War Production Board and the Office of Price Administration established a Premium Price Plan intended to set copper prices high enough to encourage domestic production while not triggering excess profit taking. The government-sponsored Metals Reserve Company entered into a *special contracts* program with smaller producers (including Quincy, Isle Royale, and Copper Range) to assure them the best rates for their copper so they could produce as much as possible during the war. As a result the Michigan mines remained in production and paid out modest dividends even while paying higher wages to their employees. The Metals Reserve Company contracted to buy all Quincy's copper, paying as much as twenty cents per pound by 1945. Under the contract, Quincy hardly thrived, but neither did it go under.[25]

The Metals Reserve Company played an important role in Quincy's future in another important way. In 1942, Quincy received a $1.2 million loan from the agency to help fund construction of a reclamation plant at Torch Lake, near Mason and Quincy's No. 1 mill. C&H had been reclaiming stamp sands for years. C&H designed and built Quincy's plant for a fee of 8 percent of its cost. Here, Quincy "mined" its tailings: it sucked them up with a dredge, piped them to the reclamation plant, reground them, captured the copper, and then sent it off to be smelted. Almost as soon as it opened, the reclamation plant proved far better than the old mine at making money for Quincy.[26]

The Metals Reserve Company's contract to buy all Quincy's copper expired on 31 August 1945. Thereafter, the copper would not bring a price high enough to cover the operating costs, so on 1 September 1945, the Quincy Mining Company closed its mine.[27] For all intents and purposes, the mine was closed for good. Some exploration work went forth on Quincy Hill, jointly undertaken by Quincy and the Homestake Copper Company. This venture resulted in the taking of a bit of rock occasionally, especially from the vicinity of the No. 8 shaft.[28] But after 1945, although the Quincy Mining Company survived, its mine—as a viable copper producer—was dead.

Quincy's reclamation plant carried the company in the postwar years and even let the old firm pay a modest dividend of twenty-five cents per share in 1948, the first time Quincy had paid a dividend since 1920. In this year of renewed dividend payments, the company also renewed operations at its smelter, which had been idle for seventeen years. The smelter handled mineral from the reclamation plant, and smelted scrap copper, just to keep going. Once finally back in the black, the company continued to pay modest dividends for many years thereafter. The reclamation plant operated with few interruptions from the end of 1943 until mid-1967, when it finally exhausted its supply of untreated stamp sands. Over that run, the reclamation plant produced about 100 million pounds of copper.[29]

The Quincy Mining Company held investments in other metals companies; it owned real estate in Michigan and on Manhattan. It kept up with its leases for all its company houses that were still occupied. But after the late 1960s, after 120 years of corporate existence, it had only a handful of employees, a quiet mine, defunct mills, a played-out reclamation plant, and a cold smelter. Entire neighborhoods of company houses had disappeared up on the hill. "Old Reliable" was through. Soon the same could be said for C&H, Copper Range's Champion mine, and the entire native-copper-mining industry.

18

❖ ❖ ❖

CALUMET AND HECLA

Down with the King

T he men who guided Calumet and Hecla (C&H), Quincy, and Copper Range through the strike of 1913–14 and the First World War still led them when they struggled to make it through the 1920s and 1930s. At Copper Range, William Schacht put in thirty-seven years with his company, including fourteen as president, before dying in 1944. At Quincy, Charles Lawton served as superintendent from 1905 until 1946, when he died. At C&H, James MacNaughton put in a total of forty years with the company (1901–41) as general manager and, ultimately, president. The mines had stable leadership, even in the most unstable of times.[1]

The mining company that James MacNaughton joined in 1901 changed greatly over his tenure. For decades, it worked just the Calumet Conglomerate lode. When that lode started to falter, MacNaughton directed the company's quest for new sources of copper. C&H opened works on the Osceola and Kearsarge amygdaloid lodes; it bought up substantial shares, usually a controlling interest, in numerous mining companies owning large tracts of mineral lands; and it planned to reclaim copper from the stamp sands washed out of its mills and into Torch Lake.

Many of C&H's expansion efforts never paid off. For every hit, there were misses. Still, from the First World War into the troubled 1920s and 1930s, C&H benefited tremendously from a few of its newer endeavors. C&H reclaimed hundreds of millions of pounds of copper from tailings at the bottom of Torch Lake. C&H also discovered that much of its future resided north of Calumet. Mines there, like Ahmeek, had started as independent operations. Then C&H bought up stock in them and finally consolidated

In 1915 C&H started reclaiming 152 acres of stamp-mill tailings from Torch Lake, which contained about one-fourth of all the copper it had hoisted from the Conglomerate lode since the 1860s. A dredge (like the one shown here in 1947) sucked up as much as 10,000 tons of tailings daily from as deep as 120 feet underwater. The dredge pumped the stamp sand back to an onshore reclamation plant through a pipeline that floated on steel pontoons. *(Michigan Technological University Archives and Copper Country Historical Collections)*

with them. Out of many companies emerged one: the Calumet and Hecla Consolidated Mining Company (C&H Consolidated).[2] This consolidation made C&H the premier producer in Michigan again (a rank it had lost for a while to Copper Range) and enabled the company to cope better with economic challenges.

About one-fourth of all copper hoisted from the Calumet Conglomerate lode since the late 1860s resided in a huge tailings deposit at Torch Lake. The tailings ran 120 feet deep and covered 152 acres. C. Harry Benedict, C&H's master mill metallurgist, was responsible for capturing as much copper as possible from the mine's stamp rock and for reclaiming copper from the stamp sands. For years, several companies had considered the prospects of reclaiming stamps sands. Under Benedict, C&H put the region's first full-scale reclamation plant into operation in 1915.[3] This plant, after regrinding the stamp sands, initially captured the liberated, fine copper particles by using improved gravity-separation machinery, such as Wilfley tables. In 1916 the plant began treating the copper chemically, too, when Benedict introduced leaching in a new facility that was about 500 feet long by 175 feet wide. Sixteen covered tanks each held up to one thousand tons of sands. An ammonium solution flowed into the sands and dissolved only one mineral—

The suction dredge delivered tailings to C&H's regrinding plant, where sixty-four Hardinge mills stood in long rows. Each mill contained hard pebbles, as well as a charge of stamp sand. As the mill rotated, the pebbles rolled around in the interior, regrinding the sand, and liberating more copper. After being discharged from the Hardinge mills, copper was separated out from the fine sand by gravity at Wilfley tables and also by the newly adopted processes of leaching and flotation. *(Michigan Technological University Archives and Copper Country Historical Collections)*

the copper. The copper-rich solution next went to a closed distillation vessel heated by steam pipes. When heated, the ammonia boiled off as a vapor, which the plant captured, condensed, and then reused in the leaching tanks. Meanwhile, the dissolved copper combined with oxygen and precipitated out of solution as a black solid: copper oxide. C&H sold some copper oxide to chemical companies; the bulk of it went to the C&H smelter to be made into ingot copper.[4]

C&H sought other means of capturing very fine copper liberated in its mills and reclamation plant. The flotation process—which used special oils to cause a heavy mineral to float rather than sink—had first been applied in the treatment of zinc sulfides in Australia in 1911. Over time, many mining districts used flotation to recover different metals. Each application called for research into what would work with a given mineral.

Critical technical details to be resolved included such things as the proper oil(s) to be used and the mesh size of the copper particles to be floated. Some early tests on conglomerate copper in 1914, conducted by an outside lab, suggested it could not be floated. But shortly thereafter, new flotation oils offered C&H promise. In the boom time of the First World War, C&H built an experimental flotation unit at its Hecla mill that proved successful. So in 1919–20 C&H phased in a full-scale flotation plant that treated slimes (fine copper, rock, and water) from the milling of mine rock, plus very fine copper coming from the reclamation plant's regrinding units. Thereafter, flotation served as an important auxiliary to gravity separation; Benedict recognized it as the most important technological change in mineral capture in fifteen years. It accounted for something less than 10 percent of the copper captured in the milling of mine rock and for something more than 10 percent of the copper reclaimed from stamp sands.[5]

At C&H the technological changes at mills and reclamation plants proved more valuable than any changes made underground. An organizational change brought great benefits, too: the creation of C&H Consolidated in 1923. In 1911, lawsuits had blocked C&H's earlier attempt to effect consolidation. But in 1923, what had been separate mines (all of C&H's various works, plus the Ahmeek, Allouez, Centennial, and Osceola mines) started operating legally as one. From 1915 until consolidation, C&H had not been the region's largest producer. That mantle had fallen to Copper Range, and C&H accounted annually for about 30 percent of the region's copper. But from 1923 through the end of the Second World War, C&H Consolidated accounted for at least half of Michigan's production and in some years for as much as 70 or 80 percent.[6]

Consolidation enabled C&H to produce copper more efficiently. While the companies remained legally separate, they had their own boundaries to honor, and all had their fully equipped shops at the mine, plus their own railroads and mills. Their underground works and surface plants included many redundant facilities, which under consolidation could be winnowed down.[7] The mines were contiguous properties running north from Calumet. Once legally consolidated, their old boundaries no longer applied. The stoping ground, once subdivided into several mines, was now one, worked by one company. C&H Consolidated operated fewer shafts and employed fewer trammers. It shut down many hoists, boilerhouses, and rockhouses that were no longer needed, plus some railroads and stamp mills. It closed underused facilities and worked those that survived at higher capacity, achieving greater economies of scale.

Like other companies, C&H virtually shut down in 1921, and over the remainder of the decade it struggled to get back on its feet. Here, reclamation proved especially important. Besides running its original reclamation plant on Torch Lake, C&H opened a second one there in 1925 to reprocess tailings from the Tamarack Mining Company's mill. In 1917, C&H had bought up all Tamarack properties, including its mill site and stamp sands. While C&H struggled to profit from mining, it was easier to make money from reclamation. While copper sold for twelve to fourteen cents per pound, it cost

C&H only five to seven cents per pound to make copper from stamp sands. At that rate, reclaimed copper accounted for 60 percent of C&H's dividends in the 1920s.[8]

Even though the downsized industry had lost so many jobs in the 1920s, the mines were labor short. C&H had too few good men for the jobs it still offered. Many men did not linger around Red Jacket and the other mine towns. Instead, they boarded trains and left. This became a perpetual problem at the mines: they had fewer jobs and still fewer men to take them. To counteract a labor shortage, C&H sought to obtain a higher output per man. C&H maintained the charade of using "contract miners," but a miner really worked for a daily wage under a bonus system. On the Osceola lode, for instance, a man was expected to drill fifty feet of holes per shift; that was the standard. But if he pushed himself and drilled beyond that standard, he received an additional ten cents per foot.[9] To cope with labor shortages, C&H, Quincy, and Copper Range started recruiting workers from England (mostly Cornwall), Germany, Canada, and even Mexico, but met with marginal success.[10] The companies often deceived their recruits. They promised them miners' jobs, but when the recruits arrived on Lake Superior, the companies often put them to work as trammers.

In the slowest of times, the companies did what they could to keep experienced men around. C&H, at its peak of profitability in the First World War, employed just over 6,000 men, including nearly 2,800 underground workers. In 1921, the bleakest year after the war, the mine averaged 794 on its payroll, including just 113 underground workers.[11] To keep a modest number of men from bolting the area, C&H put them to work at various tasks around the mine. Some of these men worked on the surface to create a park built in honor of Alexander Agassiz, who headed the company from 1871 until his death in 1910. This was a "make work" project in one sense: it gave men something to do. From that perspective, the new park symbolized the company's recent decline. Simultaneously, however, it celebrated the firm's long, paternalistic past.

Agassiz Park occupied the wedge of land between the northern end of the mine and Red Jacket village. Since the 1860s, C&H had used this land parcel in myriad ways—as pasturage, as athletic fields, as open storage for timber and other mining materiel, and, during the strike of 1913–14, as a bivouac for National Guardsmen. In 1916 C&H dressed up the twenty-five acres, making the parcel a suitable place for the company and the community to celebrate C&H's fiftieth anniversary. By the late teens, industrial growth along the old Calumet Conglomerate lode was not in the cards, so C&H made the land available for a permanent park. This park would be something new at a time when few new things were going up at the stagnant mine or its adjacent village.

Agassiz Park served as a community asset but was not a public project. C&H remained true to its way of doing things. The company decided to build a park, the company decided what to include within it, and the company decided to do it right. It turned to Warren H. Manning, a renowned landscape architect, to design the gardens, tree colonnades, paths, plantings, playgrounds, amphitheater, and other facilities. Cor-

In the early 1920s, when many miners were idle in bleak times, C&H had them build playgrounds and plant trees and gardens in a new park honoring longtime company president Alexander Agassiz. When Agassiz Park opened in 1923, the company unveiled a bronze statue of a robed, seated Agassiz, created by New York sculptor, Paul Bartlett. The statue presented Agassiz as a Harvard scholar, but the industrial background made it clear that he was also a mine boss. *(Michigan Technological University Archives and Copper Country Historical Collections)*

porate support for Agassiz Park grew out of the company's fiftieth anniversary celebration in 1916, and the time seemed right to create it, given the high profits of the First World War years. But park construction stalled during the war as C&H put more men to work producing copper instead of planting shrubs. In the postwar recession, however, the opposite happened. Copper production stalled, and mine employment dropped precipitously. C&H put some needy men (those it wanted to keep around) to work creating and landscaping Agassiz Park.[12] By the time the park opened in 1923, C&H had weathered the worst years of the postwar recession; it had accomplished its consolidation; and while things did not look particularly bright, they certainly looked better.

From about 800 employees in 1921, C&H rebounded to employ 2,600 to 2,700 in 1922–23. After the consolidation of 1923, counting employees at Ahmeek and the other mines now incorporated into the C&H fold, the company's employment climbed to as high as 4,900 men in 1928–29.[13] Then the stock market crashed and the Great Depression arrived. C&H's total employment slipped to 4,300 in 1930 and to 4,000 1931. Employment would have—and probably should have—fallen off faster, but C&H, hoping to ride out the Depression, kept producing copper as long as possible and ate up over $7.4 million of cash reserves to do so. The gambit did not work because the Great Depression outlasted C&H's financial resources.[14] The company had to rein in spending, and employment dropped steeply as C&H trimmed and closed operations. In 1932, C&H closed its reclamation plants, and halted mining on the Osceola lode and at the Allouez, Centennial, and North Kearsarge mines, and a little later (in 1933) at the Ahmeek mine. In 1932, company employment more than halved, falling to 1,700. Many of these men worked part time or shared jobs with one another. That way, C&H spread its jobs out to support more families. In 1933, employment fell to only 1,300 men. During 1932 and 1933, copper bottomed out and sold, respectively, for only 5.6 and 7 cents per pound. C&H recovered a bit after that, especially in 1937, when copper prices hit a Depression era high of 13.2 cents per pound. The price rebound encouraged C&H to reopen its reclamation facilities and the Ahmeek mine. Still, in the late 1930s, employment stayed around 1,400 to 1,800 men, and it hit 2,000 only once before the start of the Second World War. During the war, it never surpassed 2,250.[15]

The Great Depression ended mining on the once-great Calumet Conglomerate lode. For years, the company left broad pillars alongside its shafts to protect them from being crushed as they went deeper. Those pillars contained much copper that could be mined cheaply because all the development costs to reach that copper had long since been paid for. In 1933, C&H stopped the pumps at its original mine. Miners started robbing pillars at the bottom of the mine and worked up. As they retreated up the shafts, the water grew deeper in the worked-out mine. In October 1939, men closed operations and capped off the No. 12 shaft, the last one working on the Conglomerate lode. The first C&H mine was finished, after producing 3.275 billion pounds of copper.[16]

The timing of this closure could not have been much worse from the standpoint of preserving of many of the showcase buildings and machines that once stood along Mine Street or along the Calumet, Hecla, South Hecla, and Red Jacket shaft branches of the mine. The Second World War accelerated the scrapping out or selling of some of the most impressive mining machinery built in the United States in the late nineteenth century. C&H scrapped out obsolete or surplus machinery to the tune of twenty-five thousand tons of iron and steel.[17] On 20 August 1943, the *Daily Mining Gazette* carried what amounted to an obituary for C&H's once-mighty surface plant:

> All that remains [are] a few surface structures from which all the machinery and equipment has been scrapped and turned into war weapons, but eventually the buildings that housed them, along with the 250-foot smokestacks that stood as silent sentinels over what for many years was the world's richest copper mine, will be razed.

While old buildings and machines disappeared from the C&H landscape, old leadership disappeared, too: namely, Big Jim MacNaughton, who retired in 1941 after four decades of service. About the time MacNaughton retired—a bit more than a quarter century after

In 1939 the original C&H mine on the Conglomerate lode shut down after having produced 3.275 billion pounds of copper. Thereafter the company mined copper elsewhere and dismantled much of what had once been the nation's grandest copper mine. This aerial view shows the closed mine and Agassiz Park in the foreground; the village of Laurium in the middle distance; and C&H's mills, reclamation plant, and smelter in the distance, where the smoke rises along the shore of Torch Lake. *(Keweenaw National Historical Park Archives)*

he had led the copper companies to their decisive victory over the Western Federation of Miners—peacetime gave way to the Second World War, and a successful union drive took place on the Keweenaw.

The Lake mines had risen to their peaks over a long era dominated by *big business* within the American economy. The Great Depression of the 1930s, with the more liberal New Deal administration of Franklin D. Roosevelt, saw the rise of two counterbalancing forces in the economy: *big government* and *big labor.* Roosevelt's New Deal made the federal government a more active player in economic issues, and legislation passed under his administration, such as the Wagner Act, placed the weight of the government behind a push for greater recognition of unions and collective bargaining. C&H had a long history of antiunionism, but in the years on either side of 1940, workers at Copper Range, Isle Royale, Quincy, and even C&H Consolidated finally unionized.

At this time, union drives succeeded in many industries across the country, including Michigan's auto industry. In the late 1930s, wages at the downsized copper mines were about one-third lower than at Great Lakes iron mines. So, it was not surprising that organizers once again returned to the Copper Country. Led by men such as Gene Saari, the Congress of Industrial Organizations (CIO) started enrolling workers in locals of the International Union of Mine, Mill, and Smelter Workers: IUMMSW. Local 494 formed at South Range, covering Copper Range Company workers, in 1939. By early 1941, Local 515 operated at the Isle Royale mine and Local 523 represented workers at Quincy.[18] These locals turned three of the four operating companies into union shops. That left only C&H out of the union fold, but not for long.

By 1942 the IUMMSW had attracted a sufficient number of C&H employees to its ranks to call for a union election under the watchful eye of the National Labor Relations Board. To help win that vote and forestall this union's arrival, C&H granted some wage increases, provided vacations with pay, and promoted an in-house organization—the Independent Copper Workers Union—to serve as an alternative to the IUMMSW. Also, C&H strongly implied that should the union drive succeed, C&H would take back some paternal benefits long offered, such as low rents for company houses, coupled with low energy and utility costs.[19]

C&H escaped unionization in the spring of 1942 by a vote of 656 in favor of the IUMMSW and 736 against. But the company could not buck the nationwide and local upsurge in unionization for long. In November 1942, C&H workers voted again, and this time the IUMMSW-CIO won.[20] Now C&H faced collective bargaining whenever it moved from the end of one union contract to the start of the next. The bargaining dealt not only with wages, but with benefits, hours, seniority, work rules, and grievance procedures. It was hardly surprising that C&H, the largest, the most paternal of the companies, and the most vigilant in combating unions, was the last to be pushed into collective bargaining. It bargained first with the IUMMSW during the Second World War years. After 1950, when the men switched their affiliation over to another union, it bargained with the United Steelworkers.[21]

With the Calumet Conglomerate lode shut down and with the reclamation of tailings at Torch Lake nearing an end, C&H was in no position to experience a boom during the Second World War. In 1941, C&H claimed 65 percent of Michigan's total production (35 percent from reclamation and 30 percent from its mines). During the war, C&H Consolidated relied on its strongest producer, the Ahmeek mine, for much of its production but also put additional shafts into service. It unwatered shafts at its Seneca, Kearsarge, and Centennial properties and reopened them; it sank new shafts on properties over the Iroquois and Houghton lodes.

Well before the war ended, C&H looked ahead at an uncertain future and decided to diversify to sustain itself in the postwar world. Diversification was indeed a wise plan, given what was to come. During the war, a peak of 1,517 men worked underground at the Houghton County mines. From 1946 until the late 1960s, the mines of Houghton County employed only 250 to 500 men underground at any time. Although C&H operated as many as seven shafts in Houghton and Keweenaw counties in the postwar era, mining native copper became a minor part of its overall operations because of its branching out into new endeavors. C&H's diversification began in the early 1940s and continued after the war.

Throughout their history, the mining companies had largely stayed out of manufacturing businesses. Other firms drew copper into wire or fabricated it into hardware or munitions. For C&H Consolidated, that changed in mid-1942 when it acquired Detroit's Wolverine Tube Company. Wolverine, employing about a thousand people, made copper tubing for plumbing, refrigerators, air conditioners, and automotive and aircraft use. A year after the war ended, C&H expanded Wolverine by building a new manufacturing facility in Decatur, Alabama.[22] Besides going into the tube business, C&H used its shops in and around Calumet to produce new products. It manufactured detachable drill bits for the mining industry, and its foundry produced castings for outside customers. It created a small firm to manufacture buffing compounds from stamp sands. In 1944, using its knowledge of leaching and other copper-recovery technologies, C&H established a copper-scrap-recovery program (to reclaim copper from shell casings and the like); within a few years, its copper-recovery works produced more copper than its mines. In 1945 it organized the Lake Chemical Company, which produced a variety of copper-based chemicals for industrial and agricultural uses.[23]

After the war, C&H expanded its mining operations beyond copper and beyond Michigan. It developed fledgling mining ventures in search of zinc, lead, or uranium in Wisconsin, Illinois, and Nevada. Finally, C&H generated additional income from its large land holdings. By the late 1930s, it started selling off company houses. While this continued, the company also sold off timber resources from its lands and began leasing or selling lakefront property to vacationers or to locals wanting a camp.

C&H's copper mines struggled from year to year. The firm closed some shafts and reopened others as copper prices permitted. During the Korean War, for instance, C&H reopened its works on the Osceola lode. But seen as part of the big picture, C&H's copper mines were becoming less important to the firm; in fact, all its operations on the Keweenaw Peninsula were becoming less important. This could be witnessed in two highly symbolic ways. In 1952, C&H shortened its corporate name, took out the word "mining," and became "Calumet and Hecla, Inc." Three years later, the company moved away from its copper-mining heritage by moving its headquarters from Calumet to Chicago.[24]

Laborers had switched allegiance from the International Union of Mine, Mill, and Smelter Workers over to the United Steelworkers Union in 1950. As the native-copper industry wound down at C&H and Copper Range, the collective bargaining of three-year labor contracts became increasingly contentious. The men wanted more money; the companies felt they had none to give. While its men still belonged to the IUMMSW, in 1949 C&H closed its mines throughout the summer season; this act succeeded in convincing workers to accept a wage cut. Three years later, in 1952, the United Steelworkers struck C&H for two months. In 1955, workers struck early in May and did not return until near the end of August. To bring the men back to work on its terms, C&H announced it was going to close its mines and sell off its properties. C&H halted its move to liquidate once the union men agreed to company terms and returned to work. The threat of closure apparently made the men more fearful of engaging in strikes, because labor relations calmed for a period, going into the mid-1960s.[25]

While occasionally brandishing the threat of closure, C&H continued the chase for new copper. For a time in the early 1960s, it appeared that new works on the Kingston conglomerate lode would carry the company's mining operations, but that did not prove to be the case. Like so many other lodes C&H had tested and tried over the decades, the Kingston did not develop into a bonanza. In 1965, labor troubles arrived again as union workers went out on a ten-week strike over wages and select work rules.[26] In 1968, labor and management butted heads again—and for the last time.

By 1968, C&H did not even exist anymore as a separate, independent company. Universal Oil Products (UOP) out of Chicago had bought up C&H's properties and products. UOP was a conglomerate engaged in numerous activities. Its largest source of revenue was the petroleum- and petrochemical-processing market. UOP marketed itself as an advanced scientific and engineering research company. It dealt with materials science, minerals processing, catalytic converters, and plastics. It also fabricated metal parts, including parts for truck suspensions and airplanes.[27] UOP, typical of a conglomerate, entered new fields by acquiring companies already engaged in those fields. Wanting to avoid a takeover itself—wanting to avoid being swallowed up by a larger conglomerate—UOP sought in the mid-1960s to become bigger. C&H caught UOP's attention. It, too, was headquartered near Chicago. C&H's Wolverine Tube Works and its markets fitted nicely with UOP's interests in energy and metals fabrication. At the end of April 1968, UOP acquired C&H through an exchange of stock. Each share of C&H stock was traded for 0.6 shares of UOP common stock; the UOP stock traded to acquire C&H was valued at about $100 million. UOP then incorporated a new Calumet and Hecla Corporation as one of its wholly owned subsidiaries.

In August 1968, only months after UOP assumed control of C&H, the last labor contract with the United Steelworkers concluded, and employees went on strike. The strike halted mining and the production of copper, copper alloys and chemicals, and castings. Some four hundred miles south of the Keweenaw, UOP leadership did not budge in negotiations. That leadership had no history or tradition connecting it to the 1,200 C&H workers living in the vicinity of Calumet, who were now idled. It had only economic ties, and they were marginal. The deadlock continued over the long winter of 1968–69, and discussions tended to be acrimonious. In the spring, renewed negotiations failed. On 9 April 1969, UOP announced to strikers that their employment was terminated. They had no jobs to wait for, no jobs to return to.[28]

Workers in the small communities around C&H had been threatened with closure before and were not certain whether this was the real deal or only the most recent games-manship intended to get the union to drop its demands. UOP's announcement on 9 April was surely ominous. Church bells tolled in Calumet; many thought it was a death knell. Surely, if the closure happened, the loss of over a thousand jobs, the loss of a $9 million annual payroll, would devastate workers, families, businesses, and institutions of all sorts.[29] Calumet village and Calumet Township had already withered a great deal since the peak of mining, when over 37,000 persons, combined, lived in those two places. They had seen people leave as the copper-mining economy had declined. The remaining population of about 10,500 desperately needed to hold on to those jobs now threatened by UOP, and so civic and religious leaders attempted to broker a settlement, to forge a new contract between UOP and the Steelworkers Local 4312. But last-ditch negotiations failed, attempts to find new ownership for the mining and industrial plant failed, and more months passed with no resolution. Finally the mine pumps halted, UOP let the last operating shafts fill with water, and by the last months of 1970 the end was at hand. UOP liquidated the Calumet division's mining and manufacturing works and scrapped out or sold off its machinery and equipment.

The mine had been there first. It had struggled at the start but then flourished to become the great C&H, one of the richest mines in the world. The mine had given rise to adjacent communities filled with shops, stores, churches, schools, fraternal organiza-tions, houses, and families. The communities and the host mine had fed off each other for a full century—about a half century of dynamic growth and expansion, followed by a more painful half century of contraction and decline. With the industrial base gone, the communities faced hard times and a very uncertain future.

19

❖ ❖ ❖

COPPER RANGE

Staying Alive

F ew mines lived as long as Quincy and Calumet and Hecla (C&H). Copper
Range's Baltic and Trimountain mines started up at the turn of the century and
by 1910 they had already peaked. Only the Champion mine on the Baltic lode
expanded between 1910 and 1920. In 1916–17, Champion built over sixty new houses
at Painesdale to attract and keep the men it needed to push production to as high as
33.6 million pounds annually during the lucrative war years.[1] At the start of the troubled
1920s, those boom times seemed like a distant memory at Champion. No more houses
would ever be built; no more production or profit records would be set. While the times
turned bad at Champion, things were even worse at Trimountain and Baltic.

The Trimountain mine never lived up to expectations. In 1908, Horace Stevens's
Copper Handbook noted that Trimountain was "somewhat disappointing," and a few
years later it stated that "Trimountain has proven a disappointment . . . and is by no
means in the same class as the Baltic and Champion." In 1929, Copper Range acknowledged that Trimountain had "never been an important source of earnings." The mine
limped through the 1920s, when Copper Range maintained minimal production there,
just to keep men in the area and equipment in working condition. But in May 1930
Copper Range gave it up and closed Trimountain, which over its lifetime produced 140
million pounds of copper.[2] It never reopened.

Like Trimountain, the Baltic mine ran at a loss throughout the 1920s. Copper
Range's *Annual Report* for 1925 noted that, over the past six years, ownership of these
two mines had been "a liability and not an asset." Only a short stretch of lode on the
southern end of the Baltic had been a decent producer. Copper Range, while winding

down Trimountain, started finishing off Baltic, too. Starting at the bottom of the mine, miners robbed the copper-charged pillars that had been left intact to support the hanging. Besides producing copper, robbing the pillars "gave employment as long as possible" to men who Copper Range didn't want to see leave for the auto plants of Detroit. But on 31 December 1931, it gave up the effort and closed the Baltic mine, after it had produced about 270 million pounds of copper.[3] Like Trimountain, it never reopened.

As the stronger Champion mine entered the 1920s, it remained mostly profitable, paying out $600,000 in dividends in 1920, none in 1921, then $600,000 again in 1922, and up to $1.1 million by 1925. Its production dropped to 13.6 million pounds in 1920 but recovered to run from 17 to 21 million pounds annually through 1925.[4] One key problem that Champion faced was keeping men to fill its jobs. In these poor times the Copper Country lost workers at an even faster rate than the companies cut back on jobs. Unemployed workers had limited access to welfare payments that might have kept them around through the tough times. The biggest source of welfare was Houghton County's Superintendents of the Poor, with an annual budget $55,000 to $65,000. In 1920 the Superintendents of the Poor provided relief to the tune of $62,045, which provided groceries, clothes, furniture, rent, fuel, and medical service to 534 individuals: 196 widows, 110 elderly residents, 91 women with disabled husbands, 85 hospital patients, 39 women whose husbands had run off on them, 12 orphans, and 1 illegitimate child. No relief at all went to healthy, unemployed miners. This picture changed a bit when the Superintendents of the Poor offered relief to 158 unemployed persons in 1921 and to 139 out-of-work residents in 1922.[5] Still, that was a drop in the bucket compared to the number of men—about 7,500—who had lost jobs in local mines, mills, and smelters between 1918 and early 1922. For the vast majority of the region's unemployed men, the wisest thing to do was to get up and leave. The men who left were fortunate that while the mines were in decline, most of the economy, after an initial postwar dip, was booming. There were jobs out there.

The mines were labor short not only because unemployed men left the region, but because men who had jobs quit them for better jobs elsewhere. Copper Range paid wages in the early 1920s that were 50 percent higher than before World War I, and still the men left.[6] The biggest culprit in luring workers away from the Keweenaw was Detroit, and the Ford Motor Company in particular. Copper Range officials wrote back and forth on this issue, commiserating with one another. The day of the Cornish miner was over because Ford seemed particularly keen on putting those men to work at Highland Park or River Rouge. Cornishmen "were glad to go into factories and only an exorbitant advance would attract them to mining." Contract miners left, and trammer bosses and surface men drawing good salaries left. The railroads happily aided the exodus by putting extra coaches on their southbound trains. The company doubted it could ever afford wages high enough to stanch the flow to industrial cities: "Unless the increase in pay brings wages nearly up to that in cities, men will continue to voice their discontent and threaten to leave—and will leave."[7]

Failing to hold men, Copper Range recruited new men from Duluth and Minneapolis, from England (mostly Cornwall), Canada, and Germany. The federal government waived new immigration restrictions to allow for this recruitment. By 1924, Copper Range had spent $50,000 just to lure Germans to the Keweenaw. That money covered the salary of the overseas recruiter, the cost of ships' passage for workers and families, and the expense of setting them up with furniture once they arrived. Imported men over time were to pay the company back for bringing them over the Atlantic. Some did, but others ran on their new employers. By 1924, Copper Range had lost $8,500 invested in workers who fled Painesdale shortly after arriving.[8] But if some men cheated the company, the company cheated some of the men. William Schacht did not want the U.S. Department of Labor to investigate Copper Range's program for importing men, because it "could not help [but] bring out the fact that our Germans as well as the Cornish miners imported are not being used as drill machine runners."[9] In other words, many of these men, promised miners' jobs, found themselves working as trammers.

Despite its problems with labor availability and costs, the Champion mine remained profitable until the end of the 1920s. In 1929 its economic future looked brighter than it had in years because the price of copper that year jumped from twelve to fourteen cents per pound up to eighteen.[10] But along came the Great Depression, and despite the efforts of a copper producers' cartel to limit production in order to maintain high prices, the price of copper steeply declined to 13 cents in 1930, to 8 cents in 1931, and to only 5.6 cents in 1932. This decline pushed Copper Range to close Trimountain and Baltic and to limit operations at Champion. In May 1931 Champion trimmed the workweek from six days to five; in June it trimmed it to four, while reducing wages; in October it cut wages again. In March 1932 it went back to a six-day workweek but closed its No. 1 shaft and worked only the richer southern end of the mine.[11]

When deciding who to keep on the payroll and how to schedule their hours, Champion was guided by old paternalistic and patriarchal ideas:

> In order to cause the least hardship, work is given men in accordance with their needs. Those having dependents work three or more days per week, and others two days or less per week. All are rotated so as to spread employment to the greatest possible number.[12]

To help employees further to ride out troubled times, Copper Range encouraged families to garden:

> We are plowing up all of our available farm land. Each family . . . is supplied with a 50 × 100 foot lot and 2 bushels of seed potatoes and other seeds, such as rutabaga, beet, onion, carrot, pea, bean, etc., and told to go to it so that now instead of having a lot of idle miners on my hands, we have a lot of busy farmers.[13]

Copper Range had always operated a railroad; in 1925 it started a bus company. The Eckland Bus Company of Minneapolis took this builder's photo of an early Copper Range bus. Both transportation businesses ultimately ran into hard times. The bus company closed in 1955; the railroad, in 1973. *(Musser Collection, Lake States Railway Historical Association)*

When greater state and federal relief money became available, Copper Range laid off more men at Champion and let New Deal programs take care of them. For the mine and mill employees it kept after 1933, Champion worked them on a more regular schedule. It expanded the workweek to five 8-hour days, "with some increase in wages in effort to cooperate with the National Recovery Administration."[14] By selectively mining only rich ground and juggling work schedules and wages, Copper Range managed to keep the Champion mine open throughout the Great Depression, which was a rare accomplishment in the 1930s.

As its native-copper industry struggled to survive, Copper Range, like C&H, followed the strategy of diversification. It moved capital into new endeavors. In 1925 the struggling Copper Range Railroad formed a subsidiary firm: the Copper Range Motor Bus Company. The buses relieved the railroad of the burden of running an unprofitable passenger service. The buses ran from Painesdale down to Ontonagon (when the roads weren't choked with snow) and up to Houghton and over to Lake Linden.

In the late 1920s, Copper Range thought that a reviving copper industry was ready to lessen its reliance on steam power and increase its use of electricity. It joined with the Middle West Utilities Company of Chicago to form the new Copper District Power Company, which had a grand vision of harnessing all the waterpower of the Ontonagon River by building seven hydroelectric plants that would operate off a total head of 580 feet and generate 175 million kilowatt-hours per year.[15] In 1929, construction began on the first of these, the Victoria plant, at a spot where the river fell seventy feet over a run of three thousand feet. The plant, with its multiple-arch concrete dam, long penstock, and powerhouse, cost 1.6 million dollars and could generate 45 million kilowatt-hours. Starting in 1931, Victoria sent electricity northward to Painesdale, Baltic, Trimountain, and Atlantic; to Copper Range's mill towns and smelter; and to the Houghton County Light Company, which bought power from the Copper District Power Company. The utility also delivered power to Ontonagon, Victoria, Rockland, Mass, and Greenland. But as the power company started up, the region fell into the worst years of the Depression, and copper production all but halted. The demand for Victoria's power ran far short of its design capacity, and the Copper District Power Company scuttled plans for the six additional hydros.[16]

The electricity transmitted from Victoria to the Champion mill encouraged Copper Range to electrify and revolutionize some technologies there. In 1905 Copper Range had hoped to replace steam stamps with rotary crushers. At that time, the new technology failed, and the Copper Range mills continued on with steam stamps. In 1927–28, because efficient milling was so important to generating profits, Copper Range had installed state-of-the-art flotation technology at its Champion mill to capture the finest copper. With that technology in hand and with cheap power at its disposal, in 1932 Copper Range returned to the idea of rotary crushing, using a machine designed by its own William Schacht. Electric motors powered the Schacht Impact Crushers, which had impellers that turned in one direction while circumferential striking pads rotated in the opposite direction. When copper rock was fed into the machine, the impeller drove the rock against the counterrotating striking pads, breaking it. It took a few years to sort out mechanical problems and make the new crushers fully operational, but the savings in energy costs—over coal-fired boilers and steam—made the effort worthwhile. By 1935 the Champion mill became the only one on the Keweenaw to have abandoned the use of steam stamps.[17]

Inexpensive electric power also encouraged Copper Range to build a reclamation plant at its Champion mill. Because of a decrease in mine production, the mill had regrinding and flotation units sitting idle; it also had a deposit of tailings sitting in Lake Superior. The lake's waves had washed away fine sands but had left behind coarser tailings. Copper Range purchased a dredge and created an enclosed pool by building up a ridge of sands on the periphery of the tailings. The dredge lifted sands from the bottom of the pool and sent them to the reclamation plant. This operation never brought Copper Range the benefits that reclamation brought to C&H and Quincy. It operated only during parts of summers, and from 1937 to its closure in 1948, recovered just 3.5 million pounds of copper.[18]

Copper Range also invested in manufacturing. In 1931 it purchased a large interest in the C. G. Hussey Company of Pittsburgh, and in 1936 it acquired all of Hussey's assets.

The Hussey firm's ties to Lake copper mines went back to the mid-1840s. Curtis Grubb Hussey wore many hats in Pittsburgh: physician, merchant, and pork dealer. That was not enough for this entrepreneur. In 1843 he helped bankroll a trip to Lake Superior made by a friend, John Hays. Based on Hays's positive accounts of the region's copper, Hussey invested in Lake Superior's first great profit maker: the Cliff mine. In the late 1840s Hussey played a key role in developing the reverberatory furnace for smelting native copper, and he established a smelter in Pittsburgh. In 1848–49 he also formed the C. G. Hussey Company, which operated a rolling mill that produced sheet copper. Hussey rapidly became a multimillionaire, and his rolling mill did well until his death in 1893. The firm faltered after that, before being acquired by Copper Range.[19] For years C. G. Hussey had been a buyer of Copper Range's product. Now Copper Range used its own copper to fabricate sheet copper, plus copper nails, rivets, ferrules, and building materials.[20]

Copper Range also invested in mineral lands. This was nothing new for the company. In 1905 to 1909 it had taken out an option to work the Globe mine's property, just south of the Champion mine. Copper Range expended $350,000 on this exploration, but to no avail. In 1911 the company acquired the Atlantic mine.[21] Copper Range accelerated its acquisitions as the Baltic and Trimountain mines wound down. In 1928 it took another option on the Globe, and through 1937 it spent $680,000 on development there. In 1938 Copper Range purchased the Globe property outright and mined it until 1945. In some years the Globe mine accounted for 10 percent of Copper Range's production but never became profitable.

In 1929 Copper Range acquired the White Pine and Victoria mines plus control of the old National mine. All three Ontonagon County mines were idle. The National had not produced copper since 1895; White Pine and Victoria had ceased operations in 1920–21.[22] In 1931 Copper Range acquired the Naumkeag Copper Company and the St. Mary's Canal Mineral Land Company (which had been its partner in forming the Champion mine). These two purchases gave Copper Range more contiguous properties to explore from the Atlantic mine down to the Globe. In 1932 Copper Range purchased the Mohawk mine, north of Calumet, which had just closed. It also acquired the Wolverine mine. These properties, in years of steep economic decline, came cheap. Copper Range acquired the Mohawk mine for only $25,000.[23] Most of these acquisitions, however, proved dear at any price. The capital invested in them was lost, was never returned by a revival of the old mines or by any new discoveries of copper. There was, however, one major exception to this rule: the White Pine mine about seventy miles southwest of Painesdale.

The White Pine mine sat atop the Nonesuch lode, so called because it was unlike others in the copper district. The lode contained native copper, but much of its rock was a shale containing copper sulfide, called *chalcocite*. The grains of chalcocite were "so small as to be practically invisible" and so were the grains of native copper, making them hard to capture in the milling process. Another important characteristic of this large orebody, which later affected the way it was mined, was that it did not outcrop and then dip steeply into the ground but sat as an underground dome of mineral.[24]

Frank Cadotte had discovered the Nonesuch lode near the base of the Porcupine Mountains over the winter of 1865–66. Discovery led to incorporation: the Nonesuch mine formed in 1867 and exploited the lode's native copper sporadically from 1868 through 1885, producing about 390,000 pounds of ingot.[25] Other speculators purchased land in the vicinity of the Nonesuch mine, but development lagged until early in the twentieth century, when C&H—seeking new sources of copper to compensate for the faltering yield of its Calumet Conglomerate lode—stepped into the picture. After exploring the area in 1907, C&H organized the White Pine Copper Company in 1909 and opened a mine about three miles east of the old Nonesuch mine. C&H held a controlling interest in the White Pine mine and managed the company but did not own it outright.

The White Pine Copper Company sank 110 diamond-drill holes into the lode to trace its extent and find the richest portion. In 1910 Stevens's *Copper Handbook* noted the novelty of working this lode:

> The property is decidedly interesting, from a geological and mining stand-point, but it is difficult to estimate its value, for the occurrence of copper is under such entirely different conditions than those noted in the main Keweenawan series, and so little effective work has been done [on the Nonesuch lode] that the property must be considered an interesting exploration, rather than a mine.[26]

Nevertheless, White Pine did mine it, sinking four shafts into the lode, erecting requisite boilers and hoists, building a mill and a rail connection to the outside world, and erecting over thirty houses for employees. White Pine began production in 1915, when it sent 2.8 million pounds of ingot to market, all coming from the Nonesuch's native-copper deposits; the mine did not work the copper sulfide. Before ceasing operations in 1920, White Pine produced 18.2 million pounds of copper. Thanks to the high copper prices of the First World War years, it managed to pay out $33,500 in dividends over 1917–18.[27] Since the 1860s, that was the only money ever made from mining the Nonesuch lode.

By 1929 the White Pine mine had been closed for nearly a decade. It carried an indebtedness of $118,611.85 and in May the property went up for auction on the Ontonagon County Courthouse steps. Copper Range's new president, F. Ward Paine, wired William Schacht at Painesdale, "I figure we should bid. . . . I feel there is a real merit here." At the auction, C&H bid up to the level of White Pines' indebtedness; if it won, it would fully own the property, which would be in the clear financially. But Copper Range bid $388.15 higher than C&H—and its bid of $119,000 carried the day.[28] It would take a quarter century for Copper Range to realize just how much it had won, and how much C&H had lost, at the auction of the White Pine mine.

Going into the Depression, Copper Range lacked the capital to develop its new property, so White Pine sat idle. By 1937–38 the company started diamond-drill prospecting,
and it unwatered and reopened one of the old White Pine shafts. Copper Range had a
new plan for the Nonesuch lode: instead of taking native copper, it would be the first
Lake mine to exploit copper sulfide, which existed here in far greater quantities. Copper
Range contracted with the Michigan College of Mining and Technology to study ways to
mill and recover chalcocite.[29] New smelting technologies would be needed, too. By 1940
Copper Range had taken 2,800 tons of rock from White Pine for tests, "made with a
view to developing a mining method by which this deposit may be developed and mined
efficiently and economically."[30] There was no immediate return on this investment, and
Copper Range had no way of knowing whether the day would ever come when it had
both the money and the technologies needed to open a sulfide mine at White Pine.

While conducting experimental work on the Nonesuch lode, in 1940 Copper Range
still operated the Champion and the Globe mines, the Champion mill at Freda, the
smelter near Houghton, the Copper Range Railroad and the bus line, the Houghton-
Hancock water line, the Copper District Power Company, and the Hussey works in
Pittsburgh.[31] Most of these operations were doing less than a stellar business, but a few
were modestly profitable, like the power company, which paid out dividends of $100,000
to $125,000 annually from 1935 to 1940.[32] Champion hoisted rock only from its two
southernmost shafts, which were 4,615 and 5,040 feet deep, and maintaining production had become "increasingly difficult . . . because of the spotty nature of the ground."[33]
Similarly, at the Globe mine the lode was fractured and the mineralization erratic. Early in
the Great Depression, when Copper Range acquired the Mohawk mine, it also obtained
Mohawk's one-third share in the Michigan Smelting Company. Now Copper Range
claimed the facility as one of its wholly owned divisions, but mine closings and cutbacks
drastically reduced the smelter's business. It ran at a small fraction of its capacity and
sometimes intermittently.[34] The Copper Range Railroad, too, suffered from overcapac-
ity and underuse. It had little use for its northern spur after the Wolverine and Mohawk
mines closed. It no longer had the tonnage to carry between the Trimountain and Baltic
mines and mills and then on to the smelter. After cutting back on routes, rolling stock,
and employment, the railroad filed for bankruptcy in 1935. Reorganized in 1938, the
rail line benefited from a bit of an upswing in the production of copper and timber at the
end of the Depression but never come close to recovering what it had lost.[35]

Copper Range sustained its many operations and divisions through the Second
World War. It continued to mine, mill, and smelt copper. Thanks to government contracts that paid a high enough price for the metal, the company could afford the wartime
wages of its now-unionized workers, which had risen 60 percent. When the war and
government contracts ended, Copper Range could not mine without substantial losses,
so it closed the Globe mine permanently in 1945 and shut down Champion, which it
paid to keep unwatered.[36] It also closed its reclamation plant and smelter. In 1946–48,

Copper Range, rather like a terminal patient, got its house in order. It knew it would have little use for much equipment in and around the mines, so it sold or scrapped it. It removed "all useful tools and equipment" from the bottom up to the twelfth level of the Champion mine and from the Globe, bottom to top. Although it left some equipment behind at the Champion mill, it also conducted major scrap operations at its three stamp mills.

Early in 1947, Copper Range bowed out of the Copper District Power Company, and ownership passed to the new Upper Peninsula Power Company.[37] That same year, the company added two of the three diesel locomotives that it ever ran on the Copper Range Railroad. The diesels hardly represented positive growth, however; they merely replaced numerous old steam locomotives. The railroad remained marginally profitable in the late 1940s and early 1950s but had fewer customers to serve and faced growing competition from motortrucks. The railroad had scrapped a few steam locomotives in the 1920–21 recession, and from 1930 through 1955 it sold off or scrapped another twenty or so. The diesels purchased in 1947 and 1951 provided motive power to the ever-declining railroad until the Interstate Commerce Commission authorized its total abandonment in 1973, when Copper Range sold off all remaining equipment.[38]

In its 1948 *Annual Report,* Copper Range, not so much looking ahead as looking back, boasted of its lifetime achievement through that year: its mines had produced 1.23 billion pounds of copper, and the company had paid a total of $34.6 million in dividends. But in 1948 all mining was still shut down on the Baltic lode. After conducting much experimental milling of sulfide ore at Freda and drilling 239 diamond-drill holes into the Nonesuch lode, work at the White Pine mine also stopped. Copper Range's main sources of revenue were the C. G. Hussey Company, the Copper Range Railroad, the Copper Range Motor Bus Company, and the sales of timberlands.[39]

In 1949 and 1950, Copper Range's sales and profits came largely from its Hussey works. Importantly, it began mining again at Champion and milling at Freda. Hussey had something to do with that move—Champion was to supply copper to Hussey's manufacturing works in Pittsburgh. Putting Champion back into production provided employment to some miners, encouraging them to stick around. Once reopened, the Champion mine and mill limped along. The same could not be said for the underused Michigan smelter, which Copper Range scrapped out in 1952.[40] Thereafter, the company sent its mineral to the Quincy mine's smelter.

Diversified Copper Range was dying off one piece at a time. Next to go after the smelter in 1952 was the Copper Range Motor Bus Company. Just as the railroad faced new competition from trucks, the bus line suffered from growing competition from cars, as well as from reduced ridership because of local population losses. The buses logged 780,000 passengers in 1945. Ridership dipped to 595,000 in 1950 and to 340,000 in 1952. In 1955 Copper Range discontinued bus service and liquidated the company's assets.[41]

In 1958 a *Daily Mining Gazette* photographer captured a landscape of decline, closure, and waste along the shoreline of Portage Lake west of Houghton. The Atlantic stamp mill is long gone. The Michigan smelter is gone. Unneeded and unused Copper Range railcars sit stationary on their tracks. Smelter slag and stamp sands complete the picture. *(Michigan Technological University Archives and Copper Country Historical Collections)*

The Champion mine and mill remained in production during most of the 1950s, sometimes with the assistance of government copper-purchasing agreements (Mine Maintenance contracts) that paid Champion a higher price than it would have received in the normal market place. These contracts were designed to keep American copper producers afloat in an era of stiff foreign competition. Shortly after a Mine Maintenance contract expired, in mid-February 1958 Copper Range shut Champion down. By the end of the year, it reopened Champion. The company was now in an era of fitful stops and starts. Copper Range shut down Champion again and put it on standby on 1 February 1961. It resumed operations at the mine and mill on 15 May 1961, but they closed in the face of a railroad strike on 11 July 1961. Champion opened again on 6 September and made a modest profit in 1962.

In the midst of on-again, off-again mining, in 1961 Copper Range began selling off its 593 company houses in Baltic, Trimountain, and Painesdale. The company should have done this earlier, when the houses would have been more in demand, in better condition, and worth more. In 1961, 193 of the company's houses were "vacant and gener-

Painesdale once had an impressive cluster of community buildings at its center, including the high school and grade school, the Sarah Sargent Paine Memorial Library, the
Albert Paine Memorial Methodist Church, and the two structures shown here: the
Independent Order of Oddfellows Lodge, also called the Opera House, in the front,
and the Finnish Hall, behind. As mining slowed and finally ceased, many community institutions disappeared. In Painesdale, the grade school, Paine Library, I.O.O.F.
Lodge and the Finnish Hall are all gone. *(Michigan Technological University Archives
and Copper Country Historical Collections)*

ally in poor condition." Presumably, Copper Range had held onto its houses because it
still believed in paternalism. Controlling the housing stock and renting to employees—
those had been the right things for a mining company to do. When Copper Range finally
dropped this obligation, its house sales did not yield a high return. By 1 February 1962
it had sold 330 houses for a return of $255,398—or about $775 per house. Each buyer
received a ninety-nine-year ground lease for the house lot, with a $24 payment due each
year. By the end of 1962 Copper Range still retained ownership of 241 houses (only 39
were occupied and in fair condition); by the end of 1963 it still owned 197 houses, most
of which, surely, were vacant and in poor repair.[42]

In 1963 the Champion mine produced 5.8 million pounds of copper but operated
at a loss. It lost money again in 1964 when a strike closed it for three months. Copper
Range undertook a considerable amount of prospecting and development work in 1965;
it did 8,500 feet of drifting and put in 7,000 feet of diamond-drill holes. But it found
no new rich ground. In Champion's early days, its rock carried twenty-four pounds of
copper per ton, whereas in the 1950s and 1960s the rock yielded only nine to twelve
pounds per ton. The mine remained at the mercy of copper prices; it needed high prices
to be profitable.[43] In the face of continued losses, Copper Range decided to close its last
native-copper mine for good:

The decision to close was difficult for us to reach due to the human element involved. This property has materially contributed to the economic life of several small adjacent communities. Prior to making a final determination, we discussed the problem with representatives of our employees in an effort to minimize the effects of any dislocation upon this loyal group, who have served us faithfully for so many years.[44]

In January 1967 a letter sent by Copper Range to all Champion employees announced that the mine and mill would be shut down by the end of the year: "We had certainly hoped never to be in this position, but it is, we suppose, inevitable that the economic reserves, after 66 years, would be exhausted."[45] It happened to all the mines sooner or later: they closed. It happened to the Champion on 11 September 1967, when it hoisted its last skip of rock. Thereafter the only thing coming out of the mine, at the No. 4 or "D" shaft, was the water pumped up to serve several local communities. Adams township, instead of Copper Range, continued that service.

The venerable Hussey works in Pittsburgh lived on after mining ceased on the Baltic lode. In the 1960s it suffered strikes and floods and its first unprofitable years since the early 1930s, yet it moved ahead aggressively. The C. G. Hussey Company became Copper Range's Hussey Metals Division. It built a new plant on the outskirts of Pittsburgh; purchased a rolling mill in Anderson, Indiana; and built a plant in Eminence, Kentucky, to manufacture copper bus bars for collecting and distributing electrical currents.[46]

Copper Range expanded on the Keweenaw, too. In the late 1920s and early 1930s it had bought large tracts of land up and down the peninsula. In 1960 a timber inventory showed that Copper Range owned 185,000 acres of forest, containing some 270 million board feet of sawmill lumber and 1.2 million cords of pulpwood. So shortly after the Champion mine closed, Copper Range in December 1968 opened its new Northern Hardwoods Division. Its mill near South Range could produce 35,000 feet of sawn lumber during each of its two 8-hour shifts.[47]

The lands Copper Range purchased in the 1920s and 1930s eventually yielded trees, but most of the properties never produced any copper. The White Pine mine proved the exception. Copper Range started taking a closer look at this property in the late 1930s, and then during the Second World War the War Production Board asked all U.S. copper mines to report on ways they could boost production. On the basis of this request, Copper Range undertook more focused research on the possibility of putting White Pine into production. None of this work bore immediate fruit, but it eventually led to the opening of a major mine.

Besides putting in hundreds of diamond-drill holes, in 1945 Copper Range at White Pine sank the Bill Schacht Shaft, which was named in honor on the company's onetime president. The Shacht Shaft ran 1,413 feet through the Nonesuch lode. From the shaft, Copper Range drove over eight thousand feet of drifts and took out 100,000 tons of ore to test at a pilot concentrating mill built at Champion's stamp mill at Freda. Between

1945 and 1950 Copper Range, with research aid from the local mining school and the Battelle Memorial Institute, worked out efficient means of milling the Nonesuch lode's chalcocite ore.[48]

With key concentrating technologies in hand, Copper Range still lacked the capital to open a novel sulfide mine. Then, during the Korean War era, its chance came. Between 1951 and 1953, in order to boost domestic copper production, the federal government offered economic incentives that led to the start-up of six new mines: four in Arizona, one in Nevada—and one in Michigan.[49] The Reconstruction Finance Corporation provided a $57 million construction loan to Copper Range to open the White Pine mine, and the Defense Materials Procurement Agency contracted to purchase nearly 245,000 tons of White Pine copper. With loan and purchase contract in hand, Copper Range incorporated the White Pine Copper Company as a wholly owned subsidiary and invested $13 million of additional capital into the project. Construction began in March 1952, and the White Pine mine smelted its first ingot copper in January 1955.[50] Here was the long-hoped-for bonanza. All together, the native-copper mines on the Keweenaw, from the 1840s to the late 1960s, produced eleven billion pounds of copper. Over its much shorter life span, the White Pine mine, by itself, would produce 4.5 billion pounds. In its geology, production levels, technologies, and townsite, White Pine cut a sharp contrast to every Lake copper mine that came before it.

20

❖ ❖ ❖

WHITE PINE

A New Mine, a New Era

The Copper Range officials who acquired and explored the White Pine property did not live to see it into production. F. Ward Paine had followed in father William's footsteps to become company president in 1929. Soon he decided to buy White Pine at auction, risking precious capital to increase his company's ore reserves. Paine died in 1940. William Schacht tendered the winning bid at the White Pine auction. Schacht, who became company president himself, guided the diamond drilling to map the extent of the White Pine deposit and supported research into new technologies for milling chalcocite ore. In the first years of World War II, Schacht wrangled unsuccessfully with the U.S. Office of Production Management and the Reconstruction Finance Corporation for a loan to put White Pine into production. Schacht died in 1944, putting the future of White Pine into the hands of a new generation of corporate leaders led by president Morris LaCroix and Frank Ayer.[1]

LaCroix, born in Massachusetts, had a Harvard degree. Married into the Paine family, he worked at Paine, Webber, and Company before moving into oil and mining. Ayer served with the Office of Production Management to secure copper and zinc during the war, and afterward became a Copper Range vice president in charge of developing White Pine.[2] Once they had secured the $57 million loan for the project, LaCroix and Ayer turned to outside firms for help. Copper Range awarded the major White Pine contract to the Turner Construction Company, which had been around for a half century. Turner Construction had erected over 2,500 large buildings, including the Chrysler Building in New York City, and helped design the National Lead Company's open-pit mine in

Tahawus, New York. Because White Pine called for diverse technologies and structures, Turner Construction subcontracted parts of the work to specialized firms. Stone & Webster designed the power plant. The architectural and engineering firm of Pace Associates laid out the White Pine townsite and drew up plans for offices, houses, the school, and the hospital. Western-Knapp Engineering Company designed the mine, mill, smelter, and water plant. Turner Construction orchestrated all this, did much of the building, and turned over to Copper Range (or its subsidiary, the White Pine Copper Company) each part of the project as soon as it was completed.[3]

Aboveground and belowground, White Pine bore little resemblance to other Lake copper mines. Obviously, houses built at White Pine would not look like houses built at Painesdale fifty years earlier. No Lake mine had been launched with so much capital, on such a massive scale. No other mine had worked copper sulfide, which called for new technologies underground and at the mill and smelter. Even some nomenclature differed. The mine product here was *ore*, not *copper rock* or *stamp copper*. A native-copper mine had a *hanging wall* and *drifts;* White Pine had a *roof* and *entries*.[4]

Extensive diamond drilling over a three- by five-mile area gave Copper Range a good idea of what lay underground. Beneath a clay overburden, four sedimentary beds rested atop one another. A *lower sandstone* bed, several hundred feet thick, showed little copper except for its very top, which was charged with twenty-four to twenty-six pounds of copper per ton, and in some places with as much as one hundred pounds per ton. Above the lower sandstone sat the *parting shale,* seven to eight feet thick, carrying twenty-five to twenty-six pounds of copper per ton. An *upper sandstone,* three to five feet thick, topped the parting shale and assayed at only four to five pounds per ton. Above the other beds sat the three- to five-foot thick *upper shale,* with twenty-five to twenty-six pounds per ton. Copper Range initially planned to mine just the top foot or so of the lower sandstone bed plus the parting shale. It estimated that White Pine held 309 million tons of ore that would yield an average of 21.3 pounds of copper per ton. The engineers designed facilities to handle a proposed mine output of 12,500 tons of ore per day, 300 days per year. The mill, running 360 days a year, would handle 10,500 tons daily; and the smelter would produce 75 million pounds of copper ingot annually. At those rates, the known orebody would support mining for forty years.[5]

An estimated 65 percent of the orebody lay at a vertical depth of five hundred feet. The deposit lay quite flat with a dip of about 6 percent; it had "the shape of a large, low hill with gentle slopes on either side" that undulated. The engineers planning White Pine faced "the challenge of mining bedded ore too flat lying and too deep-seated to permit either block caving or open pit mining." The mining methods adopted at White Pine had precedents "more common to coal or phosphate ore" mining than to hard-rock copper mining.[6] An observer noted that White Pine's techniques were "common to the Birmingham, Alabama, iron mines; the potash mines near Carlsbad, New Mexico; and the zinc deposits of the Mississippi Valley."[7]

The main entrance to the White Pine mine was not a shaft, but a tunnel-like portal running from the surface down to the top of the orebody, about eighty feet underground. Large shovels and other excavating equipment cut the route; then crews constructed wooden forms for pouring the arched, reinforced-concrete, thirty-two-foot-wide portal. After the concrete work, heavy equipment pushed dirt back into the cut, covering the portal up. *(Ontonagon County Historical Museum)*

To open the underground, a declining portal was constructed from the surface down to the orebody, which at the entry point was only eighty feet below ground. The concrete-lined portal, thirty-two feet wide and nineteen feet high, carried two roadways for trucks and other vehicles and a manway for pedestrians. Five months in construction, the portal was fully operational by March 1953.[8] All men, equipment, and supplies (including up to 25,000 pounds of explosives per day) originally entered the mine through this "tunnel," which had a 10 percent slope; eventually, as the works expanded, men also entered the mine through shafts.[9] One early vertical shaft, fifteen feet in diameter, holed through to the mine entries some one thousand feet from the lower end of the portal. This shaft carried a man-way for emergency exits, as well as pipes and electric utilities, but primarily functioned to ventilate the mine. Two large electric fans pulled air from the mine that had been fouled by blasting gases and the fumes emitted by diesel-powered vehicles.[10] (White Pine used up to 450 vehicles underground.) Miners drove a third early and essential connection between mine and surface—a conveyor tunnel that carried ore from the mine to the aboveground mill.

Where the portal intersected the parting shale, miners pushed three parallel service entries, which were large, nearly flat passageways (separated by rock pillars) running five thousand feet through the copper-rich shale. These served as the mine's main roadways through the ore zone. The two outermost service entries were eight feet high by twenty-four feet wide; miners blasted the central service entry to an increased height of twelve feet so it could better handle the transport of large equipment. The service entries formed the core of the underground works and had to be protected. Along the outside of the service entries, miners left pillars 175 feet wide to guard "against subsidence or ground movement."[11]

To open up ground on either side of the service entries, miners cut through the protective pillars in five locations, where they drove cross-drifts out into the orebody. Once clear of the 175-wide pillars, these cross-drifts continued moving out from the service entries. Working off from these drifts, miners alternately took ground and left other ground as support pillars. They drilled and blasted out *rooms* of copper about thirty-two feet wide while leaving in place regularly spaced rock pillars, twenty by twenty feet, to support the roof.[12] As originally planned, as miners proceeded farther away from the central service entries, they would leave smaller pillars to allow for a greater extraction of the orebody. But the plan changed. As the mine went deeper and the weight of the overburden increased, miners had to leave larger pillars standing.[13]

Miners at White Pine broke considerably more ground per shift than did men at earlier mines. They drilled shot holes two at a time, using Jumbo drills mounted on a vehicle. Each percussion drill with a carbide bit could sink a shot hole one and three-eighths inches in diameter at a rate of thirty-six inches per minute. Typically, a shot round consisted of twenty-eight holes, each about twelve feet deep. After miners fired the round, White Pine used mobile heavy equipment and conveyor belts to move broken ore to the surface. A rubber-tired front-end loader pushed broken ore into a pile, and from the pile a gathering-arm loader fed the ore into a diesel-powered truck or *shuttle car.*[14] The shuttle car traveled to an underground crusher station located 3,600 feet from the mine's entrance. When dumped from the shuttle car, the ore passed onto vibrating screens that separated it into two sizes: over eight inches and under eight inches. The smaller ore, ready to go to the surface, fell through the screens and into an ore pocket below; the larger rock tailed off the screen and into a jaw crusher. After being crushed, this ore, too, fell into the ore pocket elevated over a conveyor-belt tunnel. This belt tunnel did the mine's "hoisting." The twelve-foot-high, twenty-foot wide tunnel did not run in the copper-rich parting shale, but one hundred feet beneath it in the lower sandstone. Moving toward the surface, the tunnel sloped upward on a 16 percent grade. Ore fell from the storage pocket onto a 54-inch-wide conveyor belt powered by three 250-horsepower motors, which carried it part way along the 3,585-foot tunnel at 450 feet per minute. Then the first belt passed the ore to a second belt of the same size and speed. The second belt carried the ore beyond the tunnel, up a 585-foot-long belt portal, and delivered it

This 1953 development plan for White Pine's underground shows three entries or main service drifts heading off from the mine portal. They run down past the crusher station and the ventilation shaft. Cross-drifts lead out from the service entries, opening up ground on both sides. Miners drill and blast, alternately taking ground and leaving ground. They blast out rooms, where the copper is extracted, and leave behind pillars that support the roof. *(Michigan Technological University Archives and Copper Country Historical Collections)*

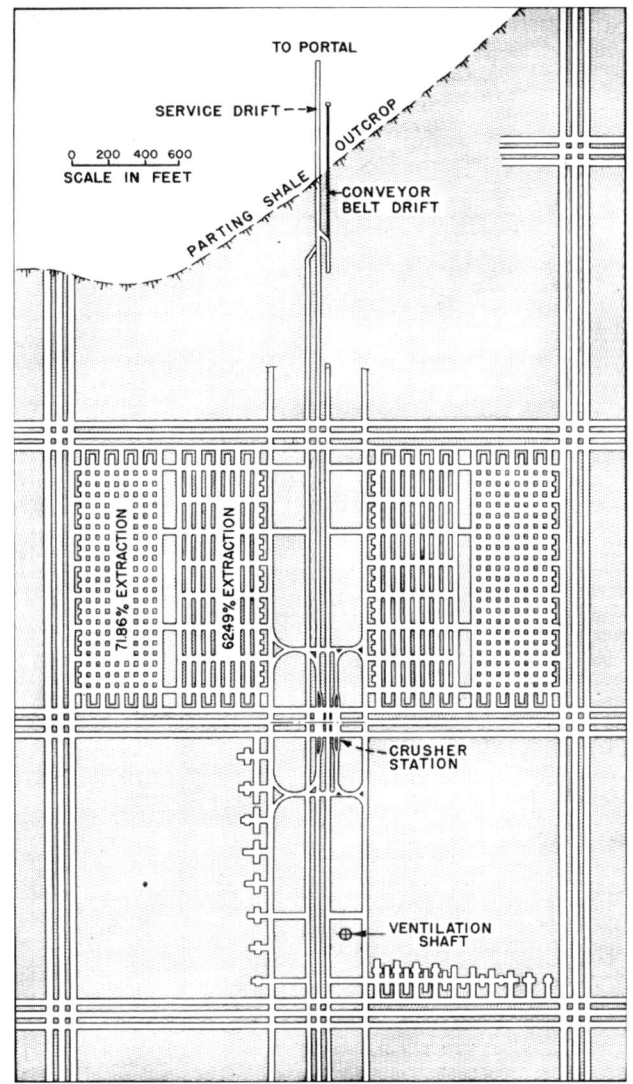

at the mill's crusher bins on the surface.[15] The two-flight conveyor system could handle about 2,400 tons of ore per hour.

White Pine's mill did not use stamps. Instead, the brittle chalcocite ore passed through two gyratory, eccentric cone crushers arranged in series. The first broke the ore to less than two inches; the second reduced the ore to less than three-eighths of an inch. The reduced ore next traveled to one of six large ball mills—perhaps the largest ball mills in the world at the time. Each mill, without its charge of ore, weighed three hundred tons and carried ninety to one hundred tons of steel balls up to two inches in diameter. Large motors rotated the mills at sixteen revolutions per minute, and the balls rolling around inside ground the ore, liberating the grains of chalcocite from the shale. From

White Pine covered a large area and had a relatively flat floor. Men moved around the mine in jeeplike vehicles, and heavy, diesel-powered vehicles did much of the work. Unlike earlier mine vehicles, such as electric haulage locomotives, these self-propelled vehicles did not run on fixed tracks, but on large rubber tires, which allowed for greater maneuverability. *(Michigan Technological University Archives and Copper Country Historical Collections)*

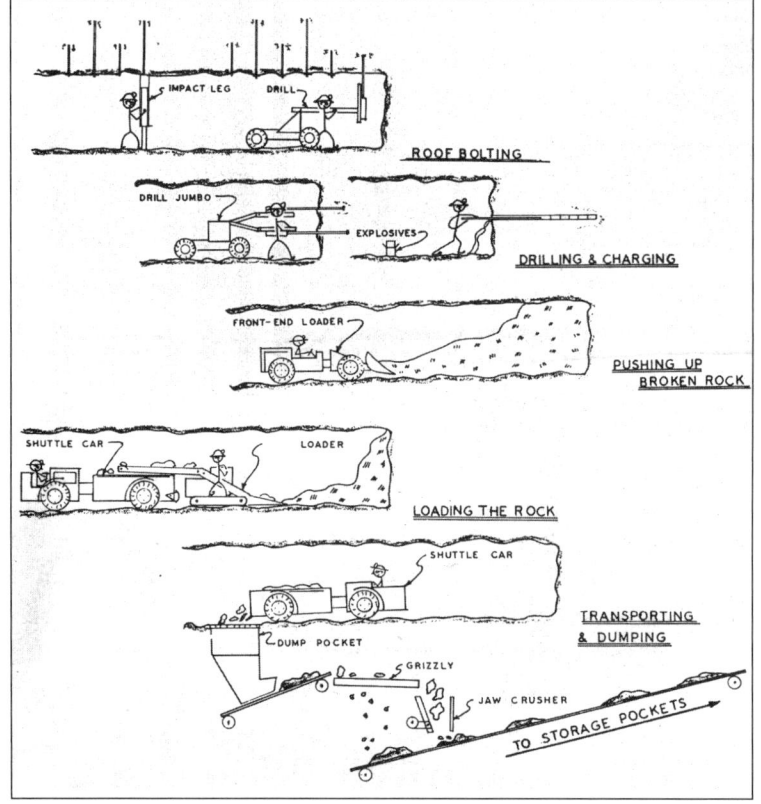

The White Pine mine used equipment underground that was unique in the copper region. To show the public what went on belowground, Copper Range created and published cartoonish sketches that delineated key technologies. *(Ontonagon County Historical Museum)*

the ball mills the mineral passed to flotation cells filled with liquid reagents, frothers, and collectors. The bubbles rising up in the cells had an affinity for the miniscule grains of chalcocite, which they carried up to the top, while much of the ground shale fell to the bottom. Skimmed from the froth atop the cells, the collected mineral (copper sulfide and silicates) ran about 30 to 35 percent copper.[16]

From the flotation cells the mineral passed to pug mills, which mixed it thoroughly with ground pyrite and limestone—two fluxes that were needed to effect a good separation of molten copper and slag in the reverberatory furnace. At the same time Copper Range was demolishing its old Michigan smelter on Portage Lake, it built a new one at White Pine. This big furnace, 110 feet long and 28 feet wide, ran almost constantly throughout the year. It was fueled by about 120 tons of powdered coal per day, and the molten copper flowed from this melting furnace to a refining furnace (by 1959 there were two), where rabbling and poling reduced impurities in the melt, leaving a product that was 99.9 percent pure—with just a trace of silver. Meanwhile, three 15-ton slag pots on railcars moved by a diesel-electric locomotive hauled away 504 tons of slag per day and dumped it nearby.[17]

In the early 1950s, Copper Range erected many ancillary facilities that supported the mine, mill, and smelter. For the men, it erected a change house connected to the portal so employees did not have to walk outside to go to or from the underground. The change house had 750 racks for clean clothes on one side, 750 racks for dirty work clothes on the other, and showers and toilets in between. It also had separate lockers, toilets, and showers for supervisors and foremen.[18] For transporting coal, fluxes, and other materials to the mine, and for transporting copper out, it constructed a 14.5-mile-long rail line that ran from White Pine to Bergland, where it connected to the main line of the Duluth, South Shore and Atlantic Railroad (DSS&A). Service along the road started in January 1953, and in 1955 Copper Range sold this spur to the DSS&A for $520,000, about what it had cost to build it.[19] The mine and its townsite also needed a water-supply system. Local streams were wholly inadequate, so White Pine drew its water from Lake Superior, near Silver City, 6.5 miles away. Copper Range sank a shaft near the shore, and from the shaft men drove a tunnel 2,600 feet long out under the lake, where it connected to an intake shaft. A plant at Silver City pumped the water to White Pine's new water tank through reinforced concrete pipe. By mid-1956, White Pine was already using fifteen million gallons daily.[20]

Much of this water carried tailings from the mill to large, landlocked tailings ponds. Early stamp mills, which sat right beside lakes that served as dumps, sent vast tonnages of stamp sands into Portage Lake, Torch Lake, or Lake Superior. In planning for White Pine, Copper Range thought about doing things the old way, just as it had done a half century earlier at its Baltic, Trimountain, and Champion mills. It could ship ore by rail to a mill near Silver City and dump all tailings straight into Lake Superior. Upon reflection, Copper Range rejected that idea as not "good practice" for the mid-twentieth century.

The idea of dumping tailings into Lake Superior had become "unthinkable," so Copper Range acquired some five thousand acres needed to impound tailings near the mine.[21] In doing so, the company reacted to the early stirrings of an environmental movement in the 1950s.[22]

Copper Range believed that it represented a new approach that maintained a good environmental balance.[23] Instead of dumping tailings into Lake Superior, it delivered them through a pipeline to an interior location north of the mine and outside of public view. In 1953, construction contractors cleared hundreds of acres of land and built an earthen dam to hold back the mill discharge. Later, Copper Range regularly raised the height of the dam to increase the tailings storage capacity. It added ten feet in 1955 and another ten feet in 1956, moving and compacting more than a million cubic yards of earth or clay to do so. In 1958 it added an extension on its first dam and built a second. The first dam, which ended up being 2.5 miles long and 50 feet high, created a 1,750-acre tailings pond (3 square miles); the second dam and basin created a tailings pond covering 2,450 acres (4 square miles). Together they could hold more than 100 million tons of tailings. By the late 1960s, when Copper Range's mill discharge had been upped to 24,000 tons of tailings daily, the company characterized its tailings ponds as "the largest man-made structures in the State of Michigan, and perhaps the most expensive."[24]

Even in 1955 Copper Range did not get to set its own environmental standards. It had to apply to the Michigan Water Resources Commission for a permit to discharge wastes into the Mineral River, which ran through the site and ultimately into Lake Superior. The permit put effluent restrictions on Copper Range's wastewater discharges. It set standards for the total amount of water to be released (sixteen thousand gallons per minute), for pH levels (above 7.5 but below 10.4), and for parts per million of suspended solids (no more than 25). The wastewater could not have "any chemical substances of a nature or in sufficient quantity to contaminate the waters of Lake Superior" and could contain no bacteria originating from human sewage.[25] A newly constructed, company-owned sewage and wastewater treatment plant helped Copper Range meet these standards, and the tailings ponds were themselves a pollution control measure. The company did not decant water from the ponds into the Mineral River until solids had settled out.[26]

From the start, Copper Range also faced issues regarding air pollution, especially from its smelter. Copper smelters were notorious polluters in Montana and elsewhere, and "there was a lot of fear in [downwind] Ontonagon that there was going to be a lot of damage done by the fumes." At first, Copper Range did not try to abate smelter pollution; it tried to disperse that pollution broadly to minimize damage to the environment. A smelter stack two hundred feet tall would have provided adequate draft for running furnaces, but in 1953 Copper Range built a smelter stack 504 feet tall.[27] By emitting smoke, fumes, and particulates farther from the ground, the stack allowed air currents to disperse them quickly over a wider area, lessening their concentrations and effects. But within a few years of opening the smelter, the tall stack was not enough, and Copper

On the surface, White Pine did not look anything like the old native copper mines. It had no line of shafts, and no great distance separated its mine, mill, and smelter. Everything clustered together. Copper ore emerged from the underground on a conveyor that fed it to the mill, shown here in the middle distance, on the right. Directly to the left of the mill, by the tall smokestack, stands the smelter and power plant. *(Ontonagon County Historical Museum)*

Range introduced a new technology to limit air pollution rather than just disperse it. In 1958 it completed a smelter dust precipitator to catch "approximately 1,000,000 to 1,200,000 pounds of copper . . . that would otherwise be lost in flue gases."[28]

Construction of the White Pine mine and townsite began in March 1952. In January 1955—thirty-four months after construction began—White Pine poured its first copper.[29] Immediately, the new mine dominated Lake copper production. Before White Pine came on line, Copper Range's copper production in the early 1950s from its old works on the Baltic lode ran from 1.9 to 4.4 million pounds per year. Thanks to White Pine, the company's production total jumped to 68 million pounds in 1955. From 1955 through 1959, Copper Range produced an average of 77 million pounds of copper per year. In the 1960s, it smelted an average of 122.5 million pounds annually. In 1969, Copper Range hit its peak to that point. With its Champion mine now closed, the company mined 8.2 million tons of White Pine ore and produced 157 million pounds of copper.[30]

White Pine's first decade or two came at a time when the American copper industry generally enjoyed economic stability and prices high enough for its product (thirty to thirty-two cents per pound) to generate healthy profits.[31] The new mine continued to swell Copper Range's employment figures throughout the 1950s and 1960s. In 1954, White Pine employed 650 workers. By 1970 it employed about 2,885, making it the second largest employer in the Upper Peninsula, behind only the Cleveland-Cliffs Iron Company in Ispheming, which employed about 3,000.[32] On a national or international scale, White Pine was by no means a dominant producer. A half-dozen western firms (Anaconda, Phelps Dodge, Kennecott, ASARCO, Newmont, and AMAX) accounted for nearly three-fourths of American production.[33] But White Pine was extremely important as a producer and employer in Upper Michigan.

Between 1955 and 1975, White Pine increased the size of its underground works and the scale of surface facilities in order to reach a mine output of twenty-five thousand tons per day. Through the early 1960s, White Pine mined out a half acre of ground per day, or 130 to 160 acres per year; in the early 1970s, it extracted 435 acres annually.[34] By the mid-1970s, White Pine had opened up seven square miles underground, and miners took out ore in twenty different places.[35] As mining expanded the underground each year, and as mining depths increased to 1,800 feet in some directions, conditions changed. The copper became more or less concentrated; several major orebody displacements, or *faults,* were encountered; and the mine roof required more attention. Underground expansion increased the cost of mining because it lengthened the distances that men, materials, water, electricity, and especially broken ore had to be transported, and it also intensified ventilation problems.[36] Copper Range responded to new conditions by changing—or attempting to change—its technologies.

A key proponent of change at White Pine in the 1960s was the Copper Range president at that time, James Boyd, who held a doctorate in mining engineering and had worked early in the Korean War era at the U.S. Bureau of Mines and then with the Defense Materials Administration. He had been involved in awarding the $57 million government loan to Copper Range and ended up heading the mine that the loan made possible.[37] Boyd believed in investing in new machines and mining methods. During Boyd's reign, he moved Copper Range to the cutting edge of change; he hoped to enhance the profitability of his underground mine as it competed in the copper market with lower-cost open-pit mines in the West. Some changes worked to the company's advantage, while others failed decisively. The failures, by the 1970s, caused Boyd's successor, Chester Ensign, to adopt to a more conservative management style.[38]

When White Pine opened, supporting its roof was a "minor problem."[39] In addition to the rock pillars left to support the roof, in places miners installed *roof bolts*—a technology not used in early native-copper mines. They drilled small-diameter holes about four feet apart and four to seven feet deep into the roof. They inserted a long bolt into each hole after first passing it through a band of steel plate that ran across several holes. When the miners spun a bolt head with an air wrench, they drew the steel plate tight against the roof to hold it up; inside each hole, a threaded wedge shell—acting as a nut—expanded against the rock to anchor the bolt in place.

Over time, as the mine opened new areas, supporting the roof became anything but a "minor" problem. In some areas highly stressed rock had to be supported by heavy steel posts and beams.[40] In other areas the rock was highly fractured. In 1970, Copper Range's *Annual Report* noted, "Swarms of small geological faults encountered in the mine . . . hampered production," and in its 1972 *Report* the company stated, "The White Pine Mine would not exist without roof bolts." Copper Range was then installing 700,000 roof bolts per year, and it began experimenting with new bolts that did not rely on a mechanical wedge, but on a resin bond that kept them cemented tightly in their holes. In 1973 the heavy use of resin bolts reduced rockfalls by 40 percent over the year before, and they "made possible the safe mining of areas today that would have been thought impossible several years ago."[41]

To better secure the roof, Copper Range enlarged its typical pillar size up to twenty by sixty-four feet. Since the large pillars contained considerable ore, after some areas were mined out, men robbed the pillars by drilling and blasting, paring them down to take out more copper. Once miners finished this work and retreated from this ground, abandoning it, if the remnant pillars deteriorated and the roof structure failed and fell, it did no harm. Other areas being actively mined required stable pillars, and Copper Range wrapped some of them with steel cables to strengthen them: "As the cables stretched due to loading, the cables would occasionally emit tones like a heavy guitar string being struck."[42] Copper Range also pumped mill tailings back into some underground mine openings: filling them helped stabilize the roof. With its increase in roof concerns, Copper Range developed considerable expertise in rock mechanics and in maintaining safe work spaces:

> A crew, referred to as "Rock Doctors," was available 24 hours a day. These men were trained, upon being informed of a potentially unsafe roof condition, to set up a device to measure the movement between the roof and floor. . . . These devices, called extensometers, could measure movements in thousandths of an inch. Experience defined at what rate of movement roof failure could be expected and upon observing these rates and the overall roof condition, work in an area of concern could be suspended. The "Rock Doctor" program was well received and was effective in assuring whether or not an area was safe to work in.[43]

As early as 1963, Copper Range had once hoped to abandon room-and-pillar mining altogether. The company experimented with switching over to the *longwall mining* method that had been used in coal mines for decades. Under the longwall system, instead of drilling and blasting, miners used machines with rotating cutters to remove the ore—and they would leave no mineral-rich pillars behind. They would drive long tunnels or galleries underground, and then

slice ore off the walls of this gallery in the direction we wish to mine. Meanwhile, we support the roof by several rows of heavy steel posts with steel caps
on top stretching over the area in which the miners work. These posts are so
designed as to support the entire weight of the overlying rock. . . . As each
slice of ore is mined the posts are moved forward.[44]

A longwall machine with its rotating cutters traversed the gallery or *panel,* which could
be six hundred feet long. The rock that was cut or sliced off the 5- to 6.5-foot-high face
fell to the floor to be dragged off with scrapers, or it fell upon a chain conveyor that carried it away. After each pass, men moved the machinery and steel props forward to repeat
the process. The roof behind the machinery, and behind the protective steel posts, was
left unsupported and could collapse later over the worked-out ground without doing any
harm.

The longwall method would increase the amount of ore extracted, put men at less
risk from roof failures, and save the expense of roof supports. Copper Range believed
that mining by the longwall method was "thoroughly feasible," and it experimented with
longwall mining from 1963 through 1972. But it struggled to find cutters that could
slice the White Pine orebody, and what was a promising new technology was abandoned.
Copper Range had to stick with room-and-pillar mining, and it kept putting in hundreds of thousands of roof bolts to secure the hanging.[45]

Because it failed with longwall mining, Copper Range for a time deferred opening
up a new Southwest Orebody discovered in the late 1950s. A fault line separated this
orebody from the original works; it had been pushed downward two thousand feet below
the main mine. Diamond drilling discovered the new deposit and indicated that its ore
was "at least 50 percent higher grade" than the original deposit. Miners sank a development shaft into this orebody, only to discover late in 1961 that its roof was unstable and
many openings had to be supported with steel beams instead of roof bolts. Copper Range
hoped to sink hoisting shafts into the Southwest Orebody and have it in full production
by 1964 but first had to find a safe way to mine it.[46]

Copper Range initially believed that room-and-pillar mining would not work here:
the rock was at greater depth and pressure, and the roof was too insecure. Longwall
mining seemed to offer a good solution because workers would be shielded from the
unstable roof by the machine's steel posts and overhead caps. When Copper Range failed
to find longwall machines that worked under local conditions, its plan for opening the
Southwest Orebody went on hold.[47] In the 1970s Copper Range eventually opened the
Southwest Orebody by using a modified room-and-pillar method of mining. To work
this valuable deposit, Copper Range connected it via entries to the main mine so that its
ore could be carried to the mill by the usual conveyors. Besides its fragile roof, the Southwest Orebody presented Copper Range with two other problems. Here, the orebody was
on a steeper slope, a 20 percent grade, which could cause difficulties with the operation

of mobile vehicles. Secondly, up to ten thousand gallons of water per minute flowed from the face of the new works. No other part of the mine had ever experienced such a large inflow. Copper Range installed numerous pumps and ran them for months until the flow reduced to a trickle.[48]

Besides discovering the Southwest Orebody, diamond drilling in the late 1950s and 1960s expanded the boundaries of known reserves held within the original White Pine deposit. Some reserves stood five miles away from the mine's portal, mill, and smelter. Copper Range, as the mine expanded, transported ore longer distances to the central crusher and the conveyor tunnel that carried it to the surface. Instead of putting more diesel-powered shuttle cars underground to make these longer runs, White Pine, as early as 1957 and 1958, began installing *lateral* conveyors that ran thousands of feet from mine faces to the underground crushing station.[49] But in the mid-1960s, when White Pine opened up more distant mineral reserves, it did not want to haul the ore underground to the crusher and the conveyor tunnel. It already had ten miles of conveyors underground. So in 1965–67 it sank and equipped a new shaft, 24 feet in diameter and 1,600 feet deep, that intersected the ore horizon five miles east of the mine's portal. Copper Range would hoist rock to the surface through this shaft and then transport the ore overland to the mill.[50] For purposes of ventilating this shaft and the ground tributary to it, Copper Range called for a long tunnel to hole through to the established works.

To drive this passage, the company used a tunnel-boring machine. In 1966 it had experimented with a tunneling machine that cut a seven-foot bore, using it (instead of drilling and blasting) to remove ore from the rooms standing between pillars in the mine. In 1968 Copper Range installed a tunnel borer, with an eighteen-foot cutter head, at the bottom of the new hoisting shaft and began tunneling back toward the older part of the mine, not through the orebody itself, but through the underlying sandstone.[51] Copper Range experimented with several machines simultaneously: the longwall machines and smaller boring machines to remove ore without using explosives, and the large tunnel-boring machine to drive the 7,500-foot-long connection between new and old parts of the mine. Copper Range could not make any of these new technologies work satisfactorily.

The maker of the tunnel-boring machine failed to provide "cutters durable enough" for working in sandstone. Copper Range shut the machine down for a half year to make modifications, but they didn't solve the problem. Cutters dulled too rapidly, and progress proved slow. To push the tunnel to completion, Copper Range angled the tunnel-boring machine upward to get it out of the sandstone and into the more easily bored copper-bearing shale. In July 1972 the tunnel finally holed through to the original works. Instead of celebrating the work of the tunneling machine, Copper Range concluded that four years of operation showed it "could not be operated economically."[52]

The company also failed with its smaller boring machines for cutting the copper-bearing shale. After starting with a seven-foot machine, by 1970 it turned to a double-headed, eight-foot boring machine. That equipment, too, needed to be debugged, but modifications didn't help. By 1972, Copper Range admitted that the smaller boring machines were "not found to be capable of economic operation."[53] Unfortunately, Copper Range's failures with new technologies did not end underground. On the surface, six years of experimentation with a new overland haulage system intended to transport ore five miles from the new hoisting shaft to the mill also ended unsuccessfully in 1972.[54]

In 1965, when Copper Range started sinking its new hoisting shaft, it formally associated with the Dashaveyor Company of Los Angeles "to develop and perfect a new transportation system for hauling bulk materials over complicated routes." The two companies installed a pilot system at White Pine in 1966, and shortly thereafter Copper Range contracted with Dashaveyor "for a complete system on the surface." Workers put down twenty-seven-thousand feet of rail in 1967 at a cost of about $2.5 million. The system, slated to begin operation in 1968, was designed to carry up to ten thousand tons of ore daily from the hoisting shaft to the mill.[55] The ore rode in individually powered modules or cars with dual electric motors. Rather like a trolley, the cars traveled on a track and picked up their electric power from a third rail. The track incorporated a rack gear, which helped control the rate of descent on downgrades and allowed heavily laden cars to climb steep hills. That was one supposed advantage of the Dashaveyor system—it followed the natural terrain. To protect the system from snow or inclement weather, a rectangular tube enclosed the cars and tracks. Theoretically, cars could travel up to fifty-two miles per hour, and they could be filled or emptied at five miles per hour without stopping. The cars could be linked to operate like trains, and a single operator in a central station controlled and monitored their motion.[56]

Unfortunately, this system, always deemed experimental, never became fully operational. "Design problems could not be solved," and in 1972 Copper Range announced the Dashaveyor as yet another failed new technology.[57] Lacking a good system on the surface to transport ore to the mill, White Pine fell back on transporting it underground by using mobile vehicles and conveyors. Copper Range had a 4,000-horsepower hoisting plant at its new shaft that sat idle because of the Dashaveyor's failure, but the shaft proved useful for ventilation purposes.

In White Pine's first twenty years of operation, production levels sometimes lagged because of labor shortages, labor stoppages, or sagging copper prices that necessitated temporary cutbacks. But on the whole, production generally exceeded initial expectations, and the mine, mill, and smelter handled tonnages greater than their original design capacities. White Pine mined about 3 million tons of ore in 1955, 3.9 million tons in 1960, 6.2 million tons in 1965, 7.6 million tons in 1970, and its all-time record high of 9 million tons in 1975.[58] Select technological successes, such as resin roof bolts and lateral belt conveyors, as well as additions to underground and surface works, made this sustained growth possible. Underground, Copper Range turned to newer generations of shuttle cars. Now called *trucks,* these four-wheel-drive, articulated vehicles hauled up to twenty tons per load; they could work in tighter spaces and on steeper grades. Copper

In the 1960s and 1970s, several innovative technologies introduced at White Pine failed. On the surface, Copper Range tried to transport ore to the mill by using the Dashaveyor system. Rockcars, traveling within a rectangular tube, were supposed to travel up to fifty-two miles per hour over irregular terrain and carry up to 10,000 tons of ore per day. Copper Range put down 27,000 feet of Dashaveyor track in 1967, but the system never became fully operational. *(Michigan Technological University Archives and Copper Country Historical Collections)*

Range introduced mobile crushing machines that could be set up wherever convenient to shorten haulage runs and reduce the tonnage that had to be handled by the central crusher.[59] On the surface, the company altered and added to its plant to reduce costs, increase production, and increase copper recovery. As early as 1959, Copper Range added a second refining furnace to the smelter.[60] In the 1960s it put a dryer between the mill and smelter to rid the mineral concentrate of moisture, which had the effect of upping the smelter's capacity to 130 million pounds per year. At the same time, Copper Range added a semicontinuous casting machine to keep pace with the furnace. In 1967, Copper Range further increased its smelter capacity by adding a second reverberatory furnace.[61]

In the early 1960s White Pine's ingot copper assayed at thirty-four to thirty-seven ounces of silver per ton. Silver was a desirable trace element in the copper, but Copper Range wanted to keep it at about twenty-five ounces per ton, so it added a silver-recovery circuit to the mill. This new technology worked well and recovered one-half to one million dollars worth of silver per year.[62] Starting in 1960 the company had less success, however, in trying to reclaim the copper left in its mill tailings. In 1960 each ton of ore received at the mill contained 22.84 pounds of copper (or just 1.14 percent), and each ton of tailings leaving the mill still contained 4 pounds of copper. The native-copper mines had used leaching processes to reclaim copper from tailings, and Copper Range tried to do the same. It used cyanide as its leaching solvent and received a U.S. patent for its process. In the closed-loop system, a cyanide solution dissolved the copper, which was then precipitated out of solution while the cyanide was regenerated to be used again and again. In theory the process could reclaim 90 percent of the copper in the tailings. Copper Range built a pilot plant and ran it for a year before abandoning the idea. The process did not recover enough copper to pay for itself.[63] While Copper Range failed to get all the copper out of its tailings, it succeeded in recovering copper from its slag dump adjacent the mine works. Between 1970 and 1973, it crushed and reprocessed a mountain of slag (about a million tons of it) that had been stockpiled on site since opening the mine. Copper Range captured an estimated 11.3 million pounds of copper from its slag. After recovering this copper, it sold the slag to highway departments and others, who used it for roadbeds or ballast.[64]

A glance at White Pine's tall smelter stack, its slag pile, and its tailings ponds showed that no large mine could function without altering, threatening, or degrading its local environment. Yet Copper Range prided itself on its stewardship of the air, land, and water. In the 1950s it chose not to dump tailings into Lake Superior, but to impound them; chose not to emit smelter gases close to the forest floor, but from atop a 504-foot-tall stack; and chose not to discharge polluted water from its mine or town without first sending that discharge through a sewage and wastewater treatment plant. But as Copper Range moved into the more environmentally conscious 1960s and 1970s, it found itself operating under new expectations, higher standards, and more regulation. Copper Range initially seemed to "go along" and "get along" with the new environmentalism, and it saw expenditures on pollution control measures as a cost of doing business. But there were limits. The company did not want to be overly dictated to by regulators or overly blamed for environmental degradation by critics. While proclaiming itself as a good corporate citizen and boasting of its environmental record, Copper Range came to believe that environmental measures could cost too much and cripple a company or an industry.

Prior to the mid-1960s, Copper Range did not publicly protest environmental standards or expectations; its relations with Michigan air-quality and water-quality officers remained generally friendly and mutually supportive. Then in 1965 and beyond, the company perceived a new wave of environmentalism sweeping over American society and politics. Corporate culture encountered a new America where protecting the environment had suddenly become a national priority, as codified in the 1963 Clean Air Act, the 1965 Water Quality Act, the 1967 Air Quality Act and, most important, the National Environmental Policy Act of 1969. This last piece of legislation—the most wide-sweeping environmental law ever passed by the U.S. Congress—created the Environmental

Protection Agency (EPA).[65] In response to such legislation, Copper Range formalized its environmental efforts and placed them all under the management of Edward Bingham, who had been hired in 1964 as the company's chief chemist. In short order, Bingham had a full-time environmental control staff of three. He was joined by an engineer about to receive a doctorate in resource management from Michigan State University and by a technician. As needed, Bingham also drew upon the expertise of engineers, scientists, and personnel from other Copper Range departments.[66]

In its first decade of service, Copper Range's environmental control office tackled several pollution issues. The company learned that a technology created to solve one problem often created a new problem. As a case in point, Copper Range had to rethink its large tailings ponds. Water carried mill waste out to the ponds, where dams impounded it. Behind the dams, the finely ground tailings settled out of the water, and then Copper Range released that water into a nearby stream. This technology checked water pollution but created an air pollution problem.[67] Winds sweeping over the dry mill tailings raised dust into the air. Winter snow kept the dust down, and watering the tailings helped control the dust at other times of the year. But Copper Range sought a more permanent solution. It studied various trees and plants to see what would grow best on the tailings so that it could reestablish a forest on top of them once they were filled in. Copper Range started to cover the tailings and planted grasses on its long, high earthen dams. The vegetation helped stabilize their banks and also provided food for animals and fowl.

Earth-moving equipment had dug pits when excavating the clay used for the dams, and these had filled with water. About 120 ponds covering from one to twenty acres stood on the margins of the tailings deposits.[68] Copper Range prided itself in that its dams, ponds, and tailings deposits nurtured wildlife:

> We have determined which trees will grow in the tailings when the ponds have been filled so that we can reestablish the forest. The ponds provide resting and feeding places for [several thousand] ducks and geese [during fall migration]; the grassy slopes are feeding grounds for deer and other wild animals in the area. Bear, coyote, fox, many kinds of aquatic animals and even an occasional wolf has [sic] been seen.[69]

Copper Range worked with the Michigan Department of Natural Resources and the Michigan State Conservation Agency in developing a wildlife management plan for its tailings area. In the mid-1970s it reported "its initial success in nesting and rearing of young giant Canada . . . geese."[70] The arrival of a half-dozen goslings would not have been news across the rest of the Upper Peninsula, but they made news when hatched in the midst of a tailings dump. Bringing wildlife back to this area helped earn White Pine the National Merit Award for 1975 from the Soil Conservation Society of America.

By the late 1960s Copper Range or its officers held memberships in numerous organizations "connected directly or indirectly with environmental control." These included the Air Pollution Control Association, the Michigan and National Water Pollution Control associations, the National Conservation Foundation, the Michigan Natural Resources Council, the Izaac Walton League, the Audubon Society, and Ducks Unlimited.[71] While networking with such organizations, the environmental control department coped with maintaining water-quality and air-quality standards that tended to be set higher as the years went by. Often, Copper Range achieved results that merited commendation, such as the "A" rating it received in 1971 from Michigan for its controls of industrial waste-water discharge. Meeting select air pollution standards became more of a problem. Copper Range, to comply with new Michigan law, agreed to install additional dust control apparatus on its power-plant boilerhouse.[72] In the late 1960s and early 1970s, smelter emissions—especially sulfur oxides—became a bigger issue.

In 1967 Copper Range set up sulfur-monitoring stations across the landscape "to ensure that no vegetation damage would be caused by excessive ground level concentrations of sulfur dioxide." The company believed that "there have never been any forests . . . that have been affected in any way by the gases from [its smelter's] operation." Nevertheless, Copper Range cooperated with the U.S. Public Health Service on researching means of trapping sulfur gases before they were emitted from the smelter's tall stack.[73] Such efforts became more important after passage of the Federal Clean Air Act of 1970. Copper Range worried that Michigan's plan to meet federally mandated ambient air-quality standards would call for a major reduction in sulfur oxide emissions. It expected to be told to meet a 90 percent emission standard, meaning that "90 percent of the elemental sulfur in the copper concentrates must be removed and cannot be dispersed into the atmosphere." Copper Range's response to this expectation set up the kind of corporation versus regulator standoff that became a hallmark of the new environmental age. The environmentalists worried about plants and wildlife and how airborne sulfur could become acidic when mixed with water; the corporations worried about costs, profits, and jobs. Copper Range opined in 1971 that, "given presently available technology, we do not believe that it would be economically feasible for us to meet a 90 percent emission standard" for sulfur oxides.[74]

The company's attitude toward environmental regulations changed greatly over five or six years. Copper Range's *Annual Report* for 1969 was the first to mention the efforts at "Environmental Control." The text made it clear that the company took pride in protecting the environment, but hinted that it was starting to feel the strain of operating in a regulatory era: "We have always conducted our operations in such a way that we improve the environment or ecology of the area rather than damage it. This was done voluntarily by the company long before the current interest in this subject, and by methods that exceed state or federal regulations."[75] The next year, Copper Range let it be known that it had already spent $14.5 million dollars on various environmental controls, and that it would invest another $5 million in the next two years. In its 1970 *Annual Report,* Copper

Range defended its environmental record: "The lengths to which your company has gone from its inception should classify it as a 'concerned' corporation." The company admitted that "operations unavoidably may upset the balance of nature in some situations" and yet claimed to be "creating an environment around White Pine that is conducive to developing an ecology in which wildlife, forests, flowers and humans can exist together more harmoniously than when we began the mine."[76]

By 1974, Copper Range found itself working in a world of "increasingly restrictive standards and rules."[77] Instead of merely defending itself, it counterattacked critics of its environmental record. Chester Ensign, president of the firm, told an audience at Michigan Technological University that "ever increasing regulations, controls and . . . over-reaction to the environmental ethic creates forces powerful enough to destroy the free enterprise system." Ensign chafed at how environmental critics viewed White Pine and the entire mining industry: "We purportedly are the spoilers of the land, we are the polluters of the nation's air and waters, we are insensitive to the environmental ethic and have no concern for ecological values." Critics, thought Ensign, saw the wrongs done by earlier robber barons (which would have included all the earlier Michigan copper mines), and in error condemned "modern-day, professional managers." In his speech, Ensign defended modern mining corporations: "The vast majority of my colleagues in industry are as concerned about the environmental legacy we leave behind for future generations as the most ardent, hard-lining, overkill-prone environmentalist in existence."[78]

Besides operating in the environmental era, White Pine lived its life in an era of organized labor. In November 1953, before the mine went into production, workers selected the United Steelworkers of America as their collective bargaining agent.[79] Over the years, Copper Range often reported that labor relations at White Pine were "satisfactory," "good," or even "excellent," and that contract negotiations were "cooperative" or conducted "in a friendly atmosphere."[80] Such statements sometimes represented wishful thinking, more than fact. Labor and management often came to loggerheads at White Pine, and work stoppages were not uncommon, both before and after the United Steelworkers arrived on the scene.

Three walkouts occurred at White Pine in 1952, and a strike by carpenters caused a work stoppage on the surface from 11 May to 5 June 1953—all before the union election. In 1954 two more work stoppages resulted in eighteen days of lost time.[81] The first major strike occurred when a labor contract terminated in 1959. A strike that began on 29 October lasted until 22 February 1960. The new thirty-month agreement that was hammered out covered not only wages but permitted the scheduling of seven-day-a-week, around-the-clock underground operations.[82] The company won the shift arrangement that it wanted, but not without cost: "During the strike a large number of men moved away and resumption of operations required the training of replacements for those men, and additional personnel to cover the swing shift."[83]

The strike taught the company that "stormy" labor relations cost them good bosses and skilled workers, some who operated large and complex machinery, and others, such as mechanics, who maintained and repaired that machinery. Lost production and labor turnover resulting from the four-month strike encouraged Copper Range to take measures to assure "that such a long interruption should not happen again": "A successful operation . . . depends as much upon the attitudes, feelings and morale of men as upon the maintenance of machines in operation."[84] Managers and employees needed to improve their understanding of each other's problems. To aid in that effort, Copper Range initiated a company magazine: the *CR News*. It also started sending employees frequent letters that dealt with future plans, staffing levels, safety measures, new technologies, rumors, and economic conditions. Copper Range also added a full-time industrial relations staff.[85] The company wanted to inculcate one lesson in particular: that it was in the employees' best interest to keep the mine open: "The only way to make our jobs secure," wrote R. C. Cole in a letter to all employees, "is to compete successfully—in other words, to make a profit."[86]

Copper Range often communicated the belief that the company and its workers shared the same self-interests, but labor stoppages nevertheless continued to crop up when labor contracts were being negotiated. In 1962, that stoppage lasted but a day, and Copper Range noted, "The negotiations generally were conducted in a friendly atmosphere with each side recognizing the problems of the other." The next time negotiations rolled around, a strike closed White Pine down from 1 September to 19 October 1964. In 1967, Copper Range sang an entirely new tune: its negotiations with union officials, instead of being friendly, "were exercises in futility," and the Steelworkers Union struck for nearly five months before signing a new contract and resuming work on 25 January 1967. When that contract expired in 1971, Copper Range found the union's position "immovable," and another strike ensued that lasted eight weeks, from 1 August until 24 September. In 1974 the Steelworkers struck again, this time for nineteen days, before signing their new contract.[87]

In its first two decades of operation, White Pine often found itself labor short. The labor pool in and around its part of the Upper Peninsula was small and growing smaller because young workers left the region to find jobs elsewhere and because Copper Range defined *worker* as male and gave no thought to hiring women. No women worked underground at White Pine until the early 1980s.[88] The company drew male workers from more than a hundred miles away but still had trouble hiring enough men, especially skilled men. Consequently, Copper Range invested heavily in job training. It started a school to train mechanics on maintaining the mine's equipment, and underground it set aside an area where newcomers learned to operate vehicles and production equipment under supervision, at a slower pace.[89]

A push-pull factor worked against the company. A generally strong American economy and the promise of greater opportunities elsewhere pulled workers away to Great Lakes cities like Detroit, Milwaukee, or Chicago. During the Vietnam War era, the military also pulled young men away.[90] Meanwhile the history of mine closings in the region, combined with the battles between Copper Range and the Steelworkers, pushed men to leave. As for those repetitive strikes, the men wanted wages and benefits higher than the company thought it could pay. The men knew what union workers received at western U.S. copper mines, and they wanted parity. But those mines, Copper Range countered, were in better economic condition. The workers at White Pine made demands that "were identical to those made upon lower cost, open-pit producers," larger firms that produced 80 to 85 percent of the nation's new copper. Copper Range sought "some concessions because it is a higher cost producer than most of the large western open pit operations."[91]

To attract and keep workers and to head off labor unrest and strikes, Copper Range did at White Pine what it had done a half century earlier at Painesdale: it built a townsite at the mine. That community's housing, services, schools, and other amenities—plus its convenient proximity to the mine—were supposed to draw men and their families to White Pine and keep them there. The town would help breed loyalty among employees who saw that life and work were closely linked, with Copper Range playing a pivotal role in each realm. But the new town never grew to become what Copper Range wanted it to be.

21

❖ ❖ ❖

SOMETHING OLD, SOMETHING NEW

The White Pine Townsite

Like mine locations populated from the 1840s to 1900, White Pine was another island of industry in a sea of trees. But aside from its remote, wooded setting, White Pine differed greatly from earlier mine locations.[1] This planned community was built after the Second World War, built in the automobile age. "Today, where there was nothing, a modern city has mushroomed," declared the *Detroit Free Press*—a city that has "spanking new, modern homes." The *New York Times,* which called White Pine "the neat new town in the wilderness," noted its distinctiveness: "White Pine, a new town rising from the forest and more nearly resembling a 'bedroom suburb' of New York than a mining community, seems out of place in the copper country of Michigan's Upper Peninsula." The *Daily Mining Gazette* expressed the view that

> the little town will be one of the most attractive in Michigan. Built with all the characteristics of a model community, it is destined to be one of the ideal residential areas in the Midwest. . . . Surely White Piners can well be said to be living in a charmed location, one of fishing, hunting, recreation, scenery, and industry.[2]

Copper Range saw the construction of the White Pine townsite as key to its successful recruitment of employees.[3] In short order, Copper Range would need over one thousand workers—especially young ones—and such men were becoming scarce in the region. In 1952, Albert Gazvoda of the Michigan Employment Security Commission stated they would have to launch a "manhunt" for workers because young couples were leaving the Copper Country for places like Detroit. Houghton, Keweenaw, and Ontonagon counties, together, had "lost 11,356 residents—some 18.1 percent [of the population]—to metropolitan areas in the last six years."[4]

Copper Range knew that White Pine had to be a new kind of mine location, one that would attract and keep American workers instead of immigrants. These workers would not crowd into company houses and take in boarders, because society now placed a premium on the nuclear family living by itself. Workers now expected paved streets, water and sewer lines, central heating, telephones, electricity, and other amenities. Because of all the complexities involved, Copper Range turned to Pace Associates of Chicago to design its townsite; this firm also planned the new iron-mining towns of Babbitt and Silver Bay in Minnesota.[5] Chicago architect Charles Booher Genther, along with nine others, formed Pace Associates in 1946. Genther gravitated toward large projects—like planning new mining towns—that required relatively inexpensive housing to be constructed in a hurry. For Genther, a client undertaking a large project wanted to hire a team of diverse professionals to complete the job expeditiously under one contract.[6] The firm he created, Pace Associates, was a diverse group. *Pace* stood for *p*lanners, *a*rchitects, and *c*onsulting *e*ngineers. Pace's broad-based team gathered data and planned White Pine with a goal to "build homes and community facilities that will sell people" on working and living at Michigan's newest copper mine.[7]

Pace conducted sociological surveys in iron and copper towns and confirmed that "adequate, economical and comfortable housing was *the* most important factor in attracting the quantity and quality of labor needed."[8] To achieve comfortable housing, Pace studied the local climate. They looked at 1940 and 1950 U.S. census data "to tabulate the probable age distribution and family size" of workers who might migrate to White Pine. They interviewed mine families to discover how big their houses were, which doors they used to go in and out, and whether they ate in a kitchen or dining room. They looked at enrollments in mine-town schools to guide them in school planning. To predict how much White Pine employees might spend annually on "groceries, clothing, or car servicing," they interviewed store owners and looked at sales figures. White Pine would hire men who were twenty-five to forty-five years old. Pace figured 85 percent of them would be married, so accommodations for family men had to predominate. Pace expected families to "be larger than average," so to give families a comfortable room to congregate in over long winters, architects drew up living rooms larger than Federal Housing Authority guidelines. And "garages were considered essential since 99 percent of mining families own a car." White Pine men would drive, not walk, the eight thousand feet that separated their homes from the mine, and cars would take families once a week to shop in Ontonagon, twenty miles away; to Ironwood, forty-five miles distant; or to Houghton, a trip of seventy miles.[9]

Pace also studied planned suburban communities like Greenbelt, Maryland, and

This ca. 1954 aerial photo shows the layout of the White Pine townsite. The Service Building, dormitories (later made into apartments), and hospital are clustered together in the middle distance, on the right. Just to the left of this cluster, the land shown as vacant is the proposed site of the Town Center, where a high school, a shopping center, and a gas station were to be built. On the left, company houses stand along curvilinear streets, backed by woods. *(Ontonagon County Historical Museum)*

Park Forest, Illinois. When the 150-member Pace team produced the 898 drawings that delineated White Pine's layout and buildings, the planners and architects incorporated bits of these other places into their plan.[10] Consequently, White Pine—sitting on a 270-acre site surrounded by a forest—became an amalgam of old and new, of industrial and suburban. Like Copper Range mine locations built a half century before, it had company houses, company-built schools and a hospital, a few churches, and limited commercial development. But, with the exception of two clusters of houses left over from the Calumet and Hecla days at the site, it all looked so different. The map showed a future town center, occupied by a shopping center, a gas station, a high school, and green space. White Pine did not have a rectilinear grid of streets named after mine companies. Its curving streets were named after trees and lined with single-story ranch houses having driveways, garages, and large unfenced lots. The mid-twentieth-century mine location also included dormitories, apartments, and a trailer park.

Pace and Copper Range provided housing options that met short-term and long-term needs. At the start, "many of the early residents of the new town were 'startup people,' those free-spirited individuals [single men] who loved to build factories, mines, roads, power plants, chimneys, and other facilities, get them up and running, and then go on to the next project."[11] To house construction workers, Copper Range erected four large dormitories that resembled two-story motels. The equivalent of nineteenth-century boardinghouses, the dormitories sheltered several hundred men. Each unit consisted of two rooms—a bedroom equipped with metal lockers instead of closets, and a living room. The men walked down the hall to use a communal bathroom with showers, urinals, and toilets. To eat they walked over to a large cafeteria with seating for five hundred in a Service Building, operated under contract by Nationwide Food Services.[12] Copper Range also erected a Staff House for personnel overseeing construction and the planning of start-up operations. Each Staff House unit had three rooms—a bedroom, living room, and small, private bath.[13] Residents of the Staff House also ate at the Service Building. After these structures met short-term needs, Copper Range converted most dormitories to apartments and built an additional apartment house for younger couples having no children. By the mid-1950s the Staff House served as a forty-four-room hotel: the White Pine Inn.[14]

About four thousand feet north of the apartment houses stood a trailer park for workers wanting to live close to work and yet save on housing costs. As drawn up in 1952, gravel roads looped around the park to create six 4-sided blocks with water and electrical hookups for two hundred trailers.[15] The plan set aside space for parking lots, a small general store, a playground, and three utility buildings with toilets, showers, and wash/dry rooms. Despite a low trailer lot rental of $7 per week or $25 per month,[16] relatively few workers selected this housing option, so Copper Range built just two of the six proposed blocks and only one utility building. As built, the trailer park took fifty trailers. In 1954 a White Pine directory of 1,109 residents showed 46 employees living in trailers.[17] No two male employees had the same trailer park address, meaning that the company welcomed families but did not allow single men to bunk together in trailers.

White Pine's *industrial suburb* of two-, three-, and-four bedroom ranch houses sat between the apartment-house complex and the trailer park. As expected in any postwar suburb, the houses had water service, sewers, and electricity. Common architectural features gave these houses a sense of unity and set them apart from older mine houses, which were two stories tall and often had strong vertical elements. Older mine houses had high foundation walls of stone (to get house sills above the snow and damp), elevated entrances, tall sash windows, and steeply pitched roofs (to help shed snow in winter). The single-story ranch houses designed by Pace had a strong horizontal element. They squatted low on concrete pads, with door sills only inches above ground. The roofs, of very slight pitch, appeared nearly flat and especially out of place on the Upper Peninsula landscape. The houses didn't look well suited for the far north, where temperatures dipped below zero and local snowfall averaged one hundred inches per year. In a snowy winter, many a resident no doubt grumbled about having to pull snow off the roof with a roof rake or about having to climb up on the roof to push snow off with a Yooper scoop.

The White Pine townsite included two clusters of company houses built almost fifty years earlier by C&H, when it mined here. Copper Range renovated these houses for its employees. This C&H house (the rear is shown) differed greatly from the houses built in 1952–53. It had more windows, more doors, a more vertical design, and a tall foundation. *(Ontonagon County Historical Museum)*

Then, when all the snow fell from the roof, it would bury or cover up the bottom half of the main picture window, set too low to the ground. A depth of cold, white evidence to the contrary, Pace did consider winter conditions when designing the houses.

Instead of building the houses on foundation walls to protect them from snow and moisture, Pace called for a vapor barrier to run halfway up all exterior walls, between the sheathing and siding. The architects designed a small overhang at roof edges to minimize the buildup of a snow load. Figuring that winter winds came from the northwest or southwest 72 percent of the time, Pace, where possible, oriented houses so that storms would not blow heavy drifts into entryways.[18] To achieve this orientation, many houses did not sit square to the street. To buffer a house from wind and cold, Pace situated the garage on the end of a house facing prevailing winds. Bathroom placement also reflected wintertime considerations. In larger houses, other rooms surrounded bathrooms so they had no cold, exterior wall. In smaller houses, the bathroom's exterior wall was sheltered by the garage and had no window to let in drafts. Thermostatically controlled oil furnaces provided heat. Water at 145 degrees flowed through half-inch copper pipes set into the

Carpenters were still building these White Pine houses late in 1952. None have been sided yet, and some are without windows or even a roof. Despite being stick-built on site, they resembled prefab houses that might have been trailered in. They sat lightly on the landscape, like so many boxes sitting on the ground. And although a forest surrounded the White Pine townsite, no trees remained standing among the houses themselves. *(Ontonagon County Historical Museum)*

concrete floor slabs, and the heat radiated upward. To keep heat in, Pace specified 3.5 inches of fiberglass insulation in exterior stud walls, and 7 inches of fiberglass between ceiling and roof.[19]

Pace architects created five basic house types for White Pine. The three for ordinary workers all had an attached one-car garage. Type A had two bedrooms and one bath; Type B had three bedrooms and a bath; and Type C had four bedrooms and one-and-a-half baths. For managers, Pace came up with two basic layouts having enhanced amenities such as two-car garages and breezeways. Type D had three bedrooms and one-and-a-half baths. Type E had four bedrooms and two full baths.[20] Pace used various tricks to create a more visually interesting streetscape and to make it look like White Pine had greater housing variety than it really did. Pace mixed different house types together on the same street. Some had their gable ends facing the street; others did not. Some houses of the same type had garages on the right, and others, reversed, had garages on the left. When carpenters from Herman Gundlach, a contractor company from Houghton, nailed up

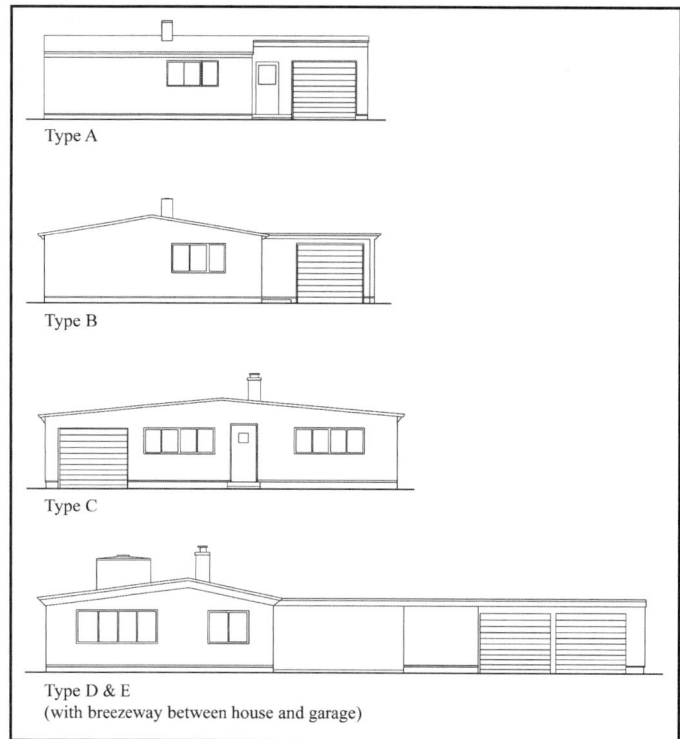

Type A

Type B

Type C

Type D & E
(with breezeway between house and garage)

These front elevations, based on original Pace architectural drawings, show the five White Pine house types. (The D and E models projected the same elevation, but differed in plan.) The houses had two, three, or four bedrooms. Some had a dining room; most did not. Some had a two-car garage; most had a single garage. Some had vertical siding; others, horizontal. They all sat low to the ground on concrete slabs. They had the same window treatments and low-pitched roofs, and they all looked out of place in Michigan's Upper Peninsula. *(Scott See)*

redwood siding, they ran the boards vertically on some houses and horizontally on others.[21] Paint or stain colors also created variety. Pace created nine exterior color schemes by mixing or matching six main colors and five trim colors.[22] An early finish schedule showed how fourteen houses, all Type A (two bedrooms) or Type B (three bedrooms), were aligned and painted along one residential street. The schedule called for five Type A houses, two Type A reversed houses (garages on the opposite end), four Type B houses, and three Type B reversed houses. Carpenters clad six houses with vertical siding and eight with horizontal. Painters applied sequoia red stain to four houses, eucalyptus gray stain to another four houses, and medium gray to two others. They painted two houses jonquil yellow, one bone white, and one tropical blue.

Journalists from the *New York Times,* the *Detroit Free Press,* and *Daily Mining Gazette* were all too kind when describing White Pine as a modern suburb. Nobody should have mistaken this mine location for a middle-class subdivision. It lacked too much. Pace's sociological study reported that most working families used the same door nearly all the time to enter and leave their houses—so the architects generally did away with a second entrance. Four of the five house types at White Pine had just one door, which wasn't even on the front but was on the side of the house by the setback garage. No brick or stone, no elaborate plantings, and no fancy lighting dressed up house exteriors. The company houses were *serviceable,* not *upscale.* This was true even for managers' houses, which were clustered along Maple Street.[23] They looked bigger, and with their detached two-car garages and breezeways (connecting garage and house), they looked different. But they did not look any *better*. All the house types sat low on concrete slabs; they all shared horizontal sliding windows and similar picture windows. Because Pace and Copper Range did not dress up the exteriors of any houses with elaborate entrances or architectural embellishments, the dwellings at White Pine looked more egalitarian than company houses of earlier eras.

Inside, the houses had half-inch plasterboard walls and ceilings. Fabric coverings selected from nine colors went up on nearly all walls. All ceilings were painted antique white. Brown, gray, or green Armstrong linoleum tile—not carpeting or hardwood—went over the concrete floors carrying the radiant heat pipes. All houses incorporated modern amenities. A washer, dryer, laundry tub, oil furnace, and electric water heater occupied a utility area, and a refrigerator and electric range stood in the kitchen, where plans left space for a dishwasher under a countertop. Exhaust fans whirred away in kitchens and bathrooms, and along one wall in the bedrooms were sliding-door closets large enough to hold both summer and winter clothing. Full bathrooms had tubs with tiled surrounds (metal tile, not ceramic) and showerheads. All kitchens came with considerable cabinetry, but of inexpensive construction and finish.

Managers' houses allotted residents more living space than did ordinary houses. The two-bedroom Type A house, not including the garage, provided only 832 square feet of living space. The three-bedroom working-class house, Type B, offered occupants 1,040 square feet of living space, whereas the three-bedroom managerial house, Type D, provided half again as much room—1,508 square feet. The Type C, four-bedroom house for ordinary workers, which in this case included the garage, enclosed 1,436 square feet of space, whereas the Type E four-bedroom house for managers enclosed 1,768 square feet. Managers' houses generally provided larger living rooms and bedrooms, but the full bathrooms in all houses were only as wide as the tub was long—a mere four feet, eleven inches.

A manager's house did display some distinct advantages besides a bigger footprint and a two-car garage. It had a breezeway, which in some plans included a hearth so residents could partake in the 1950s' barbecue craze. Inside, a manager's family found not only a refrigerator, but a chest freezer. And their kitchen had a distinct dinette area in one end, whereas in workers' houses the kitchen was simply left big enough to accommodate a table and chairs. Pace's sociological study showed that mine workers ate in their kitchens all the time, so the architects left them no place else to eat.[24] A manager's family, on the other hand, not only got a dinette, but a dining room—which was set off from the living room by a brick fireplace. The manager's family walked over the same linoleum tile and looked at the same wall coverings and cabinets that everybody else did, but the family could better entertain guests at the table and then sit in the living room, enjoying a crackling wood fire.

Employees began moving into the new White Pine houses late in February 1953. Workers have shoveled off the nearly flat metal roofs, creating piles of snow on house perimeters. Looking at this Type B (reversed) house, note the absence of a front door, the flatness of the façade, and the unsightly utility meters to the left of the windows. *(Ontonagon County Historical Museum)*

In 1952 and 1953, plans called for some platted streets to be built upon immediately; other streets where platted for lots, but no house construction was scheduled; still other streets existed on maps only and were not platted at all. Consequently, reports on how big White Pine was supposed to be varied considerably. The *Ironwood Daily Globe* wrote that Copper Range planned for 940 single-family houses; *Excavating Engineer* pegged the figure at about 400 houses. Near the end of 1952, Pace produced a map showing 465 lots platted on the site and noted there was room for another 250. In the middle of 1953, with construction under way, another Pace map showed 156 single-family houses under construction, and another 207 that were proposed. When the mine opened in 1955, White Pine families lived in a total of 183 single-family houses—156 new ones built by Gundlach and 27 rehabilitated, older houses left behind by Calumet and Hecla. Other employees who resided at the mine lived in a remaining dormitory, in one of eighty-eight apartments, or in the trailer park.[25]

It took more than living quarters to make up White Pine. In 1952 an eighteen-mile-long power transmission line reached the townsite, coming from the U.P. Power Company's hydroelectric plant at Victoria, and in 1953 Copper Range completed its pumping facilities and erected a 250,000-gallon elevated water tank to serve the mine and community.[26] That same year, while building six miles of streets, the company laid asphalt on three miles of streets and on 119 driveways. It put in twenty-seven street lights plus storm sewers, water mains, and fire hydrants.

Forty-two families resided in White Pine by April 1953. An automatic phone system with three hundred lines connected them to the outside world. A Star Route under contract to the U.S. Postal Service delivered mail to a long row of boxes put up on M-64, the Michigan highway that had the mine on one side of the road and the townsite on the other. The company's security department cruised around in patrol cars to protect residents. The company put out a mimeographed *Bulletin* to inform families of "Happenings in a Fast Growing Community." The *Bulletin* carried a schedule of free movies shown on Saturday nights at the Service Building, listed the hours of the barbershop operated by Ray Archambeau out of his house on Elm, listed available baby sitters, and carried lost-and-found notices, birth notices, and obituaries. It ran brief articles on technical subjects, like casting copper, and informed people of the dates of club meetings and youth talent shows. The *Bulletin* advised that ten days after the company laid sod around a company house, it became the tenant's responsibility to water and care for the lawn.[27]

The *Bulletin* carried notices of church doings. Catholics first worshipped in the lounge room of the Service Building at 7:30 on Sunday mornings, and a nondenominational Protestant service followed at 10:30. The "Bulletin" kept people abreast of school matters. Copper Range used a three-bedroom house at 7 Elm Street as a temporary school, but by November 1953, the permanent grade school opened. In the manner of the old paternalism, the mining company and its architects fully planned the school, without community involvement, and for their efforts Pace Associates won a design award in 1954 from the American Association of School Administrators. Copper Range built and paid for the school (valued at $665,000 plus $37,000 for contents) and then leased it to the Carp Lake Township Board of Education for one dollar a year. The board operated the school at its expense and paid for all maintenance and utilities.[28] The grade school enrolled 200 to 225 students throughout the 1950s, and 45 to 60 older children rode the bus twenty-two miles one way to Ontonagon High School. Copper Range had planned for a high school in 1952–53, but didn't erect it until 1960. Built at the mining company's expense at the Town Center, the high school added a swimming pool, auditorium, and athletic fields to White Pine's list of assets.[29]

White Pine continued the century-long tradition of providing employees with medical and hospital service. A twenty-bed hospital opened in September 1953: insurance records valued the hospital at $469,000, and its contents at $107,000. The hospital included operating and emergency rooms, a maternity ward, an outpatient clinic, a physical therapy unit, a dental clinic, and a pharmacy. Two physicians staffed the hospital, and they treated private patients, as well as company accident cases. No longer did Copper Range fund all its medical programs by charging the old dollar-per-month fee to workers. Instead it offered group hospital and surgical insurance to all employees and their dependents through the Connecticut General Life Insurance Company.[30]

Through the 1950s and into the early 1960s, the townsite received its own post office, fire station (manned by a volunteer force), motel, and restaurant. Asphalt covered all streets. Adults and children played on baseball diamonds and skated on an enclosed, natural-ice rink. A Standard Oil gas station went up out by M-64. Worshippers attended newly erected Catholic, Methodist, and Lutheran churches. In 1963, Copper Range listed all the community's assets and concluded they were "all the social and athletic activities you'd expect to find in a heads-up American town."[31] But for all the new amenities it did have, in reality White Pine wasn't a town at all—but the modern version of an old mine location. It lacked a downtown, with bars and places of entertainment. Save for a grocery operating out of the Service Building, it lacked stores and shops. Copper Range had not built the proposed shopping center at the Town Center shown on early maps, because White Pine didn't have the population to support it.

In November 1951 the *Daily Mining Gazette* wrote that some expected the proposed townsite to reach a population of five thousand. In January 1953 a company officer said that soon White Pine would have two thousand residents. In 1956, John Lally, then president of the Copper Range and White Pine mining companies, foresaw a grand future for White Pine: "Within 30 years, there'll be a city here with 15-story office buildings and homes and shops and stores. It will be one of the great copper mining and processing centers if North America." But in 1956, only 217 families and 800 people resided in White Pine, and by 1957 the numbers hadn't changed much: 236 families and 900 residents.[32]

White Pine's growth stalled in the mid-1950s. After Copper Range built its first wave of 156 houses, it didn't launch a second wave to achieve the 465 houses that planners had made room for on plat maps. The town simply failed to attract residents. Because it didn't have enough residents, White Pine couldn't support the enclosed, arcadelike shopping center proposed for the Town Center. And since White Pine lacked that shopping center and other things that a complete town would have, most workers' families eschewed living there. One thing especially curtailed White Pine's growth: the automobile. For all the planning that Pace and Copper Range did, they never foresaw how employees would use cars to maintain an independent way of life, away from White Pine.

Residents of Ontonagon village (population 2,500) appeared more prescient about the effects of the automobile on White Pine than did the planners at Pace Associates. In 1951 they had been "simply flabbergasted" by Copper Range's receipt of a $57 million loan to launch White Pine; the loan amount was six times the valuation of all real property in Ontonagon county. Those who lived in the county seat naturally wondered where Copper Range would put shipping docks, a powerhouse, and worker houses.[33] They assumed, expected, and hoped that Copper Range would locate important facilities in Ontonagon and not situate them all eighteen or twenty miles away at the proposed mine. Ontonagon residents soured on the mine project when no railhead, dock, or power plant came to their town, and they chafed at Copper Range building a whole new mine location. They pointed out "the utter folly of building a townsite . . . in the woods at terrific cost, when they had all the facilities necessary for mine personnel and their families [in Ontonagon]—which only required to be expanded with government aid."[34] Ontonagon folks said "the miners

A two-tone 1957 Ford sits in a White Pine driveway. Instead of living in White Pine, many employees used a car or bus to commute to work from other communities scattered widely across the region. Because of employees' mobility, White Pine never achieved the population that Copper Range had hoped and planned for. *(Michigan Technological University Archives and Copper Country Historical Collections)*

would commute 36 miles a day," while Copper Range insisted they would be like miners of old and "want to live on top of the shaft house."[35] They "would not travel twice 20 miles a day, often in 40-below weather and in snowfalls that occasionally set a U.S. record of 155 inches from early fall to late spring."[36] The *Daily Mining Gazette* agreed with the company. It saw automobile commutes to White Pine as a short-lived, construction-era phenomenon. In 1953 the *Gazette* noted, "It's surprising from how far the men come to put in their eight hours each day at the little, future city of the Pines. . . . One of these days they'll commute no longer. They'll hang their hats in a White Pine home."[37] But by 1956, workers had clearly shown that commuting, rather than moving to White Pine, was the preferred option. A company vice president admitted that "many are driving 90 miles a day in less than two hours and thinking nothing about it."[38]

Most workers chose to stay where they were, where housing costs were low; where they perhaps owned their own home; where working-class wives had more opportunity for paying jobs than the women at White Pine; where they lived in or near towns like Houghton or Ironwood, which offered more services, shops, and entertainment. If they stayed put, they often remained near family and continued to worship at the church they might have been baptized and married in. William Nicholls, who became a general manager at White Pine, acknowledged that times had changed:

When I was a youngster, if your father worked at a certain mine, you always lived within walking distance of work. Roads weren't plowed, and you weren't going to walk ten miles to work every day.

But at White Pine we were getting away from that thinking. Roads were better built and kept plowed. People . . . had their own social circles and didn't want to leave to risk it on a new mine. I knew it wouldn't be a huge town. The original plan was to build about 500 houses, but it was never followed through. We found that 170 houses and 40–50 apartments were sufficient.[39]

The number of auto commuters rose as employment grew at White Pine. In 1958, White Pine drew its 1,282 employees "from a surrounding area 100 miles in radius." In 1960, only 271 of 1,518 employees lived in White Pine, while 314 lived in Ontonagon; nearly 50 in Painesdale, Trimountain, Baltic, South Range, and Atlantic Mine combined; 87 in Ironwood; 22 in Houghton, 25 in Hancock, and 10 in Calumet. Those driving from 0 to 25 miles (one way) numbered 691; 503 drove 25 to 50 miles; 182 drove 50 to 75 miles; 136 traveled 75 to 100 miles; and 6 employees commuted over 100 miles each way to work. By 1968, when White Pine employed about 2,400, "the vast majority" of those employees lived in Ontonagon and Gogebic counties, but all told, they resided in sixty-seven different communities in nine counties and two states. The average employee had a sixty-seven-mile road trip each day—and from end to end, White Pine drew its workers from a line stretching 250 highway miles long. As the company noted:

This is a far cry from "the old days" when limited means of transportation forced everyone to "live over the store." The White Pine caravan includes cars, trucks, jeeps, buses and bicycles. Counting the distance traveled by each employee, total weekly mileage getting to and from the job comes to about 850,000.[40]

In 1970 the *Milwaukee Journal* reported that only 330 of White Pine's 2,700 employees lived at the mine, and the 88 percent of the workforce who commuted traveled fifty million miles a year back and forth to work.[41] During the mine's peak years, as many as thirty buses a day, operated by three different companies, carried workers to and from White Pine.[42]

Driving long distances over two-lane roads, especially in winter, carried risks. A circa 1956 issue of the *White Pine Bulletin* carried the article "Thou Shalt Not Kill," by Rev. Raymond Peterson, who campaigned against the "wanton destruction of life" caused by traffic accidents. He implored readers to "Be Your Brothers' Keeper, Not Killer" and "Drive carefully, drive as a Christian." The good reverend was not the only one concerned. White Pine's safety personnel also worried about traffic accidents, and with good reason: in some years, more employees died in car crashes than in mine accidents.[43]

For White Pine miners, vehicles posed the greatest threat to life and limb underground and on the highway. Underground, rockfalls occasionally killed men, but mobile loaders and trucks, moving in tight spaces or on inclines, posed the greatest risk to operators and pedestrians.[44] White Pine won numerous commendations for its safety record and practices, which it claimed "were equal to or even more stringent than federal law."[45] From 1958 through 1970, twelve workers died in accidents underground at White Pine, an average of just under one man per year.[46] Although the corporation's safety engineers policed the underground, they could not control highway accidents, save to warn against them. In 1965, Copper Range claimed that "our people are safer working under potentially hazardous conditions [at the mine] than they are at home. Certainly they are much safer than when commuting to work in their automobiles."[47] Underground, the worst *year* for accidents prior to 1970 was 1960, when three men died. In January 1969 a single car accident killed five carpooling men, all fathers from Bruce Crossing and Ewen. They started driving home in midafternoon after their shift. Snow was falling, and an icy crust coated the road. A half mile from the mine, their car skidded sideways into the path of an oncoming pickup. All five men died in "the worst highway accident in the memory of Ontonagon County residents."[48] White Pine barely escaped a much larger catastrophe in October 1976 when a White Pine Express bus carrying thirty-one workers to distant homes skidded off a snow-covered M-26 and rolled over. Nobody was killed, but Portage View Hospital in Hancock treated thirty passengers for injuries.[49]

Copper Range still believed in the idea that if it housed men near their work and provided them with schools, doctors, and a hospital, they would repay the company with loyalty and hard work. But it didn't happen at White Pine because of cars and commuters. Copper Range, which often incurred a strike at the expiration of every labor contract, believed that commuting workers introduced a "very complicating aspect to . . . employee relations." The company noted that the impact of long-distance commutes on "our productivity and employee relationships can easily be imagined, and the possibility of alleviating some of these problems through the development of the White Pine community is a most stimulating prospect."[50] Copper Range did not abandon its hope that by growing White Pine it could grow better employee relations. In the 1960s it improved and enlarged the community, trying to make it a more desirable place to live.

In 1960, Copper Range built a high school at White Pine, hoping it would attract and keep workers with older children. Copper Range acknowledged that, in the postwar world, Americans keenly wanted to own their own homes, so it introduced home ownership at White Pine by making building lots available to workers. Copper Range hoped to reap its own rewards while workers achieved home ownership: "If we are able to encourage the building of a substantial community with its associated shopping, service, and entertainment facilities, we shall have gone a long way toward improving the Company's community and industrial relations, and toward attracting a permanent, well-trained staff."[51]

In 1961, Copper Range sold twenty-three building lots and employees started construction on ten "attractive houses" whose plans the company had approved. By the end

of 1962, twenty-seven newly built private homes stood at White Pine, and another seven
went up in 1963.[52] By the end of the decade, Copper Range further stimulated house
construction by offering direct financial assistance. Workers borrowed a partial down
payment from their employer, and they didn't have to fret about selling their old homes,
away from White Pine, because Copper Range purchased them from relocating employees. This program resulted in twenty-one new houses going up at White Pine in 1969.[53]
These new, private houses included some of split-level design, and in their architecture
and finish they were a distinct cut above those erected by the company. Parts of White
Pine now looked like any middle-class subdivision of the time.

In April 1963, Copper Range offered for sale most of the houses it had built a decade
before. By the end of the year, it had sold 83 of its 168 houses. In 1968 it helped form
a Citizens Improvement Organization, to enhance White Pine by planting trees and
such, in order to attract more residents and businesses to the town. But no matter what
it did, Copper Range still could not get past the tipping point. It could not grow the
community large enough "to support the facilities of a good sized town," and lacking a
good sized town, it suffered an "inability to obtain and keep manpower in sufficiently
increasing numbers" to meet the requirements of a sought-after expansion of production.[54] In the second half of the 1960s, sustaining a labor force of up to 2,400 men
proved difficult.[55] In 1969, Copper Range hired and trained 507 new men, but many
soon left, and "the net addition for the year was only 149 men."[56] At this time only about
7 percent of employees lived at White Pine.[57] To increase that percentage, in 1970 Copper Range again went back to its social-engineering drawing board. It hired Unlimited
Development (a firm that it partially owned) "to develop a master plan for the growth of
the community from its present population of 1,300 people to approximately 5,000 by
1975 and 10,000 people by 1980."[58]

This plan called for investments in housing, commercial businesses, and recreational
facilities that would draw more men to White Pine, where they would enjoy "continuous employment in an ongoing and steady industry." The revamped townsite, so they
hoped, would draw others to White Pine not to work, but to lead "a pleasant life" near
Lake Superior, where they could ski, hike, hunt and fish.[59] The president of the Michigan
State Chamber of Commerce lauded the proposed development, noting it was conceived
"along good lines of planning, and avoid[ed] the evils of urban living." Copper Range
hailed the new White Pine as "the finest all-round community in the nation."[60]

Planned as a three-phase, $40 million development, on paper the new White Pine
included single-family houses, town houses, cooperative apartments, an enlarged mobile-
home court, senior citizen housing, and a nursing home—all equipped with cable tel-
evision. It had a man-made lake that was over four miles long and rimmed with boat
launches and a marina; an island in its center served as a park and campground. It had a
golf course. It had an additional elementary school and an intermediate school, professional building, government building, library, exhibition hall, new churches, and cemetery. It had a shopping center (the Mineral River Plaza mall), housing a supermarket,

bakery, and drug store; clothing, variety, hardware, and furniture stores; plus gift and book shops, beauty salon, and barber shop. Planned construction also included a bank, motel and restaurant, theater, and bowling alley.[61] Unlike the copper mines before it, White Pine now had ambitions beyond being a mine location; it aimed at being a complete town. But the effort failed, and most new developments never came to pass.

At the junction of White Pine's Main Street and Highway M-64, by the end of 1971 "a fully enclosed 58,000 square foot shopping mall" did go up, along with an "attractive" 15,000-square-foot Konteka restaurant and lounge, and an eight-lane bowling alley outfitted with Brunswick's new Astroline bowling system, with new-style seating and ball returns, plus automatic scoring. A Red Owl grocery store anchored Mineral River Plaza mall. A druggist and other stores opened there, too, but at year's end retailers occupied only 45 percent of the mall's space. Copper Range also carved out of the woods a new residential subdivision of 250 lots called Evergreen Acres, where it built twenty-five modular houses and offered them for sale at prices ranging upward from $18,500, and encouraged others to build their own houses there. But at the end of 1971, Copper Range admitted, "The rate of new residential construction . . . was disappointing." Despite the big push, White Pine's population had increased by only 9 percent during the year, bringing it to a total of 1,471 residents."[62]

No matter what it did, Copper Range could never attract enough people to White Pine. It wanted to lure in thousands of employees but attracted only hundreds, at best. Of those who chose to live in White Pine, many happily lived with that decision and loved the place. One long time employee wrote:

> I raised a family of five children there and each of them has the fondest memories of growing up in White Pine. It was a most vibrant community and one of the best places to raise a family. It had a superior school system which provided a sound education for all of my children.[63]

Another employee who resided at White Pine for thirty-three years sounded a similar note. People from many different states and countries mingled in the small community. After people could purchase their own homes at White Pine, "the mine manager lived across the street from a mill laborer, the head of engineering lived next to an underground miner, and the general manager lived across from the smelter clerk." The schools were excellent and they, along with churches and community and fraternal organizations, "made the place hum. Our kids still refer to their White Pine years as their Camelot."[64] Despite the fond memories of those who lived at White Pine, though, clearly more people in the 1960s and 1970s simply chose not to live there, next door to a copper mine and smelter, even if they worked there. And White Pine's promise of ready access to nature and outdoor recreation hardly proved a great selling point—because people could live close to trees and water all over the Upper Peninsula. White Pine had peaked by the early 1970s, and in terms of population and amenities and services, it would only decline from there.

22

❖ ❖ ❖

WHITE PINE

No Solution

The native-copper mines of Lake Superior went into permanent decline in 1920, but a few limped along for almost a half century before the final curtain came down on that industry in the late 1960s. White Pine, in its turn, encountered life-threatening times in the mid-1970s. Thereafter the mine endured major financial losses, downsizing, strikes, closures, takeovers, and environmental challenges. Still, it carried on for two more decades before shutting down twice: once in 1995 and finally in 1997.

For the American copper industry as a whole, 1975 was "the worst it had experienced in many decades," and Copper Range did not escape the downturn. Copper Range suffered the worst financial loss in the history of the company: $13.7 million. To get out from under its financial woes, stockholders in August of 1975 proposed that Copper Range become a wholly owned subsidiary of a much larger mineral and mining company: AMAX. But the U.S. Department of Justice challenged that move on antitrust grounds, and a court order enjoined any merger, sale, or consolidation putting White Pine under the control of AMAX.[1] With that deal nixed, in November 1975 Copper Range retained Paine, Webber, Jackson & Curtis to offer advice on how best to sell off all or part of the company.[2] Then it took major steps to cut production and employment.

On 4 January 1976, White Pine slashed mine output by 75 percent; its monthly output for the rest of the year ran only 25 to 60 percent of normal. Copper Range also cut two-thirds of White Pine's labor force of nearly 2,900. Chester Ensign, the chief executive officer (CEO) of Copper Range, explained that eighteen straight months of losses, caused by low copper prices of only fifty-five to fifty-eight cents per pound, made severe cuts imperative. White Pine terminated 102 workers, laid off 1,615, employed 285 on a reduced work schedule, and retained only 850 workers on a full-time basis.[3] White Pine entered a painful new era. It never again approached an employment level of 2,900. Instead, even in the best years to follow, employment ranged from 1,000 to 1,300.

By the end of 1976, things looked better because copper had climbed to 70–74 cents per pound. Also, Copper Range had found a suitor, the Louisiana Land and Exploration Company (LL&E), to bail it out of its cash-poor predicament. LL&E engaged in the gas, oil, and coal industries. It had not invested outside the energy industry since 1969, but by November 1976 it reached a tentative agreement to take over Copper Range and the White Pine mine.[4] In May 1977 the two companies hammered out the final details. Copper Range stockholders agreed to sell their company to LL&E via an exchange of stock. For each share of Copper Range stock they owned, they received 0.85 of a share of LL&E stock. To complete the deal, LL&E issued stock valued at about $51.2 million, approximately half the book value of Copper Range. In addition to acquiring White Pine, LL&E acquired from Copper Range the Hussey manufacturing facilities and about 185,000 acres of mixed hardwood timber on the Keweenaw Peninsula.[5]

LL&E expected the depressed condition of the U.S. copper industry to improve, but there was no quick turnaround. Despite "stringent cost reductions," White Pine lost money in 1978, although the last quarter of the year had been profitable. Employment remained about 1,200. Workers laid off early in 1976 exhausted their state unemployment benefits and the supplemental benefits paid by their employer, so many moved on and out.[6] By April 1978 the population of White Pine had fallen from 1,750 to 1,300. In leaving White Pine to go "where the grass looks greener," mining families took "great losses—25 percent or more," when selling their homes.[7] Social institutions suffered losses, too. The White Pine school system lost fifty-six students and faced a $65,400 budget shortfall in the year of the big layoff.[8] In 1978 LL&E got out from under a major obligation by donating LaCroix Hospital, which Copper Range had run since 1954, to the community. Citizens organized the nonprofit organization, White Pine Hospital, Inc., to keep the medical facility open.[9] At the end of 1978 the Mineral River Plaza mall was sold off to Richard Ross, who planned to add a Laundromat, beauty shop, furniture store, and Levi Outlet store. Those ideas quickly went bust, and Ross held a public auction in May 1980 to sell the mall off. Nobody showed up to bid, and the sheriff posted a foreclosure notice on the building.[10]

By 1979, copper prices had risen from the previous year's sixty-eight cents per pound to ninety cents, and LL&E increased copper production from 81.3 to 87.5 million pounds.[11] In 1980, LL&E and the White Pine mine announced a bold step intended to open up new markets for its copper, add forty years to the mine's life, and create five hundred new jobs.[12] In the past, White Pine had often jumped into new technologies, and now it did so again. Despite the severe economic problems of the recent past, and despite the cyclical nature of the copper industry, LL&E decided to erect a $78 million electrolytic refinery at White Pine that would produce a more conductive, higher-grade copper better suited to a wider range of markets. The copper from White Pine's smelter went mostly to the automobile industry to be made into radiators. Thanks to oil crises and Japanese imports, the Big Three automakers bled red ink in 1980. While they struggled and downsized, the radiator market dwindled. To reach new markets, in May 1981 LL&E broke ground on an electrolytic refinery that would make "the purest copper in the world" and also capture silver as a by-product.[13] LL&E shopped the world for the best technology for the facility. It borrowed much from a Krupp refinery in Germany but also sought out ideas from other plants in the United States, Belgium, Australia, Japan, and Finland.[14] To complete the plant's design, LL&E turned to Fluor Mining and Metals of California and the Furukawa Electric Company, based in Tokyo, Japan.[15]

Once the electrolytic refinery was running, the White Pine smelter would cast 95 percent pure copper into anodes or plates, 3 feet square and 1.5 inches thick. These anodes, along with stainless-steel cathodes, would hang in a tank filled with an acidic copper sulfate solution heated to 160 degrees. When electrically charged, the anodes dissolved, and their released copper flowed through the solution to be deposited on the cathodes—while impurities, such as silver, sank to the tank's bottom. After being stripped off the stainless-steel cathodes, the extremely pure copper could be baled for sale or melted and cast into whatever cakes, billets, or wire bars customers wanted.[16]

While the $78 million refinery was being constructed, White Pine employed 1,300, and it started taking applications for 500 new jobs expected to come online in 1982. By September 1981, the mine had received three thousand job applications. Women made up 10 percent of the applicants, and in September White Pine announced that two had been hired, "the first female hourly employees at the mine." Women, the company said, will work underground.[17] But the hopes all applicants had of obtaining good-paying blue-collar work at White Pine were soon dashed. Before the refinery opened, LL&E found that its mine was again in deep trouble. Jobs were lost instead of gained.

The 1980 decision to erect a $78 million copper refinery came at a time when copper sold for $1.50 per pound. But economic optimism faded to gloom and doom by the end of 1981, when copper was down to eighty cents per pound.[18] Again, White Pine faced large losses. Salaried employees had had their pay frozen at the start of 1981, and in January 1982 Russell Wood, president of Copper Range, asked White Pine workers to take a pay cut. Besides being troubled by current wage rates, the company faced cost-of-living adjustments soon due to United Steelworkers under their contract. White Pine's managers had often hoped that workers would sacrifice pay to preserve jobs and corporate profitability. Now they again hoped that workers would comply with managerial requests, "do the right thing," and take a pay cut. When that didn't happen, in February White Pine laid off 125 workers. In March the mine went on a four-day workweek and cut its production 20 percent "to reduce labor and materials costs." In June it laid off another 125 employees. In September 1982, White Pine really brought down the axe. It cut another seven hundred workers and suspended mine, mill, and smelter operations.[19] Hard times at the mine quickly rippled through the community. As soon as layoffs began at the mine, they started at the White Pine schools. Faced with a looming $128,000 budget deficit, in March 1982 the school board shortened work hours for secretaries and custodians and laid off four teachers.[20]

White Pine started smelting and refining copper again using stockpiled mineral but kept the mine and mill shut down for a long time. Copper Range said that "the mine can only resume production if the United Steelworkers Union agrees to a no-strike contract clause and wage or benefit concessions."[21] In March 1983, talks began on a new labor contract while one thousand men were still laid off. The company wanted wage cuts of 15, 10, and 5 percent over the next three years; a reduction in cost-of-living increases; the trimming of one week of vacation time; ten-hour workdays with no overtime; and the ability to hire new workers at lesser wages than being paid to experienced workers. Talks broke down on 29 July and, on 1 August, "answering a 17-month-long plea for major wage and benefit concessions, unionized workers . . . went out on strike."[22] Since so many men had already been idled, only 130 laborers remained to walk off the job.

Contract talks started and stopped over several months without result. Meanwhile, the price of copper declined from eighty-one cents per pound to sixty-six cents, which did not bode well for anyone.[23] LL&E voted to put White Pine up for sale, and Russell Wood, a former LL&E executive and head of Copper Range, along with two others, purchased an option to buy it in May 1984. They had three months to study White Pine and LL&E's financial situation, followed by one month to close a deal. That deadline came and went without any sale.[24] On 1 August 1984, the strike at the mine was one year old, and later that month LL&E announced it was considering a *cold shutdown* at White Pine, which would entail performing only minimal maintenance and upkeep. A cold shutdown would be just one step shy of totally closing White Pine. By mid-November, LL&E had made good on this threat: it completed its cold shutdown of mine, mill, smelter, and refinery.[25]

At this point, to try to save their jobs, members of the Steelworkers Union hired a consulting firm to see whether they might be able to purchase the mine under an employee stock ownership plan (ESOP). But LL&E killed that idea in mid-December 1984 when it sold White Pine to Echo Bay Mines of Edmonton, Alberta. This sale was bad news for White Pine employees because the new owner of the mine had no desire to operate it. Echo Bay bought Copper Range's holdings (minus the new refinery, which LL&E retained) for $55 million because they included a 50 percent interest in the Round Mountain gold mine in Nevada. Echo Bay announced, "We are not copper miners ourselves. We're gold miners."[26] Echo Bay, Canada's second largest gold company, anticipated its share of Round Mountain would amount to sixty thousand ounces of gold in 1985. As for White Pine, Echo Bay kept it shut down.[27]

Given Echo Bay's lack of interest in operating White Pine, plans surfaced for other parties to assume ownership of the property. O. E. Anderson, who had served as White Pine's general manager for eight years before Echo Bay came in, said White Pine was "the most modern underground mine in the United States" and that "it would be a tragedy to abandon it."[28] Russell Wood was still interested in acquiring the mine, and members of the still-striking Steelworkers Union again floated the idea of taking the operation over through an ESOP.[29] In short order, Wood and union members started working together on a purchasing plan that would require a profit-sharing partnership between Wood, his associates, and employees; wage and benefit givebacks on the part of workers; financial aid from Michigan; and a willingness on the part of Echo Bay to sell at an affordable price.[30]

Not only was the future of the mine at stake, so was the future of the White Pine community. In the last ten years, its population had dropped from about 1,800 to about 1,200. School enrollment had plummeted from about 600 to 285. Bernice Houtari—the Ontonagon County clerk, had seen "an economic and emotional depression." Hard, uncertain times had led to a higher incidence of alcoholism, child abuse, and spouse abuse.[31] Tensions ran high. Echo Bay didn't help matters; it threatened to turn off the mine's pumps, an act that would make reopening the mine less likely.[32] Neighbors at White Pine sometimes stood at odds with one another. Still-picketing mine workers, some reluctant to make any wage and benefit concessions, found themselves on opposite sides of the street from women brandishing signs that said, "Welcome to the town the union is killing." Noreen Bergland, speaking of the strikers, said, "They can leave and go to hell for all I care, but don't kill my town." In turn, strikers said the women were "nuts" and "union busters."[33] Despite the talk of creating an ESOP to revive White Pine, many were tired of the ups and downs of mining and ready to give up. Marge Razmus, owner of the Konteka restaurant, said, "I'm afraid we're seeing the end of an era."[34] For Pat Smith, it was the end of an era. She liquidated her White Pine Variety Store at the mall and told a reporter, "I've poured every ounce of every bit of physical and emotional strength I had into the business. Now I have nothing left to give to it. It took everything I had." Surely families who built or bought White Pine homes for $40,000 or $50,000 felt about the same as Pat Smith when they put their houses up for sale at half those amounts, or less.[35]

By the end of April 1985, Russell Wood Associates had made a deal with the employees. To get the mine reopened, the union men abandoned their twenty-one-month-long strike and agreed to a five-year contract that lowered their wages but set up a profit-sharing plan. Under the ESOP, their wages dropped from $12.50 per hour to $8.50, but to compensate for that reduction, workers would receive 70 percent of the company's annual profits.[36] Russell Wood would manage operations, and he and his close associates and senior managers would share ownership, with workers, of a newly created Northern Copper Company, which would purchase the White Pine mine from Echo Bay.

When the mine was busy, White Pine residents sometimes "would kick" about smoke drifting over the community from the smelter stack. In May 1985, Fred Smith drove a journalist around town and told him, "I don't think people will ever complain about that again. It's going to be one beautiful sight to see smoke rising from that chimney again." People prayed to see smoke again: "The Lutheran and Catholic churches in White Pine already have held special services during which just about everyone in town prayerfully gave thanks for the movement that is underway to reopen the mine."[37] If wise, they also prayed for patience because putting together the financing to complete a purchase was a complex process. Money problems stalled the White Pine purchase throughout the summer, but by mid-September, with the support of Governor James Blanchard's Democratic administration, Michigan put together a package of grants and loans amounting to about $10 million to help complete the deal. In November the ESOP bought Copper Range and the White Pine mine for $23.7 million. After the mine hadn't been worked for three years, on 25 November eleven men went 2,200 feet underground to get things started.[38] Soon, it was hoped, another thousand men would join them.

White Pine employees now understood that "the days are gone when labor and management sat down at the table and told each other to 'go to hell.' We can't survive that way any more."[39] But putting an effective ESOP together was a novel proposition, since few American companies were organized that way. As more men returned to work in January 1986, some of them planned how to "implement team-work" throughout their new company. They set lofty goals: "to have the highest paid workers, most efficient, safest, and modern copper mining/smelting company in the world." In March and April the planning team visited three facilities where employees had a voice in decision making: a Detroit Diesel Allison plant and a General Motors (GM) plant in Michigan and an Alcoa Aluminum bauxite facility in Arkansas. They "saw obvious benefits of employee involvement," such as reduced grievances, less absenteeism, and improved product quality, training, and safety.[40]

Under the ESOP at White Pine, the mine switched from traditional eight-hour shifts to ten-hour shifts; it regularly published a *Copper Range Report* that lauded workers' accomplishments, explained decisions, and attempted to squelch rumors by printing facts; and it selected Blue Cross/Blue Shield as its medical insurance provider.[41] In terms of hiring, after receiving 3,671 job applications from thirty-six states, the mine employed 588 workers by the end of January 1986 and 776 by the end of May. Workers came from many different communities in the Upper Peninsula and Wisconsin—fifty-five communities in all: 130 workers lived in White Pine, while a total of 378 resided in Ontonagon County, 244 in Gogebic County, and 103 in Houghton County.[42] By the end of the year, White Pine provided nine hundred direct jobs and had a payroll of $19.7 million. The reopened mine had the economic impact that local and state officials had hoped for. Annually, it bought $10 million worth of goods and services from Upper Peninsula vendors, and this money had a multiplier effect as it passed from hand to hand in the local economy. When the mine closed in 1983, unemployment had shot up to 31.8 percent in Ontonagon County, 16.5 percent in Gogebic County, and 14.1 percent in Houghton

County. Peak unemployment in Ontonagon County during the shutdown approached 40 percent. In 1986, those rates dropped to 10.9 percent (Ontonagon), 11.8 percent (Gogebic), and 10.6 percent (Houghton).[43] The region still suffered high unemployment compared to the national average (about 7 percent), but with a revitalized mine, the local economy definitely improved.

Not all news was good at the mine or the townsite. In April 1986, a month when seventeen workers were hurt on the job, shuttle-car driver Melvin Manniko died underground; this was the first fatal mine accident since 1975.[44] In November the LaCroix hospital closed. Opened in 1954 as a company hospital and then run independently since late in 1978, the hospital had struggled to keep its doors open. To go with its eighteen-bed acute-care program, in 1981 it added a substance-abuse program with beds for eighteen patients. A bit later it added an eating-disorders clinic. At the same time, it had to limit itself to minor, outpatient surgery only, and in 1984 it temporarily lost its accreditation. In 1986, when only one or two of its eighteen hospital beds were occupied each day, the hospital terminated forty-three employees and downgraded itself to the status of a medical clinic.[45]

The first big payout of cash and stock to workers under the ESOP came in April 1987. Copper Range distributed $5,500 apiece to eight hundred union employees who had worked at least 870 hours in the past year, and along with the cash came 175 shares of stock, valued at $24.40 each, per employee. In the following three months, Copper Range also released a total of about $1 million in incentive pay, amounting to fifty cents per hour per employee for each hour worked since 1 January 1986.[46] Despite such payouts, by early 1989 employees were open to a lucrative opportunity just coming their way. In March, members of the Steelworkers Union at Copper Range approved a wage contract "that clears the way to a buyout of the employee-owned mine." A West German firm, Metallgesellschaft, owned 63 percent of the Metall Mining Corporation of Toronto, which had interests in copper, zinc, lead, gold, and silver. Metall Mining had made a bid for White Pine in February. If they accepted the purchase offer, employees would lose their company but be handsomely rewarded for it.

The sale of White Pine to Metall went through early in June. Metall paid $83 million for the mine, mill, smelter, power plant, and refinery. Russell Wood and other top managers split 30 percent of the money from the sale, leaving 70 percent to the rank and file. Average union workers—about one thousand of them—each owned about 750 shares of White Pine stock, which they cashed in for $80 a share, or a total of $60,000 each.[47] The checks from Metall Mining came in the mail, to the delight of all. Overnight, as the *Wall Street Journal* reported, "White Pine was transformed from one of Michigan's poorest places into one of its richest." Investment and tax advisors rolled into town to tell folks what to do with their money and how to protect it. Bank deposits swelled, the GM dealer in Ontonagon sold out of sport utility vehicles (SUVs), and twenty-two of twenty-four White Pine high school grads now planned to go to college because they could afford it.

While some saw the Metall checks as a huge windfall, others saw them as fair compensation for all the hard economic times people had suffered while on strike, when the mine had been shuttered.[48]

Copper Range, now wholly owned by Metall, employed 1,000 to 1,100 workers into the early 1990s, and its facilities operated twenty-four hours a day, seven days a week. Metall created another ESOP for employees, but this one gave them a smaller a piece of the company. Employees started out owning approximately 10 percent of Copper Range; their share would increase over time, to a cap of 20 percent.[49] This time around, however, there would be no second windfall for White Pine's stockholder employees. They would not benefit from long-term employment under Metall or from any lucrative sale of the company. In less than a decade, they faced unemployment and a permanent closure of the mine.

Like many Upper Peninsula copper mines before it, White Pine did not suddenly end operations. Mines tended to die in fits and starts while hoping for the copper market to rebound, for new technologies to reduce production costs, for explorations to turn up new copper reserves, or for the return of more favorable labor conditions. The *Daily Mining Gazette* sounded a death knell for White Pine on 13 July 1995; a headline announced in bold, "**Mine Closing.**" Nearly two years later, on 30 May 1997, the *Mining Gazette* repeated it: "Copper Range closing White Pine Mine." Between these two death notices, Copper Range dealt unsuccessfully with a host of challenges, including environmental issues never faced by earlier Lake copper mines. Many Keweenaw residents—living where air, water, and woods appeared infinite but where jobs were scarce—believed that the death of White Pine was no accident. Environmentalists had killed it.

From 1990 on, White Pine was caught in an expanding web of environmental issues that were closely connected to the mine's closing even if they did not directly cause it. In 1990 the Environmental Protection Agency (EPA) named White Pine the largest toxic polluter in Michigan. Its atmospheric emissions, particularly from its smelter, amounted to about one thousand tons of particulate matter annually. In 1992 the National Wildlife Federation and the Michigan United Conservation Clubs filed a lawsuit in federal district court seeking to force White Pine to comply with federal air pollution standards and permits. The suit stated that, in the preceding two years, White Pine had released five times the allowable limits of three especially hazardous particulates: lead, mercury, and arsenic. According to the Michigan Department of Natural Resources (DNR), White Pine had emitted 1,400 pounds of mercury alone into the air each year. The lawsuit sought from Copper Range a civil penalty of $25,000 per day for past and future air pollution violations. Claiming to be sensitive to the economic uncertainties visited upon the mine by this suit, the National Wildlife Federation, perhaps disingenuously, announced, "Our intent is not to shut down a facility that greatly affects the economy of the western Upper Peninsula." Instead, they wanted to negotiate a settlement whereby White Pine would clean up its emissions, thus protecting the environment, especially nearby Lake Superior.[50]

The EPA and the states of Michigan and Wisconsin also supported the lawsuit against the mine. After two and a half years of negotiations, early in 1995 Copper Range and Metall (soon to change its name to INMET) entered an agreement with the environmental organizations. Copper Range would pay $4.8 million in penalties, shut down its smelter, and build a cleaner one. This brought the projected cost of the settlement to $205 million, "the biggest in the history of the Clean Air Act."[51] While building its new smelter, White Pine would ship its mineral by rail to be smelted in Manitoba, Canada.[52] But the ink had hardly dried on the agreement, when a White Pine spokesman admitted that the cost of building a new smelter might be too high: "There may be no economical way to continue," he said, meaning that the 1,100 jobs at White Pine might soon disappear.[53] For Ontonagon County's nine thousand residents, the specter of closure was most unsettling because 40 percent of the county's economy was tied to the mine. Adjacent counties would suffer, as well. The "forced shutdown of the mine's smelter was, as many say, the final blow"—and "no doubt the wedge between environmentalists and the working class [was] driven deeper."[54]

In mid-July 1995, John Sanders, president of Copper Range, announced the complete closure of the mine that had produced 4.4 billion pounds of copper, opened up some 13 square miles of underground works, and had reached a maximal depth of 2,700 feet. White Pine was the second-largest underground copper mine in the United States and the twelfth largest overall, but its ore was only 1 percent copper.[55] "We've turned ourselves inside and out to try and avoid [the closing]," he said, but no way out had been found. Sanders did not place all blame on the environmentalists. Surely the shutdown of the smelter had been "one of the main reasons" for the closure, but Sanders "blamed declining ore quality and the burdening expense of getting the ore out of the ground as reasons for the shutdown." "You can't create an ore deposit for us," he said.[56]

One thousand White Pine workers were slated to lose their jobs by the end of September, while about one hundred were slated to stay on as caretakers and to conduct a pilot project that might prolong White Pine's working life.[57] The mine was done—but maybe it wasn't—if a wholly new method of mining could be put in place. For about two years, Copper Range had been working with the state to get a permit for a solution-mining pilot project. Now, in 1995, solution mining held out the only hope for saving the White Pine mine, but this technology raised many new environmental issues.

The vast network of support pillars left standing in the thirteen square miles of worked-out ground in the mine held as much as three billion pounds of copper.[58] They hoped to recover 60 percent of that copper through solution mining, which would entail blasting out 90 to 120 pillars at a time, reducing them to pieces of rubble smaller than two inches. Pipes would then flood the *rubblized* area with a *lixiviant* (leaching solution)—in this case, mine water and an iron-bearing solution that would be highly acidic because of the inclusion of 150 grams per liter of 93 percent sulfuric acid. The solution would liberate the ore's ferrous iron, which would oxidize to become ferric iron—which in turn would oxidize the copper in the chalcocite. When the lixiviant contained about 2.2 grams of copper per liter, pumps would lift it to the surface and deliver it to a new solvent extraction plant. The plant would chemically treat up to 7,600 gallons of solution per minute, removing the copper. The copper-depleted solution, now called *raffinate* would be regenerated and reused underground, but the process depended on a constant infusion of additional sulfuric acid. The implementation of full-scale solution mining would require 51 million gallons of 93 percent sulfuric acid per year, or 136,000 gallons daily. To deliver that amount, four railroad tank cars, or forty over-the-road tanker trucks, would have to pull into White Pine every twenty-four hours.[59]

In September 1995, Copper Range fired off 130,000 pounds of explosives at once to rubblize, as a test, seventy-two rock pillars, each of which was sixty feet long, thirty feet wide, and nine feet high. The blast was so large that it attracted the attention of scientists at the Los Alamos National Laboratory, who traveled to White Pine and set up seismometers to record the intentional mine collapse for study and comparative purposes.[60] The test blast went well, but Copper Range could not send acid into the rubble unless and until it received the needed permit from Michigan's Department of Environment Quality (DEQ) to test solution mining. The DEQ held public hearings as part of its permitting process, and those showed that solution mining was controversial before it ever started. Individuals spoke at the hearings of the potential for great environmental damage caused by the vast volume of acid to be used. Others spoke of the need to support solution mining because it would prolong White Pine operations for fifteen to thirty years and provide sustained employment for about 350 workers. Organized advocates of solution mining wore white shirts to the public hearings and sported green and white buttons saying, "I Support Copper Range."[61]

Before a permit could be granted, Copper Range and the DEQ had to resolve a long-term environmental issue regarding the cleanup of the mine years after solution mining would end. They projected fifty-eight years into the future, when the mine would have filled with water that would begin discharging onto the surface and finding its way into Lake Superior. The discharge would be high in metal content, acidity, and salinity—the latter caused by brine flowing into the mine at its lower levels. INMET, the parent company of Copper Range, had to agree to fund a water treatment plant to be needed nearly sixty years into the future to treat the discharge from White Mine. With commitments in place, in May 1996 the DEQ announced that White Pine had a permit to begin its pilot solution-mining project.[62] The DEQ press release glossed over environmental risks and, instead, stressed the jobs and payroll that the state's first solution-mining project would support. "This is a success story in which the state, company and community can take great pride," said DEQ director Russell Harding, who served under Michigan's conservative Republican governor, John Engler. "The Engler administration's commitment to environmental protection and economic growth, coupled with the support of a determined company and community, has breathed new life into the mine. This decision will have a positive impact on the region for years to come."[63] If the technology worked and proved profitable while being tested on 50 million gallons per year, White Pine hoped to receive approval to process, by 1999, 446 million gallons of copper-rich solution annually.[64] Copper Range initiated solution mining in the summer of 1996, but in October it cut the pilot project short, and full-scale solution mining never happened.

Copper Range applied the brakes to solution mining in the face of challenges belatedly posed by the federal government's EPA. The EPA initially passed on getting involved in the permitting process; it left decision making to the state. But in August 1996, after being goaded on by environmental groups, the EPA reported that Michigan would not be allowed to regulate solution mining by itself. The EPA now saw the mine as a kind of giant well whose polluted waters needed to be contained. It inserted itself into the permitting process, having determined that its involvement was sanctioned by the Safe Drinking Water Act and by regulations covering Underground Injection Control.[65] The director of Michigan's DEQ did not welcome the EPA's late entrance into the project, which he thought was encouraged by the Democratic president, Bill Clinton. He believed that the EPA would slow things down and take up to eighteen months to do its own environmental analysis.[66] At the same time, the head of the DEQ faulted the Clinton administration for encouraging Native Americans to lodge protests over solution mining. Clinton "had issued an executive order to give 'undue consideration' to certain 'constituency groups,' such as Native Americans," who in this case protested solution mining because it posed environmental hazards to tribal lands.[67]

Approximately nine hundred Chippewas lived on the Bad River Reservation in Wisconsin, about fifty miles west of White Pine. The reservation covers 192 square miles and has an extensive network of inland rivers and streams plus a seven-thousand-acre wetland and seventeen miles of Lake Superior shoreline.[68] Copper Range and the Wisconsin Central Railroad planned to deliver most of the acid needed for solution mining on a run of tracks and over a bridge on the reservation. Walt Bresette and other Chippewa environmentalists fought against this route, fearing what a derailment and acid spill would do to the water. In July and August, Native Americans on the Bad River Reservation physically blocked Wisconsin Central Railroad freight trains from crossing their lands, which stopped the delivery of sulfuric acid to White Pine. New truck and rail routes were inaugurated to keep supplies flowing to the pilot project, which produced its first refined copper in September 1996.[69] That same month, the EPA conducted public hearings in Wisconsin and Michigan to explain its late involvement in the project and to solicit comments to help set the scope of its environmental impact study.

For the EPA, containment of water pollutants was the key issue. For many local citizens, meddling on the part of the federal government became the issue. At one of its September hearings at White Pine, "concerned citizens made the agency well aware [that] the town is at stake if it does not grant Copper Range Co. a solution mining permit." Nearly five hundred participants, again wearing white shirts to show their solidarity, told

The proposal to introduce solution mining at White Pine generated great controversy in the region in the mid-1990s. Heated public meetings pitted insiders against outsiders, and environmentalists and protectors of Lake Superior against protectors of jobs and the local economy. The controversy also pitted Governor John Engler and Michigan's Department of Environmental Quality against President Clinton and the Environmental Protection Agency. *(Ontonagon County Historical Museum)*

Tribe still blocking tracks

HIGHBRIDGE, Wis. (AP) — A federal mediator says a communications breakdown may be at the heart of a dispute between an Illinois railroad and Chippewa Indian protesters occupying its tracks.

The conflict over the safety of trains passing through the Bad River Reservation has halted shipments of sulfuric acid to a Michigan mine and prevented about 50 carloads of goods from reaching Wisconsin Central Ltd. customers.

Negotiator arrives to assist in resolving standoff over acid shipments to White Pine

wooded area four miles northeast of here July 22. They say they fear the acid tankers could derail on the old tracks and pollute the land and water on the reservation.

ing the minerals that the acid was dissolved.

The tracks on which the acid is being hauled "received only a cursory inspection" before shipments started, Dixie said.

Agency, and U.S. Transportation Secretary Federico Pena.

Wisconsin Central has said the tracks being blockaded had been inspected and were safe for the daily run of a freight train that travels no more than 10 mph on its trip to Copper Range Co. mine near White Pine in Michigan's Upper Peninsula.

Negotiations between railroad executives and the protesters broke down Monday, leading Hanson, who refused to arrest

Mine meeting in White Pine

Big turnout expected tonight

By VANESSA DIETZ
Gazette writer

WHITE PINE — Tonight, a crowd of white shirt-clad mine supporters will face the federal agency that Copper Range Company cites as the reason it may be forced to lay off up to 100 workers at White Pine.

EPA continues to contend the action by the Bad River Band of Lake Superior Chippewas and others has no connection to its re-evaluation.

Despite an appeal from Michigan Department of Environmental Quality head Russell Harding, EPA said it will continue to collect information for an environmental analysis of the planned solution mining.

Public protests mine project

Foes of Copper Range solution mining project raise a ruckus at EPA public hearing

By VANESSA DIETZ
Gazette writer

ODANAH, Wis. — The U.S. Environmental Protection Agency

Environmental Quality has already given its permission for the process and solution mining is currently being done at the mine as a small-scale pilot project.

The procedure involves pumping a slightly pulverized solution through underground rock pillars underground in the mine shafts to extract copper.

Bad River drummers chanted at the beginning of the more-

Official: President Clinton responsible for solution-mining analysis

By VANESSA DIETZ
Gazette writer

ONTONAGON — An Environmental Protection Agency analysis of solution mining at Copper Range Co. might have ties all the way to the White House.

the planned solution mining. But, on the last day of public commentary for the department's permit, the Chicago agency sent a 13-page letter to Harding, spurring him to ask, "What kind of partnership is this?"

Harding thinks EPA first decided to regulate, and had its attention

EPA defends stance on solution mining

Agency gathers public input regarding White Pine mine, says environment is its top customer

By VANESSA DIETZ
Gazette writer

CHICAGO — Even though Environmental Protection Agency recently opened the floor to White Pine residents, it could shut the door to the town's mine.

Copper Range's solution mining project, as a federal agency, it is responsible for upholding the government's treaties with the Chippewa Indians. She said consideration of the treaties "was certainly a part of" the decision

---

the EPA to "go away." They "bashed [the EPA] for butting in" and gave the EPA a good "tongue-lashing."[70] At another hearing a month later, state representative Paul Tesanovich chastised the EPA for getting involved too late and likened the agency "to a cat toying with its prey." Joan Antila, an Ontonagon County commissioner, opined that "any Yooper knows what the environmentalists are looking for. The environmentalists have long seen us as a pristine playground. They want just enough of us left around to serve their needs," and by implication, they wanted nobody left around to mine anymore.[71]

In October, Copper Range announced, "If we don't proceed with solution mining, we will shut down the business." Then the company suspended the pilot project and asked the EPA for a decision on permitting full-scale solution mining. Copper Range could not afford to sink any more capital into the new technology unless it had a federal green light to go ahead with it.[72] For its part the EPA claimed that it was sensitive both to the environmental and the economic issues raised by solution mining. The EPA said it could produce a draft economic and environmental analysis by July 1997.[73]

In preparing the analysis, the EPA noted that the counties around White Pine were disadvantaged economically. On a statewide basis, Michigan had a per capita income of $22,192, whereas Wisconsin's was $20,884. Per capita income in Houghton, Ontonagon, and Gogebic counties in Michigan, and in nearby Iron County in Wisconsin, averaged only $15,747 annually, which was 70 percent of the national average. The median value of occupied houses in counties around White Pine stood at $27,072, which was 44.7 percent of Michigan's overall median and only 34 percent of the national median. Surely these counties could use the projected 315 blue-collar jobs and 80 technical/managerial jobs that solution mining would support directly. In addition the secondary employment stimulated by the mine's payroll would create another fifty jobs in nearby communities.[74]

Besides emphasizing the economic benefits of solution mining, the EPA downplayed some of the environmental risks of transporting acid in railcars to the mine, especially through the Bad River Reservation. The worst-case scenario would be if one day's shipment of four rail tankers of acid derailed and spilled at the Bad River Bridge. The EPA set the probability of such an accident at one occurrence every 74,000 years. The probability of losing a single tanker car at a bridge would be once every five thousand years. An accident spilling 10,000 to 25,000 gallons of acid anywhere along a railroad route to White Pine would be once in eight years.[75]

The EPA did not discover sufficient reason to deny a permit for solution mining and was "leaning toward granting" the permit.[76] The EPA was set to release its draft economic and environmental analysis on 2 June 1997. But just days in advance of that date, INMET and Copper Range threw in the towel. It would cost $10 million to fully implement solution mining, and they hadn't been able to raise that money while the EPA decision was in limbo. Company officials went over economic forecasts, studied declining worldwide copper prices and rising solution-mining costs, and in May 1997 decided that solution mining "just won't make money for the company's stockholders, and the time to pull the plug is now." The EPA permit no longer mattered. Copper Range withdrew the permit application and closed the mine because it was "too expensive to try to keep the business going and put solution mining methods in place."[77] Copper Range's president said the firm was "committed to working with local and state officials in efforts to identify beneficial post-mining uses for the site." Copper Range "will ensure that all current facilities, both underground and [on the] surface, are closed in an environmentally sound manner."[78]

Earlier mining companies, when they closed, often scrapped out or sold anything of value and then just walked away from any environmental problems they had wrought, such as ruins of buildings, uncapped mine shafts, or millions of tons of tailings in some lake. Modern Michigan law did not let Copper Range off so easily. On 29 October 1997, Michigan's DEQ and Copper Range entered a consent decree concerning the cleanup of White Pine's 32,500 acres, which the DEQ would monitor and which Copper Range would complete and pay for. This cleanup carried on for over half a decade. Some tasks

White Pine's giant tailings basins stood between the mine and the shore of Lake Superior, seen in the distance. Containing about forty years of mill discharge and covering about five thousand acres, the tailings became a major environmental problem when they dried out. Winds sweeping over the basins lifted clouds of dust into the air. *(Michigan Technological University Archives and Copper Country Historical Collections)*

were relatively small, such as the cleanup of an industrial drum dump at the mine. The two biggest environmental problems to be resolved were quite opposite each other. One involved water; the other, dust. One was underground; the other on the surface.[79]

At its deepest depths, the mine had penetrated brine-containing formations. This salt water, given forty to sixty years, would rise up through the mine and flow outward to contaminate the area's shallow aquifer and local streams. Eventually it would end up in Lake Superior. To delay the flooding of the mine with brine for about two hundred years, and to dilute the brine, in June 1999, Copper Range started pumping Lake Superior water into the mine; the freshwater would serve as a cap to suppress the in-flow of brine at the bottom. Pumps at Lake Superior in Silver City sent thirty-six million gallons of water daily into the mine. Altogether the cap required sixteen billion gallons of water.[80]

On the surface, the vast tailings basins, covering five thousand acres, presented the greatest environmental problem. With the cessation of milling, they dried out. The DEQ considered the tailings to be inert and chemically nonthreatening, but the tailings created an irritating dust problem. In April and May 1998, twenty-five days without rain, coupled with a strong northwest wind, sent dust clouds billowing into the White Pine townsite. To quell this dust, Copper Range could not simply water down the tailings occasionally; over several years, the company had to turn the basins, which "looked like a lunar landscape, into something resembling a massive meadow." Copper Range started covering the nine square miles of tailings with clay, wood chips, and sludge from an Ontonagon paper mill. With an organic base laid down, Copper Range fertilized and seeded the basins, covering them with dust-suppressing vegetation.[81]

Some parts of the site could be reused or redeveloped; others could not. The mill went down in 1999, when Copper Range demolished it. Copper Range gave away other buildings and land to support the creation of a new industrial park being created by the Ontonagon Economic Development Corporation with help of a grant from the Michigan Economic Development Corporation. Small businesses could go there, such as metal-fabricating shops.[82] The relatively new and efficient electrolytic copper refinery had the greatest value. Although the White Pine mine wouldn't send it any more copper, it could still refine anodes produced elsewhere and shipped in by rail. In 1995, Copper Range entered a contract with Hudson Bay Mining and Smelting to refine its copper anodes. The anodes, 98 percent pure, were produced 1,250 miles away in Flin Flon, Manitoba. At White Pine the refinery converted them into copper cathodes over 99.9 percent pure. In January 1998, INMET and Copper Range sold the copper refinery to BHP Copper (Broken Hill Proprietary of Australia). BHP continued to operate the refinery on a toll basis, producing over 75,000 metric tons per year, the highest production rate since the plant had been first commissioned. Ownership by 2005 eventually passed to HudBay Minerals, and the refinery's main customer remained Hudson Bay Mining and Smelting, a HudBay subsidiary.[83] In June 2009, HudBay announced that it would close the refinery in August 2010.

Without doubt, the most novel reuse of the White Pine mine occurred underground in a part of the mine not flooded by the water cap that checked the inflow of brine. In the year 2000, SubTerra, a Canadian biotechnology firm, opened a three-thousand-square-foot growth chamber underground. At an abandoned Canadian mine, SubTerra, under contract to Health Canada, grew and provided "affordable, quality, standardized marijuana for medicinal and research needs." In Michigan, no marijuana growing was allowed. Instead, under strictly controlled conditions in the mine, SubTerra grew "a new tobacco plant being bred to produce a protein to be used in the treatment of bone marrow cancer."[84]

The postmining uses of White Pine never came close to supporting the employment or payroll levels of earlier days, and the townsite wound down as its supporting industry wound down. It became less a home for industrial workers and their families, who tended to move out, and more a home for retirees of modest income, looking for inexpensive houses, who tended to move in. The continued decline of White Pine as a community—a community that had already lost many residents, its hospital, and much of its trade and stores out at the Mineral River Plaza mall—was perhaps best symbolized by the decline of its schools. Local school enrollments had long been a barometer of the economic health of the mine and community. They rose as the mine grew, climbing from 300 in 1960 to 443 in 1965 and peaking at about 600 in 1972. They fell when mining declined. White Pine tried to hold on to its schools as best it could as it passed through bad years, such as much of the 1980s, troubled by cutbacks, layoffs, and strikes. In that decade, enrollment at times fell from four hundred students to two hundred. Facing school budget deficits, in hard times the White Pine school board trimmed staff and programs. In 1984 the board dropped the football team because it had too few players and was no longer competitive, having lost seventeen games in a row.[85]

The mine closing in 1995 and the decision to forego solution mining in 1997 dealt harsh blows to the school district's children, parents, and faculty and staff. In 1994 the grade and high schools together enrolled 202 students. By 1997, K-12 enrollment had fallen to 164. Enrollment dropped to 103 in 2000 and to just 92 in 2002–3, when the grade school had 60 students and the high school only 32. To trim costs, school leaders offered up the prospect of closing the grade school and incorporating it into a remodeled high school so that only one campus need be supported.[86] But by 2003, facing a $350,000 school deficit, the community faced a hard truth. It could maintain a grade school, but not a high school. In 2002 the high school principal had said, "The school is the heart of the community . . . [and] once the school is gone, the community loses." That big loss came in the spring of 2003, when White Pine faced the harsh reality that it could no longer keep a high school alive with donations, candy bar sales, pancake breakfasts, and athletic banquet fund-raisers. High school students who weren't graduating in 2003 took guided tours of the Ewen-Trout Creek and Ontonagon high schools, twenty-seven and twenty miles away, respectively, so they could pick one to be bused to in the fall.[87]

At the last high school athletic banquet, held early in May 2003, many former athletes who had worn the uniforms of the White Pine Warriors returned to the school. "It was kind of like a funeral," said one returnee, who had driven five hours from Minnesota to attend. Then early in June, the high school held its last commencement, attended by about three hundred people who crowded into the gym. Adding nostalgia and sad irony to the event, it was looked over by Catherine Shamion, the White Pine school superintendent. Shamion's father had taught at White Pine for twenty-six years. She had graduated from White Pine High School in 1974. Hers was the largest commencement ever, with fifty-two graduates. Now she officiated at the last graduation ceremony, which awarded diplomas to only fifteen students.[88] Thereafter, the abandoned high school— soon joined by a closed elementary school—would sit in the middle of a former mining community that had never lived up to the expectations people had held for it.

23

❖ ❖ ❖

LEGACY

From 1920 until the late 1960s, as several native-copper mines closed and a few limped on, much of the Keweenaw went through the painful phenomenon of deindustrialization. Thousands of workers were pushed out of jobs or left voluntarily to find better opportunities elsewhere. Local governments lost tax revenues. Businesses lost their customers, churches their parishioners, and many families their children—because sons and daughters moved away to find a better life in places where the economy was stronger. Perhaps only the natural environment benefited from economic decline. Mines, mills, and smelters discharged less industrial smoke into the air; mills and reclamation plants tailed less waste sand and chemicals into lakes; and cutover forests regenerated themselves. The steep slope running from Portage Lake to the top of Quincy Hill had been barren of trees for about seventy-five years; with the slowdown and then shutdown of mining, milling, and smelting at its base and top, the hillside grew green again.

No economic activity on the Keweenaw stepped in to take the core role once played by the mining industry, but select sectors of the economy did improve while the mines were in decline and after they closed. In the 1960s, as Calumet and Hecla (C&H) and Copper Range's Champion mine played out, right in between them, in Houghton, Michigan Technological University expanded at a time when the postwar baby boomers headed off to college. Michigan Tech was a key legacy of the copper-mining industry, and like an underground stull supporting a piece of the hanging wall, Michigan Tech propped up the local economy. This state-supported school started in 1885 as the Michigan College of Mines. The school provided a practical and technical education, at first geared toward the local copper mines, the regional iron mines of the upper Great Lakes,

In the post–Second World War era, Americans had more automobiles, more money, and longer vacations. With only a few native-copper mines still hanging on, the Copper Country tried to boost its economy by luring more tourists to the region. This sign stood on the outskirts of Houghton in the 1950s. (Michigan Technological University Archives and Copper Country Historical Collections)

and other mining districts at greater distance. But the college had to change and undergo a metamorphosis or two to remain relevant and succeed as the mining industry, education, and the economy changed. In 1927 it became the Michigan College of Mining and Technology. By the early 1960s, with fewer than fifty mining engineering students enrolled, the school dropped the word "Mining" from its name. Armed with a new name, broader curriculum, and an enlarged, modern campus, Michigan Tech expanded from an enrollment of a couple thousand in the early 1960s to one of nearly eight thousand students. Michigan Tech's growth greatly boosted a local economy that had seen its mines shut down.[1]

In its postmining days, the Keweenaw's economy was also revived in part by a local forest-products industry, by some light manufacturing, and by some small high-tech business start-ups incubated in the shadow of Michigan Tech. Tourism became an especially important contributor. In the second half of the twentieth century, with vacation by auto in vogue and with the Mackinac Bridge connecting Upper and Lower Michigan, the Copper Country became more frequented by travelers. Some came to see nature; others to see history.

Modern travelers, like early settlers, have been taken by the Keweenaw's water, woods, and winter. Tourists come this far north because of the grandeur of Superior's shoreline, because of cool summer temperatures, and because of mile after mile of green woods seen along two-lane, winding roads. They came from Lower Michigan, Wisconsin, and elsewhere to camp, hike, mountain bike, ride off-road vehicles, boat, fish, hunt, collect mineral specimens, or just to look. In the last two decades, winter tourism has grown

In the last quarter of the twentieth century, winter tourism became very important to the regional economy. Abandoned rail lines running through the woods up and down the Keweenaw came to life again as snowmobile trails. Some residents within earshot of these trails objected to snarly engine noises, but that sound was music to the ears of those tending cash registers in local motels, restaurants, bars, gas stations, and tourist shops. In 1975 this *Daily Mining Gazette* photo captured the first snowmobiler to ride a newly opened trail. *(Michigan Technological University Archives and Copper Country Historical Collections)*

to rival summer tourism, thanks to the deep blanket of lake-effect snow that covers the land. Some winter tourists come to cross-country or downhill ski, but far more arrive in pickups pulling trailers loaded with snowmobiles. They can ride their snow machines up and down the Keweenaw, along hundreds of miles of trails cutting through woods and only occasionally through a settlement of any size.

A fair number of visitors, whether in summer or winter, think the Keweenaw is a little-touched wilderness. This notion is particularly reinforced if they drive stretches of road covered by a canopy of trees or if they visit the Michigan's Porcupine Mountains Wilderness State Park, 59,000 acres big in Ontonagon County, or the 377-acre Estivant Pines Nature Sanctuary near the tip of the peninsula at Copper Harbor. Many tourists, and local residents, too, focus on the natural environment, on surviving pieces of virgin forest, but don't see it all or understand how the hand of settlement and industry changed most of it. Trees, brush, and heavy snows can hide myriad sins, and so can the blinders that people wear that accentuate the positive and eliminate the negative. Snow-mobilers zoom along what to them is a wilderness trail without realizing that a mining company's railroad first cut that route and used it to deliver thousands upon thousands of tons of copper rock or mineral from mine to mill to smelter. Tourists flock to pristine shorelines but somehow overlook stretches of dead shorelines covered with stamp sands. Gray, desolate stamp-sand beaches have long existed on the margin of Lake Superior at old stamp-mill sites like Gay and Redridge and down near Baraga. They could be found along interior ponds and streams, such as at Calumet Lake, Eagle River at the Cliff mine site, and Boston Pond. They covered many acres of Portage Lake shoreline in the vicinity of Houghton and Hancock, and they were especially prominent alongside Torch Lake, which at one time was home to wall-to-wall stamp mills, reclamation plants, smelters, and industrial dumps.

The industrial legacy left behind at Torch Lake—which in a century's time received about 200 million tons of tailings—was so bad that in 1986 the Environmental Protection Agency (EPA) designated it as a Superfund site in need of remedial action. The eight-hundred-acre Superfund site occupied mostly Torch Lake proper but included some parcels along Portage Lake and elsewhere. The EPA removed industrial drums and contaminants. The biggest part of the cleanup mimicked what was done with the dusty tailings at White Pine. The EPA contracted with local firms to truck in soil, cover the stamp sands, and then seed them. Begun in the 1999 and largely completed in 2005, the project cost $12.3 million.[2] The vegetation, besides supporting wildlife, helped halt the continued deposition of tailings into the lake by wind and erosion. Tailings for decades had kept killing life at the bottom of Torch Lake. Thanks to the cleanup, over many years the bottom will slowly return to life.

If tourists don't come to enjoy nature, they often come to see history. Some, including many genealogists, tromp over ground that grandparents or parents had walked before. Others have no family ties to the area but come to see the old mines and the old towns along a peninsula that doesn't look like the rest of modern America. The Copper Country has intriguing attributes that other places lack. Like the 150-foot-tall No. 2 shaft-rockhouse atop Quincy Hill—the most impressive icon of the mining days—that towers over Portage Lake and can be seen for miles in several directions. Like time-capsule downtowns lined with the architecture of 1890 to 1910. Like the mine locations—some abandoned, some still clutching to modest populations—that pop up every so often along highways—places like Ahmeek, Delaware, Central, Phoenix, and Trimountain. Like the haunting, grown-over Catholic and Protestant cemeteries at the Cliff mine, which was so important as the region's first successful mine yet was finished by 1870. Like the partly submerged suction dredge by Mason on Torch Lake, which used its big iron snout to reach underwater to suck up stamp sands and pump them to a reclamation plant. Like the lone steam stamp that stands vigil at the ruins of the Ahmeek mill in Tamarack City. Like the log houses standing near the Victoria mine in Ontonagon County.

In 1986 the Environmental Protection Agency (EPA) designated about eight hundred acres of stamp sands on Torch and Portage lakes as a Superfund site. When the cleanup began in 1999, it included only a small portion of the Isle Royale mine's tailings, shown here stretching out into Portage Lake east of Houghton. Long before the EPA arrived, local citizens had claimed much of this site. Home builders in search of affordable lakefront property started building houses on the Isle Royale sands in 1971. In addition to single-family dwellings, this "made land" from a stamp mill is now home to an apartment house, several businesses, and Houghton's water treatment plant. *(Michigan Technological University Archives and Copper Country Historical Collections)*

Like the National Civil Engineering Landmark, a dam made of sheet steel, of all things, for impounding water for two stamp mills at Redridge. Like the stone boat sitting on solid ground near Mohawk. This Depression-era, Works Progress Administration (WPA) make-work project captures the region's history so well, with its hull and deck laid up of mine rock and locally quarried Jacobsville sandstone, surmounted by a gun mount that is really a machine rock drill.

The region looks different not only for what it has, but for what it has not. Since 1920, most of America has had more money than the Copper Country. The rest of America busily made itself new while much of the Copper Country lay dormant. People drive north from cities surrounded by suburbs, commercial strip developments, shopping centers, shopping malls, big-box stores, interstate highways, and beltways—to a finger of land whose cities, villages, and mine locations mostly reflect the late nineteenth and early twentieth centuries. By the 1950s, in many parts of America the modern roadside was rapidly taking shape, a roadside of broad concrete thoroughfares, fast-food outlets, and chain motels; of shopping centers and standardized signage and architecture. That roadside vista arrived in a modest, limited way in the Copper Country in the late 1970s and

Icons are important to a culture and instill a sense of place. Here, two icons of American industry coexist, even though they are one hundred years apart: the modern, yet venerable, Ford Mustang, and the old and venerable Quincy No. 2 shaft-rockhouse. In its name, Copper Country Ford, and in its plate, this Houghton dealership lets the world know where its home is. *(Scott See)*

early 1980s. Thanks largely to the economic boost given Houghton by Michigan Tech's presence, that small city saw the a modest (now struggling) shopping mall open its doors in 1978 and boasted the region's first Burger King and McDonald's. The Copper Country's first regional strip development went up on M-26 south of Houghton, between the city (with its old downtown) and the new mall. This strip now very much resembles the rest of the United States in the world below.

Besides the mall, at one time anchored by a JC Penny store and a Kmart, the M-26 corridor south of Houghton now claims an Office Depot, a Wal-Mart Supercenter, and a Shopko. It has two independent supermarkets; a Blockbuster; a Dollar store; a florist/garden center; a furniture store; a bowling alley; numerous food outlets (including Pizza Hut, McDonald's, Taco Bell, Dairy Queen, Kentucky Fried Chicken, Arby's, and Applebee's); a new car dealership plus gas stations, auto-service garages, car washes, and used car dealers; and a Holiday Inn Express and a Country Inn & Suites. Only the well tutored or the observant could find anything with a local or regional look in all of this. Even here, however, are ties to the past, links between this modern landscape and the traditional mining culture.

Inside McDonald's, attractive stained-glass panels depict parts of the old mining landscape. The mall is named the Copper Country Mall. The Ford dealership is Copper Country Ford, and when a new car rolls off its lot, the plate on the front carries the

While the commercial corridor along M-26 south of Houghton looks much like the rest of modern America, it still exhibits some local color. This retail store touts its connection to the heritage of copper mining. A piece of mass copper sits outside, and the store looks like a shaft-rockhouse. The real thing, Quincy No. 2, is seen in the distance atop Quincy Hill. *(Scott See)*

dealer's name and an image of Quincy's No. 2 shaft-rockhouse. Keweenaw Gem & Gift sells copper and mineral specimens plus trinkets and crafty things made of copper. It has a few mining artifacts on its lot, and its building, in profile, is a stylized shaft-rockhouse, complete with a head sheave.[3] The bowling alley and its attendant restaurant, opened forty years after the demise of native-copper mining, are named the Mine Shaft bowling center and the Rockhouse Hardwood Grille. To create flat building sites for Wal-Mart and for Holiday Inn Express, earthmovers carved away parts of sand hills, leaving steep banks. To keep the banks from eroding, they hauled in mine spoil from poor-rock piles adjacent to abandoned mines. They landscaped the hills with a cascade of poor rock that some miner had blasted out maybe a century earlier. These days, contractors and road commissioners rob poor rock from the piles where the mines had left it, and they spread it across the Keweenaw, putting it in places it never used to be.

The modern landscape along M-26 south of Houghton is replicated in few places along the Keweenaw, but never on such a large scale. Convenience store/gas stations, true, have been sprinkled rather liberally along the outskirts of many settlements, but other parts of the modern picture are usually lacking. To this day, the city of Hancock has no chain-operated fast-food restaurant save for one sub shop, no large national merchandiser, and relatively few buildings downtown that date after 1920. Only since the mid-1990s has a modern commercial development called Mine Street Station stood in Calumet township amid abandoned mining ground. Mine Street Station is an anachronistic island of modernity with a Burger King, Pamida merchandiser, AmericInn Hotel, auto repair shop, and a Honda dealer that sells various products, but not cars. By way of contrast, streetscapes along the villages of Ontonagon, Calumet, Laurium, and Lake

Linden evidence virtually none of this. With two- or three-story brick or stone build-ings, with elaborate cornices, with some cast-iron fronts and large display windows, they look much like they did a century or so ago. In downtown Ontonagon, you can't find a Whopper or a Big Mac, but you can go to Syl's Restaurant and, like miners of old, have a Cornish pasty for lunch.

Since the closing of the mines, there has been tension on the Keweenaw between believers in the old and believers in the new, between advocates of historic preservation and advocates of growth and development. The population has not universally treas-ured its old mine, mill, and smelter sites; its old downtowns; its old neighborhoods of company-built housing; or even its old schools, churches, and cemeteries. Some of this is very understandable. It takes a certain cast of mind to see cultural value in an aban-doned industrial site with collapsed roofs and dilapidated buildings, or in a pile of waste rock or smelter slag. Before the onset of the environmental era, the mining companies did not help matters. They scrapped out things of value while leaving behind pollutants, uncapped shafts, wrecked shells of buildings, massive concrete foundations, and those vast sterile beaches of stamp sands upon which nothing grew.

Many Keweenaw residents have seen too much old and not enough new. Old mines, old houses, old stores, old overgrown cemeteries that nobody could afford to care for or even wanted to. In Hancock the original Protestant and Catholic cemeteries, sited by the Quincy mine around the time of the Civil War, disappeared to create building sites for a new Lutheran church and new Catholic church. For many the old is symbolic of decline, symbolic of a broken promise. The mining industry was here for people for 125 or even 150 years, but it's not here now to help take care of anybody, so it is time to move on: time to become less wedded to earlier centuries and a now-dead industry; time to welcome new construction of almost any kind, be it houses, schools and churches, small-scale manufacturers in industrial parks, chain motels and restaurants, a giant Wal-Mart, or minimalls that tend to siphon small businesses from old downtowns. If you can't have something totally new, then modify the old—such as a company house—to fit your modern needs. Paint your company house a color it never carried during the mining days. Add on or enclose a front porch, add on a garage or new kitchen, and change all the windows and doors.

While home to proponents of modernization and change, the Keweenaw has always had its champions of history and historic preservation. This has been rooted, perhaps, in the strong belief that the Keweenaw was a special place and that the settlers who came here were special people. The early pioneers surely sensed this. Early on, they started collecting historical documents and photographs. Many wrote published reminiscences, with titles such as "A Narrative of One Year in the Wilderness," "Some Incidents of Pioneer Life in the Upper Peninsula of Michigan," "Some Early Mining Days at Portage Lake," and "Recollections of Civil War Conditions in the Copper Country." In 1874, residents held an inaugural Old Settlers Ball in Eagle River. On 21 January 1886, about three hundred people came to Calumet to help two of the earliest settlers, Daniel and Lucena Brockway, celebrate their fiftieth wedding anniversary. Guests signed a registry and recorded just when they had arrived at this place. These people were definitely conscious of their history. At the party were 147 guests who had migrated from the world below to the Copper Country. In addition to the Brockways, two others had come in the 1840s, fourteen in the 1850s, forty-two in the 1860s, and eighty-nine in the 1870s or later.[4]

While much history was saved—by the mining companies themselves, local governments, historical societies, private collectors, and others, much was lost. The Copper Country since the 1840s had always been in the business of making and losing history at the same time, particularly when it came to the built environment. At abandoned locations, nature reclaimed sites by regenerating forests and by collapsing dilapidated buildings under heavy snow loads. As a consequence, many nineteenth-century mines— including important and well-populated ones, such as the Cliff mine—are now just archaeological sites with no standing structures. Meanwhile, mines that ran for many decades often consumed their past while taking care of the present. They buried original sites under poor rock and replaced antiquated buildings and technologies with new. And when the end finally came, companies generally did not just walk away from a site, leaving it as a kind of time capsule. They sold off machinery and equipment or salvaged it for scrap. Sometimes they used a wrecking ball to bash down much of a building so a crew could more easily yank out its air compressor or hoist.

Because of machinery sales, World War II era salvage drives, occasional fires, environmental deterioration, and sometimes the adaptive reuse of old structures, today no native-copper mine site on the Keweenaw is as it was. In 1915, when the mines were at their peak, no fewer than seventy-five shaft-rockhouses stood tall along the mineral range; now only two still stand from that era: at Champion's No. 4 shaft and Quincy's No. 2.[5] Tramroads and railroads are gone, as are boilerhouses, dry houses, and even all the outhouses once scattered around. A wooden, five-hole privy still stood next to a Quincy dry house in 1980, but a few years later, heavy snow collapsed it to the ground. Machine and blacksmith shops, railroad engine houses, a few hoisting engines and engine houses, and several tall smokestacks still survive, but no site looks just like it did when still in production. Company houses have been more prone to survive than mine structures, and substantial blocks of historic mine houses can be found in places like White Pine, Calumet, Ahmeek, and especially Painesdale, the remarkably intact turn-of-the-century location at the Champion mine.[6] But many clusters of houses have disappeared, too, such as much of the Lower Pewabic neighborhood at the Quincy mine, where all that's left are cellar holes and some fruit trees.

In terms of preservation, stamp mills have fared far worse than mines. Company houses survive at mill towns, ranging from Mason, Tamarack City, and Lake Linden on Torch Lake; to Freda and Redridge on one side of the Keweenaw; and to Gay on the other. On many of the Keweenaw's shorelines, material evidence of milling and reclamation abounds in the form of stamp-sand beaches, occasional dams and penstocks, concrete piers and machinery pads, and the ruins of walls. But no major mill building stands in good repair, no collection of washing and separating machinery exists, and no Cornish drop stamps survive. One lone steam stamp at the Ahmeek mill and a row of ball mills in Ontonagon County represent the most significant milling artifacts to be found aboveground and in situ on the Keweenaw.

Copper smelters on the Keweenaw were not nearly as numerous as mines or mills. Six of them were clustered at Portage Lake and Torch Lake at the center of Houghton County, near the biggest mines. As smelters closed, most fell under the wrecking ball or had their furnaces and casting machinery removed so that buildings could be used for something else. One smelter did remain surprisingly intact: Quincy's, built in 1898. Perhaps the Quincy smelter, of all the historic, mining-related industrial complexes on the Keweenaw, was left most like a time capsule, as though men had just left work one day, leaving their stuff behind. It was hardly in pristine or original condition when it finally shut around 1970, but it retained reverberatory and crucible furnaces, ladles, a casting wheel, two reciprocating steam engines, an electrical plant, considerable machinery and tools, most of its buildings, and a slag pile of black, glassy smelter waste. But even a time capsule diminishes over time. Decades of little maintenance, followed by decades of virtually no maintenance, accompanied by considerable vandalism and theft, have severely compromised much of the smelter's historic fabric, leaving it today in an advanced state of deterioration.[7]

Over the years, many local historical groups and museums—small and always underfunded—have stepped in to protect and preserve the threatened, and sometimes vanishing, history and landscape of the mining era. In Calumet, Coppertown USA opened a mining museum in C&H's old pattern shop, which had been a part of the company's foundry operations. Nearby, the Keweenaw Heritage Center at Calumet strives to rehabilitate the hundred-year-old French Catholic church, St. Anne's, while using it to stage exhibits. On the other end of town, friends of the Calumet Theatre, erected in 1900, oversee the structure's physical preservation and continue to stage myriad live performances.

On Quincy Hill, the Quincy Mine Hoist Association, a nonprofit corporation, dedicates itself to the preservation of the huge Nordberg steam hoist and the shaft-rockhouse at the No. 2 shaft. In Hancock, Findlandia University (which started life in 1896 as Suomi College) operates the Finnish American Heritage Center, which includes an archives and stages historical programs. The Houghton County Historical Society—located at Lake Linden in C&H's old mill office—runs a museum and collects manuscripts, photographs, artifacts, railroad equipment, and even some buildings. Although it operates in Michigan's smallest county, one with a population of only 2,300, the Keweenaw County Historical Society has over 950 members and caretakes the lighthouse and Rathbone

School in Eagle Harbor, the church and blacksmith shop in Phoenix, and much of the abandoned Central mine. The Ontonagon Historical Society runs its museum and has acquired its local lighthouse. Painesdale Mine & Shaft adopted the Champion mine's No. 4 shafthouse as its preservation project, while Old Victoria preserves a cluster of log houses at the Victoria mine. The Copper Range Historical Society runs a small museum dedicated to peoples and companies associated with the village of South Range and the Baltic, Trimountain, and Champion mines. Small businesses run underground tours of the Adventure and Delaware mines.

The state of Michigan and the federal government became involved, too. On the northern end of the mineral range, Michigan created Fort Wilkins State Park to preserve and interpret the army post built in the 1840s. On the southern end of the range, Michigan's Porcupine Mountains Wilderness State Park includes, within its boundaries, the archaeological remains of small mining operations. The state-supported university, Michigan Tech, does its part by operating the A. E. Seaman Mineral Museum, which houses copper and mineral specimens plus select photographs and artifacts. As part of its library, the university created the Michigan Technological University Archives and Copper Country Historical Collections. Those collections became extremely important as they grew to include records from individuals, local governments, and especially the surviving corporate records of the three most important companies to mine on the Keweenaw: C&H, Copper Range, and Quincy. In a manner somewhat akin to the state's Porcupine Mountains Wilderness State Park, the federal government operates the Isle Royale National Park on the island of that name in Lake Superior to the west of the Keweenaw Peninsula. The Park Service manages Isle Royale as a wilderness area, a land of moose and wolves, but it shelters archaeological features of copper-mining activity. Isle Royale and the Keweenaw share common geological features and share a history—from the 1840s through the rest of the nineteenth century—of mining. The big difference was that the industry that grew on the mainland withered on the island.

In the 1980s, some disappointing preservation losses triggered a greater interest in local history and preservation. After decades of hard times, neglect, and even abuse, two of the most significant structures in the history of C&H and the village and township of Calumet were demolished because of hazardous structural conditions, a lack of funds for stabilization or renovation, and perhaps a lack of appreciation for what their loss would really mean until they were gone. In 1984, demolition claimed Italian Hall, scene of the 1913 Christmas Eve tragedy that killed seventy-three people in the midst of the long, divisive labor strike. The Superior engine house also became rubble. A century before, showcasing a mammoth steam engine, it had stood as the premier symbol of C&H's technological supremacy and power.

Not long after these demolitions, the idea was born, especially in Calumet, that more needed to be done to hold on to what remained. Maybe the key to the struggling community's future should be its past. The village and township of Calumet had a history that could be told and sold to revitalize the local economy, grow civic pride, boost tourism, and stimulate appropriate new development amid the old, original fabric. The proponents of economic development via the paths of history, heritage, and preservation had a recent model to follow: the Lowell National Historical Park in Massachusetts. Lowell had parlayed its history of textile manufacturing into a national park designation. In Calumet, the idea gained currency that it, too, because of its history, might become a national park. The National Park Service (NPS) would step in to help take care of a place that had been abandoned by its once-great, paternalistic mining company.

In response to a congressional request, the NPS began studying this possibility in the late 1980s. The NPS studied the area to determine whether it qualified, in terms of its historical significance and its physical remains, for National Historic Landmark (NHL) status. That hurdle needed to be cleared first. Kathleen Lidfors surveyed the area for the NPS and subsequently prepared historic district studies and an NHL nomination that the NPS approved. Besides looking at Calumet, she looked at the Quincy mine site, too, because both were extremely important to the region's history. She saw they were very dissimilar and yet complementary historic landscapes. Calumet had more town left standing, while Quincy, with its No. 2 shaft-rockhouse and hoisthouse, had more mine left behind. Consequently her 1988 NHL nomination included two historic districts: one at Calumet, and a second, ten miles south, at Quincy.

Over the next several years, local boosters on the Keweenaw strove to achieve broad support for a national historical park, while the NPS conducted further investigations and in 1991 published *Study of Alternatives, Proposed Keweenaw National Historical Park.* The extended bureaucratic and political process of creating a new park culminated on 27 October 1992, when the U.S. Congress passed Public Law 102-543, "An Act to Establish the Keweenaw National Historical Park."[8] The Calumet unit of the park encloses about 750 acres and includes the village of Calumet and much historic fabric, industrial and residential, within Calumet Township. The 1,120-acre Quincy unit includes the mine site and housing atop Quincy Hill and the Quincy smelter on Portage Lake.

The act that created the Keweenaw National Historical Park gave reasons for doing so; it spelled out the area's national significance. Its geology is unique, containing the oldest and largest lava flow known on earth, as well as the world's largest deposits of native copper. The Keweenaw Peninsula provided copper to prehistoric, aboriginal peoples who traded it widely across what became the eastern part of the United States. The early mining corporations on Lake Superior pioneered in deep-shaft, hard-rock mining techniques, as well as copper-milling and copper-smelting technologies. Those same mining companies built communities, as well as mines, and practiced a paternalism unprecedented in American industry. As they became a magnet attracting European immigrants to Lake Superior during the nineteenth century, society became an "ethnic conglomerate," whose foods and traditions are preserved on the Keweenaw today. Finally, Michigan Technological University, established in 1885 to supply technologies and engineers to the mines, today possesses, in its archives, a wealth of historical documentation of the copper-mining era.[9]

At about the time White Pine, the last of the copper mines, was closing down in Ontonagon County, the new Keweenaw National Park, dedicated to preserving the heritage of copper mining, opened near the heart of the old native-copper industry in Houghton County. That was an apt coincidence for the region, which had always had its ups and downs, its boom and bust times. Mines opening and mines closing. One mine location being emptied of men, women, and children because all jobs had been lost— while along another stretch of the mineral range a new island of industry was just opened up, surrounded by a sea trees. Those trees would fall to be used as fuel, as construction material, and as timber supports underground. Then the mine itself would eventually fall, and the trees would return.

ABBREVIATIONS

❖ ❖ ❖

AA	Alexander Agassiz
AJC	A. J. Corey
A.R.	*Annual Report*
AIME Trans.	*Transactions of the American Institute of Mining Engineers*
ASME Trans.	American Society of Mechanical Engineers *Transactions*
C&H	Calumet and Hecla Mining Co.
CL	Charles Lawton
CR	Copper Range Mining Company
DEQ	Department of Environmental Quality
DMG	*Daily Mining Gazette*
E&MJ	*Engineering and Mining Journal*
EPA	U.S. Environmental Protection Agency
HAER	Historic American Engineering Record
JLH	John L. Harris
JM	James MacNaughton
LL&E	Louisiana Land and Exploration Company
LSM	*Lake Superior Miner*
LSMI Proc.	Proceedings of the Lake Superior Mining Institute
MCJ	*Mining Congress Journal*
MH	*Michigan History*

Min. Stats.	*Annual Reports of the Commissioner of Mineral Statistics*
MPC	Michigan Pioneer Collections
MTU	Michigan Technological University Archives and Copper Country Historical Collections
NMJ	*Northwestern Mining Journal*
PLMG	*Portage Lake Mining Gazette*
PMC	Pewabic Mining Company
QAS	Quincy S. Shaw Jr.
QMC	Quincy Mining Company
SBH	Samuel B. Harris
SBW	S. B. Whiting
TFM	Thomas Fales Mason
WAP	William A. Paine
WFM	Western Federation of Miners
WPT	William Parsons Todd
WRT	William Rogers Todd

NOTES

❖ ❖ ❖

CHAPTER 1. KEWEENAW COPPER: GEOLOGY, DISCOVERY, DREAMS OF WEALTH

1. Some magma eruptions were spectacularly large. The Greenstone Lava Flow ranges across fifty miles and in places is more than one thousand feet thick. Collectively, these successive lava flows are known as the Portage Lake Volcanics. For more geological information on the Keweenaw, see David J. Krause, *The Making of a Mining District: Keweenaw Native Copper, 1500–1870* (Detroit, 1992), 44–48; Susan R. Martin, *Wonderful Power: The Story of Ancient Copper Working in the Lake Superior Basin* (Detroit, 1999), 26–30; B. S. Butler and W. S. Burbank, *The Copper Deposits of Michigan,* U.S. Geological Survey Professional Paper 144 (Washington, DC, 1929); John A. Dorr Jr. and Donald F. Eschman, *Geology of Michigan* (Ann Arbor, MI, 1977), 70–77; T. M. Broderick and C. D. Hohl, "Geology and Exploration in the Michigan Copper District," *MCJ* 17 (Oct. 1931): 478–81, 486; and Theodore J. Bornhorst and Larry D. Lankton, "Copper Mining: A Billion Years of Geologic and Human History," chap. 13 in *Michigan Geography and Geology,* ed. Randall Schaetzl et al. (New York, 2009), 150–73.

2. Krause, *Making of a Mining District,* 45–46.

3. The rich Calumet Conglomerate lode was a fine example of this. C&H owned the only outcropping part of the lode that was profitable. Just north of C&H, the Schoolcraft and Centennial mines turned no profit on the same lode; and the same was true to the south, where the Osceola mine never profited from its works on the Calumet Conglomerate. See entries for Schoolcraft, Centennial mine, and Osceola mine in Michigan Commissioner of Mineral Statistics, *Min. Stats. for 1881* (Lansing, MI, 1882), 133; and *Min. Stats. for 1899,* 136. Also see Larry Lankton, *Cradle to Grave: Life, Work, and Death at the Lake Superior Copper Mines* (New York, 1991), 17.

4. Martin, *Wonderful Power,* 33; Theodore J. Bornhorst and William I. Rose, *Self-Guided Geological Field Trip to the Keweenaw Peninsula, Michigan,* Proceedings of the Institute on Lake Superior Geology, vol. 40, part 2 (Houghton, MI, 1994), 26.

5. See "Outline Map Showing the Position of the Ancient Mine Pits of Point Keweenaw, Michigan," found in Charles C. Whittlesey, "Ancient Mining on the Shores of Lake Superior," *Smithsonian Contri-*

butions to Knowledge 13, no. 4 (1863): 1–29. A redrawn version of this map can also be found in Larry D. Lankton and Charles K. Hyde, *Old Reliable: An Illustrated History of the Quincy Mining Company* (Hancock, MI, 1982), 3.

6. Martin, *Wonderful Power,* 143–44.

7. C. Harry Benedict, *Red Metal: The Calumet and Hecla Story* (Ann Arbor, MI, 1952), 22–23.

8. Martin, *Wonderful Power,* 183–214. Also see John R. Halsey, "Miskwabik–Red Metal: Lake Superior Copper and the Indians of Eastern North America," *MH,* Sept.–Oct. 1983, 32–41.

9. Krause, *Making of a Mining District,* 24.

10. Ibid., 38–43. Also see Benedict, *Red Metal,* 12–17.

11. Krause, *Making of a Mining District,* 70–77; and Robert James Hybels, "The Lake Superior Copper Fever, 1841–47," *MH* 34 (June 1950): 100–105.

12. Sydney W. Jackman and John F. Freeman, eds., *American Voyageur: The Journal of David Bates Douglass* (Marquette, MI, 1969), 58.

13. Henry Rowe Schoolcraft, *Narrative Journal of Travels through the Northwestern Regions of the United States* (Albany, NY, 1821), 175–77.

14. Krause, *Making of a Mining District,* 103–11.

15. Ibid., 105–7.

16. Ibid., 111.

17. Willis F. Dunbar and George S. May, *Michigan: A History of the Wolverine State* (Grand Rapids, MI, 1995), 211–20.

18. George N. Fuller, ed., *Geological Reports of Douglass Houghton: First State Geologist of Michigan, 1837–1845* (Lansing, MI, 1928), 557. In Houghton's "Annual Report of the State Geologist, 1841" (528–59), he gives his conservative views on the commercial value of the copper deposits.

19. Krause, *Making of a Mining District,* 93–94, 134–35.

20. Donald Chaput, *The Cliff: America's First Great Copper Mine* (Kalamazoo, MI, 1971), 15, 81; Krause, *Making of a Mining District,* 137; and Lankton, *Cradle to Grave,* 8.

21. Krause, *Making of a Mining District,* 92–94.

22. Ibid., 136–38.

23. Ibid., 144, 146–48.

24. For a history and description of the fort, see Thomas Friggens, "Fort Wilkins: Army Life on the Frontier," *MH* 61 (Fall 1977): 221–50; and Preservation Urban Design Incorporated, *Architectural Analysis: Fort Wilkins Historic Complex, Fort Wilkins State Park, Copper Harbor, Michigan* (Ann Arbor, MI, 1976).

25. Hybels, "Lake Superior Copper Fever," 36–37; John H. Forster, "Life in the Copper Mines of Lake Superior," *Michigan Pioneer and Historical Collections* 11 (1887): 177; Charles Lanman, *A Summer in the Wilderness Embracing a Canoe Voyage up the Mississippi and around Lake Superior* (New York, 1847), 141–42; and John R. St. John, *A True Description of the Lake Superior Country* (New York, 1846), 13, 15.

26. William B. Gates, *Michigan Copper and Boston Dollars: An Economic History of the Michigan Copper Mining Industry* (Cambridge, MA, 1951), 228. The appendix in Gates's book contains numerous very useful tables regarding population, employment, production, dividend payments, and other statistics covering the copper district through 1950.

27. Charles K. Hyde, *Copper for America: The United States Copper Industry from Colonial Times to the 1990s* (Tucson, AZ, 1998), 7–16, 21–22, 25–26.

28. Ibid., 3–28. Charles K. Hyde outlines the history of American copper mines that predated Michigan.

29. The various costs of opening and developing a mine are often detailed in mining-company annual reports, especially those submitted before about 1875. Early annual reports may provide labor costs plus expenditures made to erect mine buildings, install technologies, and build company houses.

30. For the early history of investment in the region, see Charles K. Hyde, "From 'Subterranean Lotteries' to Orderly Investment: Michigan Copper and Eastern Dollars, 1841–1865," *Mid-America: An Historical Review* 66 (Jan. 1984): 3–20.

31. Gates, *Michigan Copper,* 33–35; and Lankton, *Cradle to Grave,* 20–21.

32. Chaput, *The Cliff*, 36.

33. Hyde, "Subterranean Lotteries," 9. For production and dividend records of early companies, also see Butler and Burbank, *Copper Deposits*, 64–98; and Gates, *Michigan Copper*, 215.

34. See the graph of copper production from various lodes in Butler and Burbank, *Copper Deposits*, 65.

35. Gates, *Michigan Copper*, 14; and Larry Lankton, *Keweenaw Copper: Mines, Mills, Smelters, and Communities*, Guidebook for the 26th Annual Conference of the Society for Industrial Archeology (Houghton, MI, 1997), 29–32. The early histories or descriptions of many of these individual mines are found in *Min. Stats. for 1881*.

36. Lankton and Hyde, *Old Reliable*, 18, 57, 152–53.

37. See table 6 in Gates, *Michigan Copper*, 197–200. For comparative production figures for Michigan, United States, and the world, see Hyde, *Copper for America*, 65–68.

38. Gates, *Michigan Copper*, 197, 203; and Lankton and Hyde, *Old Reliable*, 152.

39. Lankton, *Beyond the Boundaries*, 195; and *Annual Reports of the Adjutant General of the State of Michigan for the Years 1862, 1863, 1864, 1965–1866*, 3 vols. (Lansing, MI). The author is indebted to Dick Rupley for sharing his expertise on Keweenaw soldiers in the Civil War.

40. Gates, *Michigan Copper*, 207.

41. Hyde, *Copper for America*, 42; and Benedict, *Red Metal*, 6–10.

CHAPTER 2. GETTING THE COPPER OUT: EXPLORATION, DEVELOPMENT, AND THE TOOLS OF PRODUCTION

1. Hyde, *Copper for America*, 33–35; and Hybels, "Lake Superior Copper Fever," 234–43, 317–18.

2. The best history of the quest to understand the Keweenaw's geology is Krause, *Making of a Mining District*.

3. The quotations, as well as the number of companies discovering ancient works and tools, are from John R. Halsey, "'Ancient Diggings': A Review of Nineteenth-Century Observations in the Prehistoric Copper Mining Pits of the Lake Superior Basin," paper presented at the 54th Annual Meeting, Midwest Archaeological Conference, 17 Oct. 2008, Milwaukee, WI, 34–47. Also see John R. Halsey, "Ancient Pits of the Copper Country," *MH*, May–June 2009, 40–47.

4. Early mining-company annual reports make it clear when a company has moved from pure exploration into development, and they show the requisite increase in employment and investment in physical plant. Also, data on production, technologies, and employment at early mines in their prospecting or development stages are found in J. W. Foster and J. D. Whitney, *Report on the Geology and Topography of a Portion of the Lake Superior Land District in the State of Michigan, Part I: Copper Lands*, U.S. House, 31st Cong., 1st Sess., 1850: Ex. Doc. 69.

5. Good examples of companies that tried and tried again would be the North-West/Pennsylvania/Delaware mining property on the northern end of the mineral range, and the Norwich mine on the southern end (see Lankton, *Cradle to Grave*, 15–17).

6. See details of the stock assessment system in Hyde, *Copper for America*, 38; and Gates, *Michigan Copper*, 32–33.

7. Nathan Rosenberg, *Technology and American Economic Growth* (New York, 1972), 59–116.

8. See Louis C. Hunter, *A History of Industrial Power in the United States, 1780–1930*, vol. 2: *Steam Power* (Charlottesville, NC, 1985); and Louis C. Hunter and Lynwood Bryant, *A History of Industrial Power in the United States, 1780–1930*, vol. 3: *The Transmission of Power* (Cambridge, MA, 1991), which has a section on "The Use of Power in Nineteenth-Century Mining" (375–512).

9. Drifts, shafts, stopes, and adits are shown in numerous longitudinal sections of mine works, often included in early mining-company annual reports. For an example of a longitudinal section, see Lankton and Hyde, *Old Reliable*, 30. Cornish mines used the fathom (six feet) as a linear measurement, and the ten-fathom spacing of drifts in Michigan is an indicator of the importance of Cornish influence at the Lake mines.

10. For hand-drilling techniques, see George G. Andre, *Rock Blasting: A Practical Treatise on the Means Employed in Blasting Rocks* (London, 1878), 128–34. A good description of early technologies at the

copper mines is found in Thomas Egleston, "Copper Mining on Lake Superior," *AIME Trans.* 6 (1879): 275–312.

11. Larry D. Lankton, "The Machine *under* the Garden: Rock Drills Arrive at the Lake Superior Copper Mines, 1868–1883," *Technology and Culture* 24, no. 1 (Jan. 1983): 6–9; and Lankton, *Cradle to Grave,* 30–31. The sizes of individual teams performing work underground is well documented in QMC, *Contract Books,* held at MTU.

12. Henry S. Drinker, *Tunneling, Explosive Compounds and Rock Drills* (New York, 1878), 56, 59.

13. Andre, *Rock Blasting,* 124; Claude T. Rice, "Copper Mining at Lake Superior," *E&MJ* 94 (17 Aug. 1912): 308; and Robert E. Clarke, "Notes from the Copper Region," *Harper's New Monthly Magazine* 4 (March 1853): 448.

14. Egleston, "Copper Mining on Lake Superior," 285–87; Clarke, "Notes from the Copper Region," 4 (April 1853): 577; and William P. Blake, "The Mass Copper of the Lake Superior Mines, and the Method of Mining It," *AIME Trans.* 4 (1875–76): 110–11.

15. Kraus, *Making of a Mining District,* 212, 214, 216, 222.

16. Rice, "Copper Mining at Lake Superior," *E&MJ* 94 (3 Aug. 1912): 217.

17. Lankton and Hyde, *Old Reliable,* 25. The QMC *Invoice Book, 1860–63,* at MTU has an invoice dated 29 May 1860 for the purchase of forty-eight wheelbarrows.

18. Lankton and Hyde, *Old Reliable,* 25, 109. The *PLMG* of 3 Dec. 1864 discusses the liabilities of kibbles and the introduction of rock skips.

19. The first steam engine on the Keweenaw arrived at the Lake Superior mine near Eagle River in 1845 (see *LSM,* 24 Jan. 1857).

20. M. C. Ihlseng and Eugene B. Wilson, *A Manual of Mining* (New York, 1911), 127–29; Lankton and Hyde, *Old Reliable,* 25–26; and Chaput, *The Cliff,* 25, 35.

21. The *LSM* of 25 Jan. 1857 lists all steam engines then used in the district. The same information is also printed in the *Detroit Daily Free Press,* 21 Feb. 1857, and in *Mining Magazine* 8 (1857): 289.

22. To see an example of such a hoisting layout, see the photograph and map of the Quincy mine location in Lankton and Hyde, *Old Reliable,* 28–29.

23. *PLMG,* 13 Dec. 1866.

24. J. W. Rawlings, "Recollection of a Long Life," in *Copper Country Tales,* ed. Roy Drier (Calumet, MI, 1967), 1:115. Also see QMC, *A.R. for 1866,* 15; and PMC, *A.R. for 1868,* 29.

25. *PLMG,* 22 July 1869.

26. Egleston, "Copper Mining on Lake Superior," 290.

27. Lankton, *Cradle to Grave,* 37–39; and G. E. McElroy, "Natural Ventilation of Michigan Copper Mines," U.S. Bureau of Mines Technical Paper 516 (Washington, DC, 1932), 1.

28. Arthur W. Thurner, *Calumet Copper and People: A History of a Michigan Mining Community* (by the author, 1974), 40.

CHAPTER 3. ISLANDS OF INDUSTRY IN A SEA OF TREES

1. Schoolcraft, *Narrative Journal,* 178.

2. Larry Lankton, *Beyond the Boundaries: Life and Landscape at the Lake Superior Copper Mines, 1840–1875* (New York, 1997), 12, 16, 43–44.

3. See John Rowe, *The Hard Rock Men: Cornish Immigrants and the North American Mining Frontier* (New York, 1974), 62–95; and A. L. Rowse, *The Cousin Jacks: The Cornish in America* (New York, 1969), 161–95.

4. LeRoy Barnett, "What the Sam Hill!" in *MH,* May–June 2004, 14–18; Larry Lankton, "One Family's Journey to 'Earthly Paradise,'" in *MH,* Nov.–Dec. 2001, 26–30; Lankton and Hyde, *Old Reliable,* 5–6, 19; Western Historical Society, *History of the Upper Peninsula of Michigan Containing a Full Account of Its Early Settlement: Its Growth, Development and Resources* (Chicago, 1883), 285–86; and Alvah H. Sawyer, *A History of the Northern Peninsula and Its People* (Chicago, 1911), 2:1012–14. For a broad treatment on the migration of New Englanders to lower Michigan, see Susan E. Gray, *The Yankee West: Community Life on the Michigan Frontier* (Chapel Hill, NC, 1996).

5. For instance, Ransom Shelden, who established Houghton, married a cousin of geologist Douglass Houghton, who was also a sister of C. C. Douglass (see Lankton and Hyde, *Old Reliable,* 5–6). Also, Daniel D. Brockway received his job on Lake Superior courtesy of his older brother's influence: William Brockway was the head of Methodist missions on the lake. And when Daniel migrated to Lake Superior, he brought his younger brother along with him (see Lankton, "One Family's Journey," 28–29).

6. Lankton, *Beyond the Boundaries,* 23–36.

7. Ibid., 9, 26, 33, 50–51.

8. For lighthouses on the Keweenaw, see Charles K. Hyde, *The Northern Lights: Lighthouses of the Upper Great Lakes* (Lansing, MI, 1986), 17, 158–84.

9. Philip P. Mason, ed., *Copper Country Journal: The Diary of Schoolmaster Henry Hobart, 1863–1864* (Detroit, 1991), 65, 30, 97, 99, 106–7.

10. For the story of James Paul and Ontonagon, see James K. Jamison, *This Ontonagon Country: The Story of An American Frontier* (Calumet, MI, 1965), 1–27.

11. Gates, *Michigan Copper,* 43; Chaput, *The Cliff,* 46–56; and Lankton, *Cradle to Grave,* 10.

12. Arthur W. Thurner, *Strangers and Sojourners: A History of Michigan's Keweenaw Peninsula* (Detroit, 1994), 76.

13. For the Clay quote and legislative history of the canal, see John N. Dickinson, *To Build a Canal: Sault Ste. Marie, 1853–54 and After* (Miami, OH, 1981), xiii, 28–29, 48.

14. The mining companies and merchants involved are found in Portage Lake and River Improvement Company, *Articles of Association and By-laws* (Detroit, 1863). Also see material on navigation in "Letter from the Secretary of War, in Response to the Deposit of Silt and Sand in Portage Lake, Michigan," in *Executive Documents of the House of Representatives for the Second Session of the Forty-Seventh Congress, 1881. No. 85* (Washington, DC, 1884).

15. Graham Pope, "Some Early Mining Days at Portage Lake," *LSMI Proc.* 7 (1901): 26–28; *PLMG,* 2 April 1864; and *LSM,* 14 Nov. 1857.

16. See Portage Lake and Lake Superior Ship Canal Company, *Articles of Incorporation* (Detroit, 1864). The company was originally intended to receive just 200,000 acres of public land, but that later was deemed insufficient compensation for the work. After the canal was built, the company collected tolls on ships using it.

17. Western Historical Society, *History of the Upper Peninsula,* 253; and Thurner, *Strangers and Sojourners,* 81–82.

18. Gates, *Michigan Copper,* 61.

19. *Lake Superior Journal,* 9 July 1853.

20. Gates, *Michigan Copper,* 20–21; Orrin W. Robinson, "Recollections of Civil War Conditions in the Copper Country," *MH* 3 (Oct. 1919): 606–7; and *PLMG,* 24 Sept. 1864.

21. Gates, *Michigan Copper,* 20; Toltec Mining Company, *A.R. for 1852,* and *A.R. for 1855,* 12–14.

22. Lankton, *Beyond the Boundaries,* 40–41.

23. John H. Forster, "Some Incidents of Pioneer Life in the Upper Peninsula of Michigan," *Michigan Pioneer and Historical Collections* 17 (1892): 336; Alfred P. Swineford, *History and Review of the Material Resources of the South Shore of Lake Superior* (Marquette, MI, 1877), 24; and George Cannon, *A Narrative of One Year in the Wilderness* (Ann Arbor, MI, 1982), 54, 58–59, 63, 67.

24. Pope, "Some Early Mining Days," 21; PMC, *Report* (March 1858), 7; Minesota Mining Company, *A.R. for 1858,* 13; and Phoenix Mining Company, *Report* (1865), n.p.

25. Lankton, *Beyond the Boundaries,* 55.

26. Ibid., 41, 113; and Lankton, *Cradle to Grave,* 48.

27. Clarke, "Notes from the Copper Region," 578; and Robinson, "Recollections of Civil War Conditions," 598–99.

28. Lankton, *Cradle to Grave,* 55–56. The most thorough study of stamp-milling technology is C. Harry Benedict, *Lake Superior Milling Practice: A Technical History of a Century of Copper Milling* (Houghton, MI, 1955).

29. John F. Blandy, "Stamp Mills of Lake Superior," *AIME Trans.* 2 (1874): 208. Numerous companies' early annual reports document their initial use of Cornish stamps. Also see Frederick G. Coggin, "Notes

on the Steam Stamp," *E&MJ,* 20 March 1886, 210–11; 27 March 1886, 232–33; and 3 April 1886, 248–50. In *Lake Superior Milling Practice,* Benedict devotes a chapter to "The Steam Stamp," 55–59.

30. Chaput, *The Cliff,* 40–41; and Benedict, *Lake Superior Milling Practice,* 27.

31. See Thomas Egleston, "Copper Dressing on Lake Superior," *Metallurgical Review* 2 (May 1878): 227–36; (June 1878): 285–300; and (July 1878): 389–409.

32. Hyde, *Copper for America,* 22–25.

33. Ibid., 25.

34. Thomas Egleston, "Copper Refining in the United States," *AIME Trans.* 9 (1880–81): 678–730; and James B. Cooper, "Historical Sketch of Smelting and Refining Lake Copper," *LSMI Proc.* 7 (1901): 44–49.

35. See the Pewabic and Franklin mining companies' annual reports for the early 1860s, which discuss their short-rail line and the operations of the tramroads down to their mills.

CHAPTER 4. OUT AT THE LOCATIONS: FROM CAMPS TO COMMUNITIES

1. John H. Forster, "Early Settlement of the Copper Regions of Lake Superior," *Michigan Pioneer Collections* 7 (1884): 191–92.

2. Lankton, *Beyond the Boundaries,* 72. For instance, the Minesota mine contracted to have two hundred acres of land cultivated in 1855, three hundred in 1858, and seven hundred in 1861.

3. Although mining-company officials did not leave a written record of what they *thought* a mining company was supposed to do, they did leave behind all the *products* of their mental template in terms of housing, churches, hospitals, farms, stores, and so on. The mental template did not vary much from one end of the mineral range to the other, or from ca. 1850 until 1900.

4. *Min. Stats. for 1880,* 46, 58, 77, 82.

5. Phoenix Copper Company, *A.R. for 1860,* 10.

6. See "Tracing of a Geological Diagram of Quincy and Hancock Locations," a drawing by S. W. Hill, Nov. 1859, in Historic American Engineering Record documentation of the Quincy Mining Company, Library of Congress (Washington, DC). Also see Nancy Beth Fisher, "Quincy Mining Company Housing, 1840s–1920s" (unpublished master's thesis, Michigan Technological University, 1997), 41–44.

7. Lankton, *Beyond the Boundaries,* 57.

8. Many photographs exist, taken of mine locations between the 1860s and the early 1900s, that show these characteristics. In the twentieth century, virtually all the fences that once bounded nineteenth-century houses were removed.

9. See the chapter "Homes on the Range," in Lankton, *Cradle to Grave,* 142–62; Lankton, *Beyond the Boundaries,* 59–60; and Lankton and Hyde, *Old Reliable,* 35.

10. For a source on the evolution of the medical profession in the nineteenth century, see Paul Starr, *The Social Transformation of American Medicine* (New York, 1982).

11. Lankton, *Cradle to Grave,* 181–82.

12. Ibid., 182–83; and Lankton, *Beyond the Boundaries,* 72–73.

13. Lankton, *Cradle to Grave,* 15–16, 163–64; and Lankton, *Beyond the Boundaries,* 72–73.

14. For the interconnectedness of early churches and schools, see the "Saints and Scholars" chapter in Lankton, *Beyond the Boundaries,* 130–44.

15. Ibid., 131–34. Also see Regis M. Walling and Rev. N. Daniel Rupp, eds., *The Diary of Bishop Frederic Baraga* (Detroit, 1990); and Rev. John H. Pitezel, *Lights and Shades of Missionary Life: Containing Travels, Sketches, Incidents and Missionary Efforts during the Nine Years Spent in the Region of Lake Superior* (1857; repr., Cincinnati, 1882).

16. Lankton, *Beyond the Boundaries,* 131, 134.

17. Thurner, *Calumet Copper and People,* 21.

18. Michigan, *Statistics of the State of Michigan Collected for the Ninth Census of the United States, June 1, 1870* (Lansing, MI, 1873), 636, 639, 645.

19. Lankton, *Cradle to Grave,* 168–71; and Lankton, *Beyond the Boundaries,* 138–39.

20. Lankton, *Beyond the Boundaries,* 139; and Lankton, *Cradle to Grave,* 168.

21. Lankton, *Beyond the Boundaries,* 140–41. For a teacher's view of schooling at the Cliff Mine, see Henry Hobart's 1863–64 journal, published as Mason, *Copper Country Journal.*

22. Lankton, *Beyond the Boundaries,* 170. See PMC, *A.R. for 1861,* 19; Mason, *Copper Country Journal,* 199; and *PLMG,* 8 Oct. 1864.

23. For a discussion of the paternalism practiced at the mines, see the "Cradle to Grave" chapter in Lankton, *Cradle to Grave,* 163–80.

CHAPTER 5. THE QUINCY MINE: TAKING THE LONG ROAD TO SUCCESS

1. Lankton and Hyde, *Old Reliable,* 4–5.

2. Quincy Mining Company, in its early history, was headquartered in Marshall, Michigan, and then Detroit. When a new slate of company directors assumed control, company headquarters moved to Philadelphia, where they stayed from 1851 to 1856. Then the company moved back to Detroit, again following the election of a new board of directors. In 1858, when Quincy's largest stockholder, Thomas F. Mason, was elected president, he moved the corporate headquarters to New York City, where he resided (see Lankton and Hyde, *Old Reliable,* 5, 7, 18–19, 151; the last page cited lists all QMC officials from the 1840s through 1980).

3. Dividends were reported in QMC's annual reports; for a table of dividends, see Lankton and Hyde, *Old Reliable,* 152–53.

4. Lankton and Hyde, *Old Reliable,* 6, 8, 10.

5. Ibid., 10–11.

6. Ibid., 15–16, 20–21.

7. Ibid., 152.

8. Work on the early Pewabic lode shafts is documented in QMC, *Contract Book, 1856–60;* and QMC, *Return of Labor, 1857–64,* QMC Collection, MTU.

9. Egleston, "Copper Mining on Lake Superior," 290.

10. QMC, *A.R. for 1872,* 18; *A.R. for 1877,* 15; and *A.R. for 1878,* 2.

11. QMC, *A.R. for 1864,* 5; *A.R. for 1865,* 5–6; *A.R. for 1866,* 4; *A.R. for 1867,* 11; and *A.R. for 1868,* 14.

12. QMC, *Invoice Book, 1860–63,* invoices for forty-eight wheelbarrows, 29 May 1860; and, for rails, QMC, "Inventory," 1 Jan. 1862. The QMC *Contract Book, 1856–60,* carries entries not only for miners, but for men with "wheeling contracts."

13. See windlass contracts in QMC, *Contract Book, 1856–60.*

14. QMC, "Inventory of Buildings, 1 March 1859," and letter, S. S. Robinson to TFM, 14 May 1861.

15. Larry Lankton, "Technological Change at the Quincy Mine, ca. 1846–1931," unpublished HAER report, Library of Congress (Washington, DC, 1978), 289–90.

16. Numerous entries for purchases of hoisting chain and then wire rope are found in QMC, *Invoice Books, 1860–63* and *1864–73.* Hoisting chain weighed 5.7 to 6.6 pounds per foot; wire rope, only 3.6 to 3.8 pounds. The QMC *Invoice Book for 1860–63,* entries for June–Sept. 1863, document Quincy's first purchases of rock skips. Purchases of "bell wire rope" are found in QMC, *Day Book, 1859–66,* 324, 346–47, 586.

17. Lankton and Hyde, *Old Reliable,* 29. QMC *Contract Book, 1856–60,* and *Kiln-House Time Books, 1861–62* and *1862–66,* document much of this labor.

18. *Min. Stats. for 1881,* 110.

19. *Min. Stats. for 1880,* 26; Blandy, "Stamp Mills of Lake Superior," 208–15; and Gates, *Michigan Copper,* 26–27. See Benedict, *Lake Superior Milling Practice,* for a full description of stamping, washing, and separating machinery. For particular details on Quincy's mills, see Charles F. O'Connell Jr., "Quincy Mining Company: Stamp Mills and Milling Technologies, ca. 1860–1931," unpublished HAER report, Library of Congress (Washington, DC, 1978).

20. Lankton and Hyde, *Old Reliable,* 16, 35. Far less is known about these early company houses on the hillside than is known about the houses built in 1864 and later on top of Quincy Hill.

21. See the ca. 1865 map of the Quincy location in Lankton and Hyde, *Old Reliable,* 29.

22. Fisher, "Quincy Mining Company Housing," 62–62, 69–74; and Sarah McNear, "Quincy Mining Company: Housing and Community Services, ca. 1860–1931," unpublished HAER report, Library of Congress (Washington, DC, 1978), 519–21.

23. Fisher, "Quincy Mining Company Housing," 62–63, 73, 75; and Efstathios I. Pappas, "Swedetown Location: Hope and Failure in a Company Neighborhood" (unpublished master's thesis, Michigan Technological University, 2002), 5–58.

24. Lankton and Hyde, *Old Reliable,* 36; and A. F. Fischer, "Medical Reminiscence," *MH* 7 (Jan.–April 1923): 27–33. Also see McNear, "Quincy Mining Company," section on medical service, 539–46.

25. Lankton and Hyde, *Old Reliable,* 36–38.

26. Lankton, *Beyond the Boundaries,* 138; and School District No. 1, Quincy Township, *Minutes of Meetings, 1867–,* QMC Collection, MTU.

27. Fisher, "Quincy Mining Company Housing," 41–44, 49; and Lankton and Hyde, *Old Reliable,* 35.

28. The list of Hancock businesses and trades was compiled from many sources: newspaper stories and advertisements, QMC invoices from Hancock firms, and census records listing occupation.

29. Lankton and Hyde, *Old Reliable,* 5, 6, 19; and Robert L. Root Jr., *"Time by Moments Steals Away": The 1848 Journal of Ruth Douglass* (Detroit, 1998), 22–25.

30. Lankton, *Beyond the Boundaries,* 116–18.

31. To see how ethnicity at Quincy in 1870 compared to all of Houghton County, see Lankton and Hyde, *Old Reliable,* 39.

32. Charles K. Hyde, "Economic and Business History of the Quincy Mining Company," unpublished report, Historic American Engineering Record, Library of Congress (Washington, DC, 1978), 83–84; and Lankton and Hyde, *Old Reliable,* 20, 151.

33. German workers' names are found in QMC, *Time Book, 1851–55,* entries for July 1851, July 1852, and Feb. 1855.

34. See contracts to sort and break rock in QMC, *Contract Book, 1861–63,* entries for Sept.–Nov. 1863.

35. See names of German workers at the stamp mill in QMC, *Time Book, 1872.* See ethnically aligned mining teams in QMC's *Contract Book* covering 1872.

36. For discussions of the Cornish contract system, see Roger Burt, ed., *Cornish Mining: Essays on the Organization of Cornish Mines and the Cornish Mining Economy* (Newton Abbot, UK, 1969).

37. Ibid., introduction, 10–11; and Lankton, *Cradle to Grave,* 63–64.

38. Quincy's many extant *Contract Books* at MTU document its extensive use of contract labor at one time.

39. Gates, *Michigan Copper,* 197, 203; and Lankton and Hyde, *Old Reliable,* 152.

40. Pappas, "Swedetown Location," 20–58; and Hyde and Lankton, *Old Reliable,* 17.

41. Lankton, *Beyond the Boundaries,* 187; and Lankton and Hyde, *Old Reliable,* 41–42.

CHAPTER 6. THE ERA OF MICHIGAN DOMINATION: 1865–1890

1. Gates, *Michigan Copper,* 209, 230; and Lankton and Hyde, *Old Reliable,* 152.

2. Gates, *Michigan Copper,* 228.

3. Ibid., 212.

4. Ibid., 39–63, 203–4, 216–18; and Lankton and Hyde, *Old Reliable,* 152. Also see Hyde, *Copper for America,* 49–67.

5. Lankton, "Machine *under* the Garden," 9.

6. Ibid., 10–17.

7. Ibid., 17–23.

8. Lankton, *Cradle to Grave,* 94.

9. *PLMG,* 5 Aug. 1869 and 24 April 1870; and Hyde, *Copper for America,* 62–63. Also, for nitro accidents, see Arthur Pine Van Gelder and Hugo Schlatter, *History of the Explosives Industry in America* (1927; repr., New York, 1972), 325–26.

10. *NMJ,* 18 Feb. 1874.

11. Lankton, *Cradle to Grave,* 96–97.

12. Ibid., 35, 47.

13. Photographs of mine locations taken through the early 1880s sometimes show large stockpiles of cordwood located near engine houses. In addition to burning wood in boilers at their mines and mills, the companies cut wood for mine supports, kilnhouse fuel, and lumber (Lankton, *Cradle to Grave,* 32, 42–43).

14. Mining companies could set up a standard horizontal steam engine to perform myriad tasks over time, as needed. At the crank end of the engine, the driveshaft produced rotary motion. By affixing to that shaft a belt pulley, spur gear, or a friction drive, power could be transmitted to some piece of machinery in need of a prime mover.

15. W. P. Blake, "The Blake Stone- and Ore-Breaker: Its Invention, Forms and Modifications, and Its Importance in Engineering Industries," *AIME Trans.* 33 (1902): 988–1031.

16. Lankton, *Cradle to Grave,* 49–50.

17. Benedict, *Red Metal,* 67; Gates, *Michigan Copper,* 60–63; Willis Dunbar, *All Aboard! A History of the Railroads in Michigan* (Grand Rapids, MI, 1969); and Lankton, *Cradle to Grave,* 54.

18. Lankton and Hyde, *Old Reliable,* 85–89; Lankton, *Cradle to Grave,* 151–53; Fisher, "Quincy Mining Company Housing," 47–179; and McNear, "Quincy Mining Company," 521–24.

19. Michigan Tech students Kathleen Dravillas, Eric Durkee, Keri Ellis, and Kim Wilmers collected data on Calumet establishments as a class research project for L. Lankton. Their two main sources of information, both at MTU, were *Polk's City Directory, 1895–96,* and Sanborn Fire Insurance maps, 1888.

20. Lankton, *Beyond the Boundaries,* 120–21; and Lankton, *Cradle to Grave,* 204.

21. Lankton, *Cradle to Grave,* 203–4; and Gates, *Michigan Copper,* 113–14.

22. Lankton, *Cradle to Grave,* 205–6; and Gates, *Michigan Copper,* 114.

23. "The Upper Peninsula of Michigan," *Harper's New Monthly Magazine,* May 1882, 898–99; and James North Wright, "The Development of the Copper Industry of Northern Michigan," *Publications of the Michigan Political Science Association* 3 (Jan. 1899): 139–40.

CHAPTER 7. THE LARGEST AND BEST COPPER MINE IN THE WORLD: CALUMET AND HECLA

1. *Min. Stats for 1899,* 147.

2. Edwin J. Hulbert, *"Calumet Conglomerate," an Exploration and Discovery Made by Edwin J. Hulbert, 1854 to 1864* (Ontonagon, MI, 1893), 25; and C. Harry Benedict, *Red Metal,* 40–59.

3. *Credit Ledger,* Michigan—Houghton County, I, 266Q, 274a, and 327; and R. B. Dun & Company Credit Collection, Baker Library, Harvard University.

4. Hulbert, *"Calumet Conglomerate,"* 105–6.

5. Ibid., 25, 49, 108–20.

6. The Calumet Mine developed the Calumet Conglomerate lode in the southeast quarter of Section 14; the Hecla Mine developed the same lode where it traversed the northeast quarter of Section 23 (see the Boston *Sunday Globe,* 13 Sept. 1885; Hulbert, *"Calumet Conglomerate,"* 28; Benedict, *Red Metal,* 69–70; and *Min. Stats. for 1882,* 99).

7. Benedict, *Red Metal,* 46, 50, 52–53, 60, 66.

8. *Boston Sunday Globe,* 13 Sept. 1885; and Benedict, *Red Metal,* 45–60.

9. Benedict, *Red Metal,* 59.

10. Ibid., 58.

11. Gates, *Michigan Copper,* 197, 217, 230; Benedict, *Red Metal,* 69; and Wright, "Development of the Copper Industry," 131.

12. Broderick and Hohl, "Geology and Exploration," 479; and *Min. Stats. for 1895,* 120.

13. *Min. Stats. for 1881,* 120, 13, 136; and Harry Vivian, "Mining Methods used on the Calumet Conglomerate Lode," *MCJ* 17 (Oct. 1931): 496.

14. *Min. Stats. for 1882,* 96; and *Min. Stats for 1895,* 121.

15. Records that passed from C&H to Universal Oil Products are now held at MTU.

16. *Min. Stats. for 1899,* 266. Also see R. G. Dun & Company, *Credit Ledger,* 327, Baker University, Harvard University.

17. Benedict, *Red Metal,* 66, 81.

18. Vivian, "Mining Methods," 497; Benedict, *Red Metal,* 81; and *Min. Stats. for 1897,* 177–78.
19. Benedict, *Red Metal,* 83.
20. Lankton, "Machine *under* the Garden," 20; *PLMG,* 7 March and 4 April 1872; and *Min. Stats. for 1882,* 97.
21. The employment figures are from Gates, *Michigan Copper,* 208. The stamp-rock tonnages are from Benedict, *Red Metal,* 80.
22. *Min. Stats. for 1880,* 148; C&H, *A.R. for 1892–93,* 15; and Benedict, *Red Metal,* 104.
23. *Min. Stats. for 1883,* 98–99; *Min. Stats. for 1886,* 217; *Min. Stats. for 1895,* 122; and *Min. Stats. for 1897,* 179.
24. *Min. Stats. for 1885,* 238; *Min. Stats. for 1895,* 126; and C&H, *A.R. for 1892–93,* 9.
25. *Min. Stats. for 1887,* 199.
26. Ibid., 199–200; *Min. Stats. for 1888,* 67–68; and Benedict, *Red Metal,* 87.
27. *Min. Stats. for 1888,* 67–68.
28. C&H, *A.R. for 1891–92,* n.p.
29. *Min. Stats. for 1886,* 217; *Min. Stats. for 1895,* 122; *Min. Stats. for 1896,* 123–24; and *Min. Stats. for 1899,* 271.
30. *PLMG,* 6 July 1871; and *Min. Stats. for 1880,* 149.
31. *Min. Stats. for 1880,* 149; and *Min. Stats. for 1885,* 237.
32. C&H, *A.R. for 1890–91; A.R. for 1891–92; A.R. for 1892–93,* 12–13; and *Min. Stats for 1895,* 121–22.

CHAPTER 8. BEFITTING A COPPER KING: C&H'S VISIBLE EMPIRE

1. *Min. Stats. for 1880,* 147.
2. *Min. Stats. for 1885,* 237.
3. *Min. Stats. for 1880,* 148.
4. Benedict, *Red Metal,* 43, 84.
5. Ibid., 41–42, 5, 63.
6. Ibid., 65, 69; Coggin, "Notes on the Steam Stamp," 210; and John F. Blandy, "Stamp Mills of Lake Superior," *E&MJ,* 27 June 1874, 401.
7. Benedict, *Red Metal,* 65.
8. Ibid., 69, 83; and Blandy, "Stamp Mills," *E&MJ,* 401.
9. Benedict, *Red Metal,* 67, 83; *Min. Stats. for 1880,* 149; Frank G. Haller, "Transportation System and Coal Dock," *MCJ* 17, no. 10 (Oct. 1931): 515; and C&H, *A.R. for 1892–93,* 17.
10. The most important gravity-separating machines in the Calumet and Hecla mills were jigs and Ellenbecker tables. At the jigs, plungers agitated the copper, rock, and water in a box over a sieve plate. Within the box, the copper settled to the bottom first, with the lighter rock above it. The Ellenbecker tables treated *slimes,* the watery mixture of extremely fine particles of rock and copper. The slimes flowed out onto an inclined, shaking table. The heavier copper flowed off one side of the table while the lighter rock tailed off another (see Benedict, *Lake Superior Milling Practice,* 1; and H. S. Munroe, "The Losses in Copper Dressing at Lake Superior," *AIME Trans.* 8 [1879–80]: 410–11).
11. Benedict, *Red Metal,* 85–86; and Munroe, "Losses in Copper Dressing," 410–11.
12. Benedict, *Red Metal,* 60; and Munroe, "Losses in Copper Dressing," 422–23.
13. *Min. Stats. for 1885,* 238; *Min. Stats. for 1895,* 125; *Min. Stats. for 1896,* 126; C&H, *A.R. for 1892–93,* 15; Haller, "Transportation System," 516; F. F. Sharpless, "Ore Dressing on Lake Superior," *LSMI Proc.* 2 (1894): 99; and Benedict, *Red Metal,* 84.
14. In part, the Leavitt stamp achieved higher output because its mortar box had discharge screens on four sides instead of the usual two. The machine's most important component was its *differential* steam cylinder. The Ball machine had one piston that drove the stamp down and lifted it up. The Leavitt stamp had two pistons atop the machine, which were mounted on a common rod and yet worked in cylinders of different size. The upper cylinder had a diameter of 21.5 inches. High-pressure steam acting on the top of this cylinder's piston drove the stamp down through a power stroke of twenty-four inches. The lower

cylinder's piston measured fourteen inches. A uniform steam pressure constantly maintained against the bottom of this piston powered the machine on its upstroke. One cycle of the stamp functioned like this: The intake valve on the upper cylinder opened, high-pressure steam entered, and both pistons, their common rod, and the stamp head descended. When the top cylinder's intake valve closed, its exhaust valve opened, and the constant steam pressure on the bottom of the lower piston caused the machine to reverse its stroke. The Leavitt stamp applied greater force than the Ball did on its downstroke, allowing it to do more work while it consumed less steam on the essential, yet unproductive, upward stroke (see Sharpless, "Ore Dressing," 100; Coggin, "Notes on the Steam Stamp," 210, 232; and Benedict, *Red Metal*, 83).

15. *Min. Stats. for 1886,* 216; *Min. Stats. for 1895,* 125; *Min. Stats. for 1896,* 126; and C&H, *A.R. for 1892–93,* 15–16.

16. Benedict, *Red Metal,* 98–99.

17. *Min. Stats. for 1895,* 126; and C&H, *A.R. for 1892–93,* 18.

18. H. D. Conant, "The Historical Development of Smelting and Refining Native Copper," *MCJ* 17 (Oct. 1931): 531; and James MacNaughton, "History of the Calumet and Hecla Since 1900," *MCJ* 17 (Oct. 1931): 474.

19. Benedict, *Red Metal,* 94; C&H, *A.R. for 1891–92; Min. Stats. for 1891,* 24; *Min. Stats for 1896,* 126; and *Min. Stats. for 1899,* 275. C&H continued to run its smelter at Buffalo until 1913–14, when it closed it down (see C&H, *A.R. for 1913,* 7; and *A.R. for 1914,* 7).

20. *Min. Stats. for 1881,* 132; and *Min. Stats. for 1885,* 237.

21. Lankton, *Cradle to Grave,* 19–20; and Benedict, *Red Metal,* 40–41, 46–50, 52, 54–57, 59–60, 62–65, 67–69.

22. This summary of Leavitt's life and career is based largely on his obituary in *ASME Trans.* 38 (1916): 1347–49. Engineering drawings for much of his machinery designed for C&H can be found in the C&H Collection at MTU and in the Leavitt Collection at the National Museum of American History, Smithsonian Institution, Washington, DC.

23. For a definition of the *shop culture* school, see Monte A. Calvert, *The Mechanical Engineer in America: 1830–1910* (Baltimore, 1967), 6–8.

24. J. B. Francis to AA, 19 Jan. 1874, C&H Collection, MTU.

25. *Min. Stats. for 1880,* 148.

26. *Min. Stats. for 1899,* 275. Also see an illustration of a "Giant Lifting Wheel for the Copper Mines" on the cover of *Scientific American,* n.s., 50, no. 10 (8 March 1884).

27. Erasmus D. Leavitt, "The Superior," *ASME Trans.* 2 (1881): 107; and *Min. Stats. for 1880,* 148.

28. *Min. Stats. for 1881,* 131, 133; Leavitt, "The Superior," 106–12; *Min. Stats. for 1882,* 97; and C&H, *A.R. for 1892–93,* 10.

29. *Min. Stats. for 1895,* 124; *Min. Stats. for 1899,* 271–72; and C&H, *A.R. for 1892–93,* 10–12.

30. Thurner, *Calumet Copper and People,* 8.

31. Lynn Bjorkman, "Calumet Village, Laurium Village, Calumet Township: Historic and Architectural Survey—Phase I," Report by the Western Upper Peninsula Planning and Development Regional Commission (Houghton, MI, 31 Aug. 1995), 30, 33.

32. Thurner, *Calumet Copper and People,* 9–10; and Bjorkman, "Calumet Village," 34.

33. Thurner, *Calumet Copper and People,* 106.

34. Ibid., 9–10.

35. Based on a 1919 map, the Keweenaw National Historical Park generated a series of maps of individual neighborhoods or locations. These maps indicate whether houses were company or privately owned, and whether they were single-family or double or triple houses.

36. Lankton, *Cradle to Grave,* 149–50, 158–59.

CHAPTER 9. A FAR MORE TYPICAL MINE: QUINCY, 1865–1890

1. *Min. Stats. for 1881,* 107.

2. QMC, *A.R. for 1866,* 15. The man-engine's initial cost was about $17,500.

3. Lankton, *Cradle to Grave,* 33–35; and Rawlings, "Recollections of a Long Life," 115.

4. Lankton, *Cradle to Grave*, 70.

5. QMC, *A.R. for 1871*, 7, 18; *A.R. for 1873*, 8; and *A.R. for 1872*, 16.

6. AJC to Burleigh Rock Drill Company, 7 Dec. 1872, QMC Collection, MTU.

7. QMC, *A.R. for 1872*, 16; and *A.R. for 1873*, 15, 20; and *E&MJ* 16 (12 Aug. 1873): 107. No Burleigh drill contracts for miners can be found in the QMC *Contract Books* after 1873, and no compressor runners show up in the QMC *Time Books*.

8. Henry S. Drinker, *A Treatise on Explosive Compounds, Machine Rock Drills and Blasting* (New York, 1883), 260–74; and AJC to WRT, 3 March, 10 April, and 12 Aug. 1873, QMC Collection, MTU.

9. Lankton, "Machine *under* the Garden," 20–23.

10. Ibid., 24–26.

11. Ibid., 26–28.

12. Ibid., 33–34.

13. *Min. Stats. for 1885*, 241–42.

14. Larry D. Lankton and Jack K. Martin, "Technological Advance, Organizational Structure, and Underground Fatalities in the Upper Michigan Copper Mines, 1860–1929," *Technology and Culture* 28, no. 1 (Jan. 1987): 48–49.

15. AJC to WRT, 15 Feb. and 21 April 1874; and AJC to Austin, 1 April 1874, QMC Collection, MTU. Saltpeter powder had potassium nitrate as a main ingredient; the less expensive powder used sodium nitrate.

16. Quantities of the various explosives were calculated by summing the invoices for these materials found in QMC *Invoice Books, 1879–81* and *1881–82*. Also see SBH to TFM, 16 Feb. 1884.

17. SBH to TFM, 16 Feb. 1884.

18. QMC, *A.R. for 1889*, 11; and *A.R. for 1890*, 11. Also see SBH to TFM, 24 Oct. 1889 and 9 Aug. 1890; and SBH to WRT, 3 July 1890, QMC Collection, MTU.

19. QMC's annual reports throughout this period tell which shafts are in operation, as well as their depths and production.

20. QMC, *A.R.* for 1876, 20; *A.R. 1877*, 14; and *A.R. for 1873*, 20.

21. AJC to WRT, 12 March 1873; and QMC, *A.R. for 1873*, 20.

22. QMC, *A.R. for 1882*, 12; and *Weekly Mining Journal*, 21 July 1883, 4.

23. QMC, *A.R. for 1870*, 14; *A.R. for 1871*, 18; *A.R. for 1872*, 16; and *A.R. for 1873*, 20. Also see Jackson & Wiley invoice, 29 June 1872, in QMC *Invoice Book, 1863–73;* and AJC to WRT, 10 April 1873.

24. Daniels to SBH, 28 Jan. and 22 May 1884.

25. Daniels to SBH, 22 May 1884.

26. SBH to TFM, 16 May 1884 and 13 Aug. 1885; and QMC, *A.R. for 1885*, 12.

27. QMC, *A.R. for 1872*, 21; *A.R. for 1875*, 22; and *A.R. for 1881*, 7, 12.

28. QMC, *A.R. for 1881*, 12; and *A.R. for 1882*, 11.

29. QMC, *A.R. for 1862*, n.p.; and *A.R. for 1886*, 12; and SBH to TFM, 19 Jan. 1885.

30. SBH to WRT, 11 June 1891 and 5 April 1892; and WRT to SBH, 15 June 1891.

31. QMC, *A.R. for 1863*, 14; *A.R. for 1864*, 13; and *A.R. for 1871*, 18.

32. QMC, *A.R. for 1872*, 16; and *A.R. for 1873*, 15. Also see QMC invoices from Portage Lake Foundry and Blake Crusher Company for rockhouse machinery (*Invoice Book, 1872–73*). Records of rockhouse labor are found in QMC, *Time and Day Book, 1873–77*, entries for Jan. 1874. Quincy spent nearly $60,000 on the rockhouse and tramroad.

33. QMC, *A.R. for 1873*, 21; and QMC, *Invoice Book, 1872–74*, invoice from Portage Lake Foundry, 1 May 1873.

34. All data were taken from QMC annual reports, 1871–75.

35. QMC, *A.R. for 1879*, 3; *A.R. for 1880*, 3; and *A.R. for 1887*, 11–12. Also see SBH to TFM, 23 Feb. and 16 May 1884; and SBH to N. Daniels, 24 Aug. and 6 Oct. 1884.

36. QMC annual reports, ca. 1865–90. Besides documenting operations and production at the mine, these reports also record machinery used, operations, and production at the stamp mill on Portage Lake. QMC, *A.R. for 1884*, 13, discusses the company's retention of drop stamps. Also see O'Connell, "Quincy Mining Company," 572–601.

37. See U.S. Engineers Office, "History of the Keweenaw Waterway, Michigan" (Duluth, MN, 1940).

38. Lankton and Hyde, *Old Reliable,* 77; and Charles F. O'Connell Jr., "A History of the Quincy and Torch Lake Railroad Company, 1888–1927," unpublished HAER report, Library of Congress (Washington, DC, 1978), 649–88.

39. H. G. Wright, "The Quincy Amygdaloid Mills," *E&MJ,* 22 July 1911, 166; T. A. Rickard, *The Copper Mines of Lake Superior* (New York, 1905), 122, 129; and O'Connell, "Quincy Mining Company," 601ff.

40. Fisher, "Quincy Mining Company Housing," 141, 143–48; and McNear, "Quincy Mining Company," 524.

41. Pappas, "Swedetown Location," 54ff.

42. McNear, "Quincy Mining Company," 521, 523; and Fisher, "Quincy Mining Company Housing," 115–19, 138–39.

43. Lankton and Hyde, *Old Reliable,* 89; and QMC, *A.R. for 1880,* 3. For costs associated with this house, see QMC, *Journals, 1879–82,* 303, 413; and *1882–87,* 32. In a letter to SBH, TFM called this dwelling "our extravagantly large house." In 1880–81 the company lavished some $25,000 on a house that it provided rent free and well furnished to its agent.

44. Hyde, "Economic and Business History," 83, 133.

CHAPTER 10. QUINCY MAKES ITSELF OVER: 1890–1912

1. Gates, *Michigan Copper,* 198–99; and Hyde, *Copper for America,* 82.

2. Lankton and Hyde, *Old Reliable,* 52, 54, 101, and map, 102.

3. Lankton, "Technological Change," 360, 370, 376–77, and 441–45.

4. Lankton and Hyde, *Old Reliable,* 68, 70–72; and Lankton, "Technological Change, 363–65, 371–72. The old No. 4 shaft was the only one not to get a new shaft-rockhouse; it continued to use the tramroad and separate rockhouse.

5. Sharpless, "Ore Dressing," 98.

6. Lankton and Hyde, *Old Reliable,* 113–14, 116–19.

7. L. Hall Goodwin, "Shaft-Rockhouse Practice in the Copper Country—IV," *E&MJ* 100 (10 July 1915): 53–56; and T. C. DeSollar, "Rockhouse Practice of the Quincy Mining Company," *LSMI Proc.* 13 (1912): 217–18.

8. Lankton, "Technological Change," 366–67, 372–73, 376; Lankton and Hyde, *Old Reliable,* 63–64; QMC, *A.R. for 1894,* 14; and *Engineering News,* 18 April 1895, 255.

9. Lankton, "Technological Change," 379–80. The No. 8 hoist cost about $30,000. Each of the horizontal duplex engines had a thirty-two-inch bore and seventy-two-inch stroke.

10. Lankton and Hyde, *Old Reliable,* 57, 74–76, 93; and McNear, "Quincy Mining Company," 552–54. Also see QMC's annual reports for 1890–1900 and the entry for the Quincy Mine in Horace J. Stevens's *The Copper Handbook: A Manual of the Copper Industry of the World,* vol. 1 (1900).

11. SBH to W. Hart Smith, 18 Feb. 1893, QMC Collection, MTU.

12. Lankton, *Cradle to Grave,* 104–6.

13. Ibid., 107–8.

14. Ibid., 97–99.

15. Lankton and Hyde, *Old Reliable,* 62, 108, 113.

16. SBH to TFM, 29 April 1896.

17. Lankton, "Technological Change," 386–88; and Stevens, *Copper Handbook,* 2 (1902):242. The company's progress in installing electric haulage is also covered in its annual reports for 1901–3.

18. JLH to J. H. Wilson, 26 Dec. 1903, and to WRT, 19 June 1902. Also see Lankton, *Cradle to Grave,* 102–3.

19. U.S. House Committee on Mines and Mining, *Conditions in Copper Mines of Michigan,* hearings before subcommittee pursuant to House Resolution 387, 63rd Cong., 2nd sess. 7 pts., 1914, 112; and JLH to WRT, 17 Jan. 1904.

20. QMC, *A.R. for 1893,* 14; and *A.R. for 1895,* 14; and Stevens, *Copper Handbook,* 2 (1902):244.

21. QMC, *A.R. for 1891,* 13; *A.R. for 1900,* 12; and *A.R. for 1905,* 9; and Lankton and Hyde, *Old Reliable,* 78–80, 83.

22. Lankton and Hyde, *Old Reliable,* 80, 82, 84, 128.

23. Ibid., 106–8; and Lankton, "Technological Change," 405–11.

24. QMC table, "Air Blast Data from March, 1914 to April, 1920," QMC Collection, MTU.

25. Lankton, "Technological Change," 411–13.

26. Lankton and Hyde, *Old Reliable,* 152.

CHAPTER 11. CALUMET AND HECLA: PROFITS NOW, PROBLEMS LATER

1. AA to SBW, 28 Dec. 1893, quoted in Benedict, *Red Metal,* 93; and AA to J. P. Channing, 21 Feb. 1894, and to SBW, 2 April 1894, C&H Collection, MTU.

2. *Min. Stats. for 1895,* 122.

3. U.S. House, *Conditions in Copper Mines of Michigan,* 1423.

4. *Min. Stats. for 1895,* 122.

5. *Min. Stats. for 1897,* 179; and *Min. Stats. for 1899,* 265.

6. C&H, *A.R. for 1890–91; Min. Stats for 1891,* 23; *Min. Stats. for 1895,* 122; and *Min. Stats. for 1899,* 273.

7. Benedict, *Red Metal,* 90–91; and *Min. Stats. for 1895,* 123.

8. Thomas P. Soddy and Allan Cameron, "Hoisting Equipment and Methods," *MCJ* 17, no. 10 (Oct. 1931): 509; *Min. Stats. for 1891,* 23; *Min. Stats for 1895,* 124; *Min. Stats. for 1896,* 124–25; and Benedict, *Red Metal,* 89–90.

9. Benedict, *Red Metal,* 92.

10. Stevens, *Copper Handbook,* 10 (1910–11):527.

11. *Min. Stats. for 1899,* 273–74; and C&H, *A.R. for 1900–1901.*

12. See C&H annual reports for these years. Also see Benedict, *Red Metal,* 93.

13. Benedict, *Red Metal,* 88; *Min. Stats. for 1881,* 132; *Min. Stats. for 1891,* 22; C&H, *A.R. for 1892–93,* 13; and Carl L. Fichtel, "Underground Power Distribution and Haulage," *MCJ* 17, no. 10 (Oct. 1931): 503.

14. *Min. Stats. for 1899,* 271.

15. *Min. Stats. for 1891,* 23; *Min. Stats. for 1894,* 124; *Min. Stats. for 1896,* 124–25; and *Min. Stats. for 1899,* 269.

16. Soddy and Cameron, "Hoisting Equipment," 509.

17. MacNaughton, "History of the Calumet and Hecla," 474–77.

18. Benedict, *Red Metal,* 122–23. Also see C&H, annual reports for years 1896–1905; and *Min. Stats for 1896,* 125–26.

19. Benedict, *Red Metal,* 123.

20. Calumet and Hecla, "To the Stockholders of the Calumet and Hecla Mining Company" (Dec. 1910), 1.

21. MacNaughton, "History of the Calumet and Hecla," 474–77; and Benedict, *Red Metal,* 125–32. Also see C&H annual reports, 1905–11.

22. MacNaughton, "History of the Calumet and Hecla," 475; and Benedict, *Red Metal,* 132–36.

23. Gates, *Michigan Copper,* 123–24.

24. C&H, *A.R. for 1892–93,* 15–16; *Min. Stats. for 1895,* 125; *Min. Stats. for 1896,* 126; and *Min. Stats. for 1899,* 274–75.

25. Benedict, *Red Metal,* 80, 87.

26. Benedict, *Lake Superior Milling Practice,* 43, 49, 51; and Lankton, *Cradle to Grave,* 248–49.

27. Benedict, *Lake Superior Milling Practice,* 118–19; and Stevens, *Copper Handbook,* 10 (1910–11):532. For another source on the evolution of C&H mills, see David A. Vago, "An Interpretive Plan for the Calumet and Hecla Mill Site" (unpublished master's thesis in Industrial Archaeology, Michigan Technological University, 2005).

28. Benedict, *Lake Superior Milling Practice,* 79–82.

29. Conant, "Historical Development of Smelting," 532.

30. C&H, *A.R. for 1899–1900;* and Conant, "Historical Development of Smelting," 532.

31. C&H, *A.R. for 1894–95;* and *A.R. for 1899–1900;* and Stevens, *Copper Handbook,* 3 (1903):61–64; and 7 (1907):107–15.

32. *E&MJ,* 31 March 1900, 373; Conant, "Historical Development of Smelting," 532; and *Min. Stats. for 1899,* 276.

33. Gates, *Michigan Copper,* 208–9, 230.

34. Stevens, *Copper Handbook,* 10 (1910–11):525.

35. Hyde, *Copper for America,* 92–94.

CHAPTER 12. AN IMPORTANT FIND: THE COPPER RANGE MINES OPEN THE BALTIC LODE

1. Gates, *Michigan Copper,* 71–72.

2. For biographical material on Stanton and Gay, and for information on the Stanton-Gay mines, see Sandra Hollingsworth, *The Atlantic: Copper and Community South of Portage Lake* (Hancock, MI, 1978), 36–41, 129; Gates, *Michigan Copper,* 71; Lankton, *Cradle to Grave,* 25; and *DMG,* 25 Sept. 1941.

3. Frederick L. Collins, "Paine's Career Is a Triumph of Early American Virtues," *American Magazine,* June 1928, 40–41, 139–140; and Paine, Webber, *Paine, Webber & Company, 1880–1930: A National Institution* (Boston, 1930), 9–11.

4. Gates, *Michigan Copper,* 72.

5. Stevens, *Copper Handbook,* 1 (1900):249; and Richard A. Fields, *Range of Opportunity: An Historic Study of the Copper Range Company* (Hancock, MI, 1997), 9–10.

6. Fields, *Range of Opportunity,* 6–9; and Stevens, *Copper Handbook,* 3 (1903):279–81.

7. Stevens, *Copper Handbook,* 3 (1903):250; Copper Range (CR), *A.R. for 1900,* 12; and Fields, *Range of Opportunity,* 10–12.

8. Stevens, *Copper Handbook,* 3 (1903):515–16; and Fields, *Range of Opportunity,* 12.

9. Fields, *Range of Opportunity,* 13–14. The $850,000 debt payment to CR Consolidated did not sit well with some Trimountain investors, especially Albert Burrage, who unsuccessfully waged a legal battle with William Paine over this issue.

10. Stevens, *Copper Handbook,* 10 (1910–11):1, 166.

11. Butler and Burbank, *Copper Deposits,* 218.

12. W. H. Schacht, "The Mining Methods of the Copper Range Company, Houghton County, Michigan," *LSMI Proc.* 22 (1922): 76–77; and Albert Mendelsohn, "Mining Methods and Costs at the Champion Copper Mine, Painesdale, Michigan," U.S. Bureau of Mines, Information Circular 6515 (Washington, DC, 1931), 2.

13. The description that follows of the "Baltic method" is based largely on Schacht, "Mining Methods," 78–99; Mendelsohn, "Mining Methods and Costs," 2–12; Fields, *Range of Opportunity,* 22–24; and W. R. Crane, *Mining Methods and Practice in the Michigan Copper Mines,* U.S. Bureau of Mines Bulletin 306 (Washington, DC, 1929), 42–54.

14. Schacht, "Mining Methods," 77.

15. Ibid., 80–81.

16. Ibid., 84–85.

17. Fields, *Range of Opportunity,* 52. Also see Baltic Mining Company, *A.R. for 1907.*

18. Stevens, *Copper Handbook,* 3 (1903):194, 516; and 6 (1906):358.

19. The Baltic, Trimountain, and Champion annual reports for these years detail shaft-rockhouse construction.

20. This discussion of changes made to the Champion shaft-rockhouses is based on Harry T. Mercer, "Rock House Practice of the Copper Range Consolidated Company," *LSMI Proc.* 17 (1912): 283–89.

21. Ibid., 289.

22. Hollingsworth, *The Atlantic,* 5–7, 10, 25–26, 75–78. The Atlantic took over a mine and stamp mill previously worked in the 1860s by the South Pewabic Mining Company.

23. Terry S. Reynolds, "A Narrow Window of Opportunity: The Rise and Fall of the Fixed Steel Dam," *Industrial Archeology* 15 (1989): 3–5.

24. Fields, *Range of Opportunity,* 15–17; and Benedict, *Lake Superior Milling Practice,* 52.

25. Stevens, *Copper Handbook,* 3 (1903):517–18; Fields, *Range of Opportunity,* 17; and Benedict, *Lake Superior Milling Practice,* 52.

26. Stevens, *Copper Handbook,* 3 (1903):253; 6 (1906):359–60; and 8 (1908):515. Also see Benedict, *Lake Superior Milling Practice,* 52, 63–65; and Fields, *Range of Opportunity,* 17–18.

27. Stevens, *Copper Handbook,* 8 (1908):516.

28. George P. Schubert, "Development of Copper Smelting Practice at Michigan Smelter," *LSMI Proc.* 27 (1929): 231.

29. The descriptions of the smelter's structures and operations are drawn from Schubert, "Development of Copper Smelting Practice," 247; Stevens, *Copper Handbook,* 6 (1906):683–84; and Fields, *Range of Opportunity,* 18–20.

30. George A. Forero Jr., "The Copper Range Railroad and the Copper Country," *Mid-Continent Railway Gazette* 37, no. 4 (Dec. 2004): 3–42; and Stevens, *Copper Handbook,* 10 (1910–11):692–93.

31. Butler and Burbank, *Copper Deposits,* 75, 97; and Stevens, *Copper Handbook,* 10 (1910–11):1697–98.

32. Butler and Burbank, *Copper Deposits,* 73, 79; and Stevens, *Copper Handbook,* 10 (1910–11):394.

33. Butler and Burbank, *Copper Deposits,* 74, 82; and Stevens, *Copper Handbook,* 10 (1910–11):589.

34. Gates, *Michigan Copper,* 198.

35. Butler and Burbank, *Copper Deposits,* 76, 87, 89, 90, 98.

36. Hollingsworth, *The Atlantic,* 94–95.

37. Gates, *Michigan Copper,* 207.

38. Ibid., 208–9; and Lankton and Hyde, *Old Reliable,* 152.

39. Gates, *Michigan Copper,* 204, 218–19; and Lankton and Hyde, *Old Reliable,* 152.

40. Hyde, *Copper for America,* 81; and Gates, *Michigan Copper,* 198.

41. Lankton, *Cradle to Grave,* 72–73.

CHAPTER 13. PATERNALISM REVISITED: THE BALTIC, TRIMOUNTAIN, AND CHAMPION MINES

1   Fields, *Range of Opportunity,* 26.

2. Champion Copper Company, *A.R. for 1900,* 11–13.

3. Julia Hubbard Adams, *Memories of a Copper Country Childhood* (by the author, n.d.), unpaginated. A copy of this booklet can be found in the MTU.

4. Trimountain Mining Company, *A.R. for 1899.*

5. Data on company housing at Baltic was compiled by James A. Rudkin, graduate student in Industrial Archaeology at Michigan Tech. Rudkin took house construction data from the Baltic Mining Company annual reports through 1902 and then from the CR annual reports. In their first years of operation, the Baltic, Trimountain, and Champion mines published annual reports that documented the building of their physical plants in considerable detail.

6. Shannon Bennett, a graduate student in Industrial Archaeology at Michigan Tech, compiled much of the data used here on the Trimountain mine. Data on house construction came from Trimountain Mining Company, annual reports for 1900–1903, and thereafter in CR annual reports.

7. See "Property Map of the Trimountain Mining Company," 1 Jan. 1908, CR Collection, MTU.

8. See "Surface Map of the Champion Copper Co.," June 1913, CR Collection, MTU.

9. CR, *A.R. for 1900,* 20.

10. Stephanie Atwood, graduate student in Industrial Archaeology at Michigan Tech, helped compile the housing numbers for the Champion Copper Company at Painesdale. Key sources were Champion Copper Company annual reports and Steven's *Copper Handbook,* volumes covering the years from 1900 to 1911. Company houses are also indicated on, and counted off of, "Surface Map of the Champion Copper Co.," found at MTU.

11. "Surface Map of the Champion Copper Co."

12. Stevens, *Copper Handbook,* 3 (1903):252.
13. The annual reports of Baltic, Trimountain, and Champion first note (through about 1902) the practice of building houses on posts and then later note the need to return and found them on masonry walls.
14. These various lots sizes were scaled off the "Surface Map of Champion Copper Co." Lots of all these sizes were found in Painesdale's various neighborhoods.
15. For the construction history and description of this house and its occupants, including several photographs, see Alison K. Hoagland, "The Boardinghouse Murders: Housing and American Ideals in Michigan's Copper Country in 1913," *Perspectives in Vernacular Architecture: Journal of the Vernacular Architecture Forum* 11 (2004): 1–19.
16. The descriptions of the housing types discussed here are based on their floor plans found in CR, "Plans of Principle Dwellings," Collection 28, box 105, folder 1, MTU.
17. For plans and some elevations (and to see how some houses were founded on cedar posts), see Claude T. Rice, "Labor Conditions at Copper Range," *E&MJ* 94, no. 26 (28 Dec. 1912): 1231.
18. For data on water service in Baltic, Trimountain, and Painesdale, see Harry T. Mercer, "Fifty Years on the South Range," unpublished manuscript, n.d., MTU, 39–40; Clarence T. Monette, *Painesdale, Michigan: Old and New* (Lake Linden, MI, 1983) 26–28; Clarence T. Monette, *Baltic, Michigan* (Lake Linden, MI, 1996) 12, 14; and Clarence T. Monette, *Trimountain and Its Copper Mines* (Lake Linden, MI, 1991), 16.
19. Adams, *Memories,* no page.
20. CR, *A.R. for 1900,* 20.
21. See drawings for "7 Room House, E Location," in CR, "Plans of Principal Dwellings."
22. Kathryn Bishop Eckert, *Buildings of Michigan* (New York, 1993), 479–80.
23. See drawings in "Residence for Dr. L. L. Hubbard," file 01-791, Wisconsin Architectural Archive, Milwaukee Public Library.
24. Floor plans for the discussed managers' dwellings are in CR, "Plans of Principal Dwellings."
25. CR, *A.R. for 1902,* 25, 34–35.
26. F. W. Denton to Michigan Railroad Commission, 2 June 1913, Van Pelt Collection, box 1, MTU.
27. Committee of the Copper Country Commercial Club, *Strike Investigation* (Chicago, 1913), 24–29.
28. F. W. Denton to W. A. Paine, 14 April 1915, Van Pelt Collection, box 1, MTU.
29. F. W. Denton to Judge Murphy, 21 Aug. 1913, Van Pelt Collection, MTU.
30. Hoagland, "Boardinghouse Murders," 11.
31. U.S. Bureau of Labor Statistics, *Michigan Copper District Strike,* Bulletin 139 (Washington, DC, 1914), 113.
32. Champion Copper Company, "Statement of Married and Single Employees," 27 Nov. 1912, and "Statement of Employees" 30 Sept. 1905 and 1 Feb. 1912, CR Collection, accession 30 Aug. 79, box 9, MTU.
33. F. W. Denton to W. A. Paine, 14 April 1916, Van Pelt Collection, box 1, MTU. This letter, written in a slightly later era, states explicitly that some ethnic groups do not need housing as good as more favored groups.
34. Champion Copper Company, "Statement of Employees," 30 Sept. 1905 and 1 Feb. 1912.
35. Simon Waitanis to F. W. Denton, 1 July 1912, and Denton to Waitanis, 3 July 1912, Van Pelt Collection, box 1, MTU.
36. Rice, "Labor Conditions at Copper Range," 1230.
37. An extensive article on the Trimountain Hospital, detailing its layout and equipment, is found in *DMG,* 20 April 1906.
38. For a discussion of other company hospitals, see Lankton, *Cradle to Grave,* 182–84.
39. Rice, "Labor Conditions at Copper Range," 1229–30.
40. Baltic Mining Company, *A.R. for 1905.*
41. Monette, "Baltic, Michigan," 32; and Monette, "Trimountain," 25–28.
42. *DMG,* 14 Nov. 1903.
43. Kathryn Bishop Eckert, *The Sandstone Architecture of the Lake Superior Region* (Detroit, 2000), 175–76. For original drawings of the library, see file 01-1148, Wisconsin Architectural Archive.

44. Rice, "Labor Conditions at Copper Range," 1229.

45. Fields, *Range of Opportunity*, 32.

46. Ibid., 31.

47. *DMG*, 21 March 1912. For original drawings of the high school, see file 01-816, Wisconsin Architectural Archive.

48. Eckert, *Sandstone Architecture*, 174–75.

49. Dorothy Klingbeil, "Copper Mining and the Sandwich Kids of Painesdale" (unpublished graduate paper, Northern Michigan University) 18, 20, located in MTU.

50. *DMG*, 21 March 1912.

51. Monette, "Baltic, Michigan," 45–50; Monette, "Trimountain," 20–25; and Monette, "Painesdale," 52–54, 62, 64–66.

52. Mercer, "Fifty Years," 13; Monette, "Baltic, Michigan," 18–21, 40–44; Monette, "Trimountain," 44–45, 47–48; and Monette, "Painesdale," 18, 78–82.

53. Siller's Hotel is shown on Champion Copper Company, "Surface Map." An ad for the Trimountain Hotel and Boarding House ran in Calumet's *Evening Journal*, 2 July 1906. The Baltic Hotel, opened in 1905, is discussed in Monette, "Baltic, Michigan," 29–30.

54. Houghton County, "Articles of Association," Libre 4, 2 Jan. 1901, and Libre 5 (1909): 170–71, MTU.

55. Adams, "Memories." All three investors are also discussed in the *Polk Directory* for Houghton (1901–2). Also see the Goodell Family biographical file at MTU.

56. Denton to Mrs. Antonia Stimach, 11 Sept. 1914, CR Collection, accession no. 564, box 81, folder 22, MTU; and Denton to Frank Santori, 26 Aug. 1908, Van Pelt Collection, box 1, MTU.

57. *Biographical Record of Houghton, Baraga and Marquette Counties* (Chicago, 1903), 92–93.

58. Houghton County, *Articles of Association*, 4, 229–30, located in MTU.

59. Stevens, *Copper Handbook*, 3 (1903):549.

60. "Resolution of Incorporation of the Village of South Range. Houghton County, Michigan," accession no. 89-464, box 21, file 1, MTU.

61. *Polk and Company's Houghton County Directory*, 7 (1907–8):1067–86.

62. Rice, "Labor Conditions at Copper Range," 1229.

CHAPTER 14. HOLDING ON: CORPORATE POWER IN AN AGE OF SOCIAL CHANGE, 1890–1912

1. See, for instance, "Upper Peninsula of Michigan," *Harper's*, May 1882, 892–902; Wright, "Development of the Copper Industry," 135, 139–40; and Rice, "Copper Mining at Lake Superior," 217.

2. Melvyn Dubofsky, *Industrialism and the American Worker, 1865–1920* (Arlington Heights, IL, 1975), 34, 41–44.

3. *PLMG*, 6 Dec. 1888; *Torch Lake Times*, 11 Dec. 1888; and *Min. Stats. for 1888*, 67–68.

4. Gates, *Michigan Copper*, 228–29.

5. For a treatment of ethnicity on the Keweenaw, see Thurner's chapter, "Ethnicity and Singularity," in *Strangers and Sojourners*, 123–57.

6. Arthur Edwin Puotinen, "Finnish Radicals and Religion in Midwestern Mining Towns, 1865–1914" (unpublished PhD diss., University of Chicago, 1973), 20–33, 63, 88, 178–79; and Al Gedicks, "The Social Origins of Radicalism among Finnish Immigrants in Midwestern Mining Communities," *Review of Radical Political Economics* 8 (Fall 1976): 2–5, 12–19. Also see Gedicks, "Ethnicity, Class Solidarity, and Labor Radicalism among Finnish Immigrants in Michigan Copper Country," *Politics and Society* 7, no. 2 (1977): 127–56; and essays by Timo Orta, A. William Hoglund, Matti Kaups, and Arthur Puotinen in *The Finnish Experience in the Western Great Lakes Region: New Perspectives*, ed. Michael G. Karni et al. (Tortu, Finland, 1975).

7. Lankton, *Cradle to Grave*, 74–75.

8. For the uses of sandstone in the Copper Country generally and for information on sandstone buildings in Calumet, Hancock, and Houghton, see Eckert, *Sandstone Architecture*, 137–79, 243–47.

9. To see photos of many turn-of-the-turn century downtown buildings and to read about the architects who designed them, see *Biographical Dictionary of Copper Country Architects,* at http://www.social.mtu.edu/CopperCountryArchitects/. This source was assembled by Kim Hoagland of Michigan Technological University and put online in April 2008.

10. R. L. Polk & Companies' *Directories* from the mid-1890s and early 1900s, covering Houghton, Hancock, Calumet, and Laurium, capture many of the commercial changes made to these towns, as do the Sanborn fire insurance maps at MTU. For Calumet, also see Bjorkman, "Calumet Village," 34–55; and Thurner, *Calumet Copper and People,* 66–88.

11. Articles announcing new communication services and utilities such as water, sewer, gas, and electricity can be found in local newspapers from the 1870s on, such as the *Portage Lake Mining Gazette,* the *Copper Country Evening News,* and the *Evening Journal,* among others. For instance, on the introduction of the telephone, see *PLMG,* 21 March, 18 April, and 2 May 1878; for electric lighting, see *PLMG,* 17 and 31 March, 14 April, 21 and 28 July, and 11 Aug. and 1 Sept. 1887.

12. Hoagland, "Boardinghouse Murders," 12–14.

13. James Reed, *From Private Vice to Public Virtue: The Birth Control Movement and American Society since 1830* (New York, 1978), 4.

14. See Ruth Schwartz Cowan, "Twentieth-Century Changes in Household Technology," chap. 4 in *More Work for Mother: The Ironies of Household Technology from the Open Hearth to the Microwave* (New York, 1983), 69–101; and Thomas J. Schlereth, *Victorian America: Transformations in Everyday Life* (New York, 1991), esp. chap. 3, "Housing," 87–139, and chap. 4, "Consuming," 141–67. Also see Susan Strasser, *Never Done: A History of American Housework* (New York, 1982); and Suellen Hoy, *Chasing Dirt: The American Pursuit of Cleanliness* (New York, 1995).

15. For a brief treatment of the Progressive Era, accompanied by a bibliographical essay, see Arthur S. Link and Richard L. McCormick, *Progressivism* (Arlington Heights, IL, 1983).

16. Lankton, *Cradle to Grave,* 130–32.

17. Ibid., 110, 124–25. The data are drawn from mine inspector reports and from Albert H. Fay, comp., *Metal-Mine Accidents in the United States during the Calendar Year 1911,* U.S. Bureau of Mines Technical Paper 40 (Washington, DC, 1913), 21. Also see Lankton and Martin, "Technological Advance," 45ff. A total of 1,900 men are known to have died underground at the mines, half of them between 1900 and 1920. After 1910 the Bureau of Mines reported that one of three working in the copper mines suffered an injury resulting in lost work hours.

18. See entries for Houghton County in Michigan, *Annual Reports of the Secretary of State on the Registration of Births and Deaths, Marriages and Divorces in Michigan* (Lansing). Houghton, Keweenaw, and Ontonagon counties maintain local death records at their courthouses.

19. Lankton, *Cradle to Grave,* 131–33, 135.

20. Lawrence M. Friedman, *A History of American Law* (New York, 1973), 409, 412–13. Also see Joseph S. Harrison, "Labor Law and the Michigan Supreme Court, 1890–1930" (unpublished master's thesis, Wayne State University, June 1980), 23–31.

21. Michigan, *Public Acts, 1887* (No. 213), 252–54.

22. Lankton, *Cradle to Grave,* 133–34.

23. Ibid., 131.

24. Ibid., 122–24; and *Mine Inspector's Report for Houghton County for 1895* (Houghton, MI), 13–16, 18.

25. William Graebner, *Coal-Mining Safety in the Progressive Period* (Lexington, KY, 1976), 1–3.

26. Michigan, *Public Acts, 1911* (No. 163), 263–67. See also *Public Acts, 1897* (No. 123), 140–41.

27. Lankton, *Cradle to Grave,* 135–37.

28. Ibid., 136–38.

29. Ibid., 138–39; and Michigan, *Public Acts, 1912,* 25–26.

30. Lankton, *Cradle to Grave,* 139–40.

31. Ibid., 188–90; and *Min. Stats. for 1899,* 277. Also see C&H's annual reports for 1898–1901.

32. Lankton, *Cradle to Grave,* 194–95.

33. U.S. Bureau of Labor Statistics, *Michigan Copper District Strike,* 117.

34. Lankton, *Cradle to Grave,* 172–73.

35. Ibid., 157–58.
36. U.S. Bureau of Labor Statistics, *Michigan Copper District Strike,* 114–15.
37. Lankton, *Cradle to Grave,* 210, 216–17.
38. Arthur W. Thurner, *Rebels on the Range: The Michigan Copper Miners' Strike of 1913–14* (Lake Linden, MI, 1984), 28–35; and Lankton, *Cradle to Grave,* 205, 206–8, 219–20.
39. Charles K. Hyde, "Undercover and Underground: Labor Spies and Mine Management in the Early Twentieth Century," *Business History Review* 60 (Spring, 1986): 1–27.
40. Lankton, *Cradle to Grave,* 206, 208–9.
41. Ibid., 213–14.
42. Ibid., 211–12.
43. Ibid., 214–16.

CHAPTER 15. SHOW THEM WHO'S BOSS: THE STRIKE OF 1913-1914

1. Hyde, "Undercover and Underground," 11–14; Puotinen, "Finnish Radicals," 203; and Hyde, *Copper for America,* 62–63, 68–70.
2. The most detailed account of the long strike is Thurner's *Rebels on the Range.* Thurner narrates the history of the strike from its origins to its aftermath, from the perspective of one who thinks that the strike was a tragic mistake (see p. xi) perpetrated by a minority of workers in the district. Another detailed account is Steve Lehto, *Death's Door: The Truth behind Michigan's Largest Mass Murder* (Troy, MI, 2006). Lehto focuses on the Italian Hall disaster but covers strike events leading up to it. His perspective is nearly the opposite of Thurner's and is very sympathetic to the WFM and its members.
3. Lankton, *Cradle to Grave,* 221.
4. Ibid., 219–21, 229; and Thurner, *Rebels on the Range,* 35–39.
5. Lankton, *Cradle to Grave,* 219–21.
6. U.S. Bureau of Labor Statistics, *Copper District Strike,* 38; and Thurner, *Rebels on the Range,* 63.
7. U.S. Bureau of Labor Statistics, *Copper District Strike,* 39; and Dan Sullivan and C. E. Hietal to JM, 14 July 1913, C&H Collection, MTU.
8. *Miners' Bulletin,* 9 Aug. 1913.
9. JM to QAS, 6 Aug. 1913 and 18 March 1914.
10. Thurner, *Rebels on the Range,* 83; and U.S. House, *Conditions in Copper Mines of Michigan,* 1482.
11. JM to QAS, 10 Aug. 1913; QAS to JM, 13 Aug. 1913; and Thurner, *Rebels on the Range,* 63.
12. Thurner, *Rebels on the Range,* 1–10.
13. JM to QAS, 24 July 1913, C&H Collection, MTU.
14. JM to QAS, telegram, 31 July, and letter, 22 Aug. 1913, C&H Collection, MTU; and Lankton, *Cradle to Grave,* 222–23. Lehto (in *Death's Door,* 29–30) reports that C&H deputies were armed with cheap handguns right from the start. He asserts that MacNaughton was telling "his superiors in Boston little white lies" when he wrote them that he was not arming deputies. Instead of believing MacNaughton's internal company correspondence from the period, Lehto rests his case on Lee Arten, "Three Guns Mementoes of Strike," in *Copper Island Sentinel,* 30 Aug. 1981.
15. Thurner, *Rebels on the Range,* 7–8.
16. Ibid., 47–48; and Lankton, *Cradle to Grave,* 228.
17. Timothy O'Neil, "Patrick H. O'Brien—The Working Man's Advocate: The Copper Country Years," in *New Perspectives on Michigan's Copper Country,* ed. Alison K. Hoagland et al. (Hancock, MI, 2007), 91.
18. Thurner, *Rebels on the Range,* 68–73; U.S. Bureau of Labor Statistics, *Copper District Strike,* 69; Hoagland, "Boardinghouse Murders," 1–3; and Lehto, *Death's Door,* 49–51.
19. WFM, *Michigan Defense Fund Ledger* 1 (Aug.–Dec. 1913), WFM Collection, Western Historical Collection, University of Colorado.
20. JM to QAS, 2 and 12 Sept. 1913. Also see Thurner, *Rebels on the Range,* 53, 88–89.
21. Clarence A. Andrew, "'Big Annie' and the 1913 Michigan Copper Strike," *MH* 57 (1973): 53–68; and U.S. Bureau of Labor Statistics, *Copper District Strike,* 70.

22. Lankton, *Cradle to Grave*, 232–33; Thurner, *Rebels on the Range*, 112–14; and C&H, *Employment Ledger, 1894–1918*, entries for "Deputies and Hotel Men," 1913, C&H Collection, MTU.

23. For O'Brien's career as a personal injury lawyer, and his service as a judge during the 1913–14 strike, see O'Neil, "Patrick H. O'Brien," 87–98.

24. Ibid., 92–93.

25. Lankton, *Cradle to Grave*, 231–32.

26. Thurner, *Rebels on the Range*, 101–2; and O'Neil, "Patrick H. O'Brien," 94–95.

27. Partial texts of the Citizens' Alliance "Membership Pledge" (1913) are found in Lehto, *Death's Door* (77), and Thurner, *Rebels on the Range* (122). Also see JM to QAS, 9 Nov. 1913.

28. Lankton, *Cradle to Grave*, 233.

29. Thurner, *Rebels on the Range*, 118–19.

30. Ibid., 120–21, 245–47. For a different assessment of the murders, see Lehto, *Death's Door*, 75–77.

31. FWD to WAP, 9 Dec. 1913, Van Pelt Collection, MTU; A. F. Rees to WRT, 18 Oct. 1915, C&H Collection, MTU; and Thurner, *Rebels on the Range*, 127–28.

32. Lehto, *Death's Door*, 81–82.

33. This letter is in the C&H Collection, MTU.

34. Thurner, *Rebels on the Range*, 137.

35. JM to QAS, 17 Dec. 1913, C&H Collection, MTU.

36. Lankton, *Cradle to Grave*, 236–37; Thurner, *Rebels on the Range*, 138–53; and Lehto, *Death's Door*, 87–94.

37. *Työmies*, 26 Dec. 1913.

38. For lengthier looks at the stories, reports, and testimonies that recounted what happened at Italian Hall, see Thurner, *Rebels on the Range*, 154–75; Larry Molloy, "Were You at Italian Hall: A Review of the Coroner's Inquest," in Hoagland et al., *New Perspectives*, 99–114; and Lehto, *Death's Door*, 99–113, 124–52. It seems that Thurner would like to see Italian Hall explained away as a "sheer accident" (174). Molloy, after examining testimony at the coroner's inquest, believes that "the tragedy was caused by someone near the front of the hall who yelled fire at least once." Molloy also notes, "The truth about what really happened at Italian Hall remains shrouded in mystery and my never be known" (113). Lehto shares no doubt and sees no shroud of mystery. He believes the mine managers "who ran the area . . . were responsible for the calamity," and that, "simply put, what happened at Italian Hall was this: A man wearing a Citizens Alliance pin ran in, cried 'Fire,' and ran out. His actions killed seventy-three people" (8, 206). Lehto, an attorney of Finnish heritage with ties to Copper Country, basically tries the mine managers and the Citizens' Alliance in his book and finds them guilty of "Michigan's Largest Mass Murder."

39. *DMG*, 25 Dec. 1913; and Thurner, *Rebels on the Range*, 155.

40. Thurner, *Rebels on the Range*, 159–61.

41. The investigation started on 9 Feb. 1914 (see Thurner, *Rebels on the Range*, 191–207). For the official record on the investigation, see U.S. House, *Conditions in Copper Mines of Michigan* (1914).

42. WFM, *Executive Board Minutes*, meeting of 3 April 1914, WFM Collection; Thurner, *Rebels on the Range*, 229, 255–56; and Thurner, "The Western Federation of Miners in Two Copper Camps: The Impact of the Michigan Copper Miners' Strike on Butte's Local No. 1," *Montana: The Magazine of Western History* (Spring 1983): 30–45.

43. Lankton, *Cradle to Grave*, 241.

44. Ibid., 240–41; and U.S. House, *Conditions in the Copper Mines*, 11.

CHAPTER 16. MAKING THE HARD TURN: FROM GROWTH TO DECLINE

1. Gates, *Michigan Copper*, 199, 205, 219.

2. Ibid., 216–20; and Lankton, *Cradle to Grave*, 244–45.

3. Gates, *Michigan Copper*, 198–99; and Hyde, *Copper for America*, 81.

4. Production statistics from Gates, *Michigan Copper*, 199–200. Also see his chap. 6, "Michigan Copper Mining in Decline, 1919–1938," 143–69.

5. Ibid., 205, 220–21.
6. Ibid., 209–10.
7. Ibid., 229; and Thurner, *Calumet Copper and People,* 67. In addition, see U.S. Census records for Houghton County.
8. Gates, *Michigan Copper,* 232.
9. *Michigan Manual (Red Book): Official Directory and Legislative Manual* (Lansing) for 1901, 1911, 1921, 1931, 1941, 1951–52, 1959–60, and 1969–70.
10. For instance, in the vicinity of Calumet, it appears that approximately twenty-seven churches held regular services in 1910. Using several volumes of the *Polk Directory,* telephone books, and local newspapers as sources, that number declined to about twenty-four in 1920, and then stood at twenty-six and twenty-five in 1930 and 1940. It still stood at twenty-two in 1950, falling to about nineteen by 1960 and to fourteen by 1970.
11. Lankton, *Cradle to Grave,* 248.
12. Ibid., 247, 250–51.
13. Lankton, *Cradle to Grave,* 253–55.
14. Ibid., 255; Lankton and Hyde, *Old Reliable,* 141–42; and Thurner, "The Truly Great Depression," chap. 8 in *Strangers and Sojourners,* 226–57.
15. Butler and Burbank, *Copper Deposits of Michigan,* 1.
16. Fields, *Range of Opportunity,* 69–70; Richard F. Moe, "White Pine Mine Development," *Mining Engineering* 6 (April 1954): 381–86; Harold B. Ewoldt, "Mining and Milling at White Pine," *MCJ* 41, no. 3 (March 1955): 25; "White Pine Uses New Methods and Equipment for Mine Development," *Mining World* 16 (April 1954): 35–38; and "Constructing $70 Million Michigan Copper Mine," *Excavating Engineer* 48 (Jan. 1954): 16–23.
17. Benedict, *Lake Superior Milling Practice,* 79–97.
18. Lankton, *Cradle to Grave,* 258–59.

CHAPTER 17. THE QUINCY MINE: FROM STRUGGLE TO SHUTDOWN

1. See "Air Blast Repairs and Expenses," table included in CL to WRT, 11 May 1921, QMC Collection, MTU; also see QMC table, "Air Blast Data from March 1914 to April 1920." Air blasts closed shafts, tore up skip roads, shattered compressed-air pipes, and caused production losses. As shown in QMC annual reports, air blasts did extensive damage at Quincy in 1909, 1914–16, 1923–24, and 1927.
2. CL to WRT, 6 Nov. 1911 and 5 April 1913, QMC Collection, MTU.
3. For much more detail on the long debate between Lawton and Todd regarding how to combat air blasts, see Lankton, "Technological Change," 408ff.
4. QMC report, "Cost of Producing Copper . . . Should be Less Regardless of . . . Increasing Depth," ca. 1921, QMC Collection, MTU.
5. QMC, *A.R. for 1906,* 13; *A.R. for 1914,* 14; and *A.R. for 1915,* 14; and CL to WRT, 28 Oct. 1919.
6. Lankton, "Technological Change," 475.
7. Ibid., 445–49.
8. QMC, *A.R. for 1922,* 13; and *A.R. for 1924,* 11–12.
9. *DMG,* 22 July and 1 and 2 Nov. 1927; QMC, *A.R. for 1927,* 12–13; and *Mine Inspector's Report for Houghton County* (1927–28), 5–7.
10. QMC, *A.R. for 1927,* 13–14; and *A.R. for 1928,* 6.
11. Lankton and Hyde, *Old Reliable,* 152.
12. Lankton, "Technological Change," 495–96.
13. Crane, *Mining Methods,* 130–31; and Arthur C. Vivian, "Cutting Mass Copper," *E&MJ* 98 (7 Nov. 1914): 825.
14. QMC, *A.R. for 1918,* 15–16; *A.R. for 1925,* 13; and *A.R. for 1930,* 13.
15. QMC, *A.R. for 1916,* 14; and *A.R. for 1926,* 6, 12.
16. McNear, "Quincy Mining Company," 562–66; QMC, *A.R. for 1916,* 16; and *A.R. for 1918,* 18; and Lankton and Hyde, *Old Reliable,* 135–36.

17. Lankton and Hyde, *Old Reliable,* 132; and Fisher, "Quincy Mining Company Housing," 253–68.

18. For the hoist, see Thomas Wilson, "Quincy Hoist—Largest in World," *Power* 53, no. 3 (18 Jan. 1921): 90–95; "The World's Largest Compound Steam Hoisting Engine," *LSMI Proc.* 27 (1929): 18–20; Ray W. Armstrong, "Compound Steam Hoist Installation of the Quincy Mining Company," *LSMI Proc.* 22 (1922): 39–41; "Nordberg Compound Steam Hosting Engine, Quincy Mining Company, No. 2 Shaft, Hancock, Michigan," *LSMI Proc.* 22 (1922): 192–94; and "The Quincy Hoist," *E&MJ,* 11 Dec. 1920, 11, 126.

19. Lankton and Hyde, *Old Reliable,* 125.

20. QMC, *A.R. for 1923,* 15–16; and *A.R. for 1924,* 13–14.

21. QMC, *A.R. for 1928,* 7, 13–14; and Benedict, *Red Metal,* 176–77.

22. Lankton and Hyde, *Old Reliable,* 127; and QMC, *A.R. for 1919,* 17–18.

23. Lankton and Hyde, *Old Reliable,* 99, 106, 141–43.

24. Ibid., 143.

25. Gates, *Michigan Copper,* 177–80. Also see QMC's annual reports for 1943–45.

26. Lankton and Hyde, *Old Reliable,* 144–46; QMC, *A.R. for 1966,* 5; and Benedict, *Lake Superior Milling Practice,* 90–94.

27. QMC, *A.R. for 1945,* 4.

28. Lankton and Hyde, *Old Reliable,* 147.

29. QMC, *A.R. for 1948,* 10; and *A.R. for 1966,* 5.

CHAPTER 18. CALUMET AND HECLA: DOWN WITH THE KING

1. Lankton, *Cradle to Grave,* 257.

2. Gates, *Michigan Copper,* 251–52.

3. Benedict, *Lake Superior Milling Practice,* 82ff.; and Benedict, *Red Metal,* 188–93.

4. Benedict, *Lake Superior Milling Practice,* 111ff.

5. Ibid., 98–102.

6. Gates, *Michigan Copper,* 230.

7. Benedict, *Red Metal,* 137–41; and Gates, *Michigan Copper,* 151–52.

8. Lankton, *Cradle to Grave,* 250; Gates, *Michigan Copper,* 145–46; and Benedict, *Lake Superior Milling Practice,* 89–90.

9. Ocha Potter and Samuel Richards, "Osceola Lode Operations," *MCJ* 17 (Oct. 1931): 490.

10. Transcript of Heritage Line broadcast, "Quincy No. 2," 4518, Oral History Collection, Finnish American Historical Archive, Finlandia University (formerly Suomi College); "Statement of German Importation Costs," n.d., CR Collection, MTU; and Schacht to WAP, 27 Dec. 1923, Van Pelt Collection, MTU.

11. Gates, *Michigan Copper,* 210.

12. Arnold R. Alanen and Lynn Bjorkman, "Plats, Parks, Playgrounds, and Plants: Warren G. Manning's Landscape Designs for the Mining Districts of Michigan's Upper Peninsula," *IA: Journal of the Society for Industrial Archeology* 24, no. 1 (1998): 49–50.

13. Gates, *Michigan Copper,* 210.

14. Ibid., 161.

15. C&H, "Operations: Past, Present and Future" (25 Nov. 1950), 2–3, C&H Collection, MTU; C&H Consolidated, *A.R. for 1931,* 7; Lankton, *Cradle to Grave,* 254; and Gates, *Michigan Copper,* 210.

16. C&H Consolidated, *A.R. for 1939,* 4–5.

17. "Scrap Recovery Campaign in Michigan Iron and Copper Country a Model," *Mining and Metallurgy* 24 (July 1943): 317.

18. Gates, *Michigan Copper,* 171–72; and Thurner, *Strangers and Sojourners,* 261–64.

19. Thurner, *Strangers and Sojourners,* 263–64.

20. C&H Consolidated, *A.R. for 1942,* 5.

21. Thurner, *Strangers and Sojourners,* 277–78.

22. Lankton, *Cradle to Grave,* 258; C&H Consolidated, *A.R. for 1942,* 3; and *A.R. for 1943,* 3; and Benedict, *Red Metal,* 239–42.

23. C&H, "Operations: Past Present and Future," 3–4; and Benedict, *Red Metal,* 236–38.

24. Ibid., 261.

25. Ibid., 261–62; and Thurner, *Strangers and Sojourners,* 279–81.

26. Thurner, *Strangers and Sojourners,* 287.

27. Universal Oil Products, *A.R. for 1968,* 5.

28. *DMG,* 9 April 1969; and Universal Oil Products, *A.R. for 1969,* 4.

29. Lankton, *Cradle to Grave,* 263.

CHAPTER 19. COPPER RANGE: STAYING ALIVE

1. Lankton, *Cradle to Grave,* 154.

2. Stevens, *Copper Handbook,* 8 (1908):1337; and 10 (1910–11):1699; and CR, *A.R. for 1929,* 7; and *A.R. for 1930,* 4.

3. CR, *A.R. for 1930,* 4; and *A.R. for 1931,* 4; and Butler and Burbank, *Copper Deposits,* 79.

4. Butler and Burbank, *Copper Deposits,* 74, 82.

5. Superintendents of the Poor, Houghton Country, MI, *Annual Reports* for 1920–22.

6. See the section titled "Decline of Michigan Copper Mining in a Period of General Business Prosperity, 1919–1929," in Gates, *Michigan Copper,* 147–56.

7. Ward Paine to Schacht, no date, and Schacht to Ward Paine, 12 June 1922; and Mendelsohn, untitled reflections on wages and labor at CR, ca. 1922. All are in the Van Pelt Collection, box 3, MTU.

8. "Statement of German Importation Costs," n.d., CR Collection, accession 30 Aug. 1979, MTU.

9. Schacht to Ward Paine, 27 Dec. 1923, Van Pelt Collection, box 3, MTU.

10. Gates, *Michigan Copper,* 154, 220.

11. CR, *A.R. for 1931,* 4–5; and *A.R. for 1932,* 3, 6.

12. CR, *A.R. for 1931,* 5.

13. Telegram, W. Schacht to Ward Pain, 20 May 1931, MTU.

14. CR, *A.R. for 1932,* 6; and *A.R. for 1933,* 7.

15. See "Power Co.'s file, box 1, Van Pelt Collection, MTU.

16. CR, *A.R. for 1930,* 6–7. For a history of CR's involvement in the Copper District Power Company, see Larry Lankton, "Victoria Dam," Historic American Engineering Record report (HAER No. MI-49), HAER collection, Library of Congress (Washington, DC). The utility generated modest profits in the late 1930s, but demand dropped off in the post–WWII years.

17. Benedict, *Lake Superior Milling Practice,* 66–69; and Fields, *Range of Opportunity,* 54–55.

18. Benedict, *Lake Superior Milling Practice,* 95–97.

19. Chaput, *The Cliff,* 17–18, 65–67, 71–72.

20. CR, *A.R. for 1931,* 6–7.

21. Hollingsworth, *The Atlantic,* 96–97; and Fields, *Range of Opportunity,* 14.

22. Butler and Burbank, *Copper Deposits,* 90, 97.

23. CR, *A.R. for 1932,* 6; Gates, *Michigan Copper,* 162; and Fields, *Range of Opportunity,* 50.

24. Butler and Burbank, *Copper Deposits,* 169–72; and Ewoldt, "Mining and Milling," 25.

25. Butler and Burbank, *Copper Deposits,* 91; and Commissioner of Mineral Statistics, *A.R. for 1880,* 111–14. A fairly detailed history of the Nonesuch mine is presented in John R. Griebel, "Cultural Landscape Report: Nonesuch Mine & Village" (unpublished master's thesis in Industrial Archaeology, Michigan Technological University, 2008).

26. Stevens, *Copper Handbook,* 10 (1910–11):1820.

27. Butler and Burbank, *Copper Deposits,* 75, 97. Also see Calumet and Hecla's annual reports for 1910–18 for work done at White Pine. For information on White Pine, the author is indebted to Vanessa McLean, graduate student in Industrial Archaeology at Michigan Tech, for her report "White Pine, Michigan: A 'Modern' Industrial Community," 2005.

28. Fields, *Range of Opportunity,* 46–47.

29. Ibid., 52.

30. CR, *A.R. for 1940,* 5.

31. Ibid., 4–6.
32. CR, *A.R. for 1946,* 6–7.
33. CR, *A.R. for 1938,* 6.
34. Fields, *Range of Opportunity,* 48–49.
35. Ibid., 49–50.
36. CR, *A.R. for 1945,* 3–4.
37. CR, *A.R. for 1946,* 6–7.
38. Forero, "Copper Range Railroad," 30; and Fields, *Range of Opportunity,* 79–80.
39. CR, *A.R. for 1946,* 3–4; *A.R. for 1947,* 4–5, 11; and *A.R. for 1948,* 4.
40. The CR annual reports for 1949–53 detailed which operations were open and which were shut down.
41. Fields, *Range of Opportunity,* 65.
42. CR, *A.R. for 1961,* 9; *A.R. for 1962,* 9; and *A.R. for 1963,* 14. Also see Fields, *Range of Opportunity,* 92.
43. The declining fortunes, the closings and reopenings of the Champion, are discussed in CR's annual reports throughout the 1950s and into the mid-1960s. For prospecting work in 1965, see *DMG,* 28 Jan. 1967. For the declining yield of the Baltic lode at Champion, see Fields, *Range of Opportunity,* 78.
44. CR, *A.R. for 1966,* 4.
45. *DMG,* 28 Jan. 1967.
46. Fields, *Range of Opportunity,* 82–84. As of 2009 the firm still survives as Hussey Copper Ltd.
47. Fields, *Range of Opportunity,* 81.
48. Frank A. Ayer, "White Pine Mine . . . Potential Major Copper Producer," *MCJ* 36, no. 12 (Dec. 1950): 27–28, 30; and Ewoldt, "Mining and Milling," 25.
49. Hyde, *Copper for America,* 183.
50. Lankton, *Cradle to Grave,* 259–60; and Fields, *Range of Opportunity,* 69–71.

CHAPTER 20. WHITE PINE: A NEW MINE, A NEW ERA

1. Fields, *Range of Opportunity,* 46–47, 67–68.
2. Ibid., 67–69; Chris Chabot, *Tales of White Pine: An Illustrated Oral History of White Pine, Michigan & Environs* (Ontonagon, MI, 1979), 82.
3. R. H. Ramsey, "White Pine Copper: Mine, Mill and Smelter Construction Stepped Up as New Michigan Project Nears Production Stage," *E&MJ* 154, no. 1 (Jan. 1953): 75–76.
4. The author thanks Ron Whiton for pointing out nomenclature differences in correspondence of 24 Dec. 2008.
5. Ramsey, "White Pine Copper," 77; "White Pine Uses New Methods," 35–36; and David Skillings, "White Pine Copper Company, Ontonagon County, Michigan," *Skilling's Mining Review* 41, no. 35 (6 Dec. 1952): 1.
6. Ewoldt, "Mining and Milling," 25–26; Moe, "White Pine Mine Development," 381; and David N. Skillings, "White Pine Copper Co.: Huge Construction Program Nears End," *Skilling's Mining Review* 43 (25 Dec. 1954): 1.
7. "White Pine Uses New Methods," 34.
8. Ramsey, "White Pine Copper," 78; "Constructing $70 Million Michigan Copper Mine," 18; and Ewoldt, "Mining and Milling," 26.
9. Ron Whiton to author, 24 Dec. 2009. In 1961 a second small shaft, 450 feet deep, was sunk two miles from the portal so as to provide an additional man shaft and to expedite the movement of crews into and out of the expanding mine. This shaft also had hoisting facilities.
10. Moe, "White Pine Mine Development," 383–84; and transcript of J. R. Van Pelt interview of Robert M. Neil, 3 Feb. 1971, Van Pelt Collection, box 12, folder 9, 5–6, MTU.
11. Ewoldt, "Mining and Milling," 26. For a good drawing of the room and pillar system used at White Pine, see Ramsey, "White Pine Copper," 76.
12. White Pine Copper Company, "A Visit to the White Pine Copper Company, a Division of the Copper Range Company" (White Pine, MI, 1963), 4.

13. Ramsey, "White Pine Copper," 76. Also, on larger pillars, see James Bekkala to author, Dec. 2008.

14. Moe, "White Pine Mine Development," 384–85; and White Pine Copper Company, "Visit to the White Pine," 4.

15. CR, *A.R. for 1953*, 24; Moe, "White Pine Mine Development," 383–84; and Ewoldt, "Mining and Milling," 26.

16. Ramsey, "White Pine Copper," 79, 81, 84.

17. Ibid., 85; Fields, *Range of Opportunity*, 72; and CR, *A.R. for 1959*, 7.

18. Ramsey, "White Pine Copper," 87.

19. CR, *A.R. for 1952*, 20. Also see the *A.R. for 1955*.

20. CR, *A.R. for 1952*, 20; and *A.R. for 1953*, 26–27; Skillings, "White Pine Copper Co.," 3; and *Detroit News*, 30 July 1956.

21. Van Pelt interview of Bingham, 3 Feb. 1971, Van Pelt Collection, box 12, folder 13, pp. 4, 5, 21, MTU.

22. See Duane A. Smith, *Mining America: The Industry and the Environment, 1800–1980* (Lawrence, KS, 1987), 133–34. Smith says one of the opening shots of the battle between environmentalists and the mining industry was a dispute over whether copper mining should be permitted in the Porcupine Mountains Wilderness State Park; he cites an editorial and an article in *Nature Magazine*, issues of Oct. and Nov. 1958. One opponent wrote, "I thought it would be a sacrilege indeed to build a mine on the doorstep of such a cathedral."

23. Van Pelt interview of Bingham, 3.

24. CR, *A.R. for 1953*, 25; *A.R. for 1955*, 12; *A.R. for 1956*, 11–12; *A.R for 1957*, 12; and *A.R. for 1969*, 9; and John R. Suffron and Edward R. Bingham, "Half a Dozen Goslings," *Michigan Natural Resources* 43, no. 6 (Nov.–Dec. 1974): 23.

25. Joseph Bal to J. R. Van Pelt, 28 Jan. 1970, Van Pelt Collection, box 12, folder 1, MTU. Bal was a District Engineer for the Water Resources Commission.

26. *Copper Range News* 8, no. 7 (July 1968): 2; and E. R. Bingham, "Reclaiming Mill Tailings Areas," 16 April 1969, paper in Van Pelt Collection, box 12, folder 2, MTU.

27. Van Pelt interview of Bingham, 27–28; and CR, "Controlling the White Pine Environment," *Copper Range News* 8, no. 7 (July 1968): 3.

28. CR, *A.R. for 1958*, 13.

29. Fields, *Range of Opportunity*, 70.

30. Average and annual production rates of copper taken from table in Fields, *Range of Opportunity*, 116–17. The mine tonnage per year figure is taken from an unpublished paper: R. C. Johnson, R. A. Andrews, W. S. Nelson, T. Suszek and K. Sikkila, "Geology and Mineralization of the White Pine Copper Deposit: 1995," 17.

31. Hyde, *Copper for America*, 183.

32. See CR, *A.R. for 1954*, 28. CR claimed that White Pine was the second largest employer in the Upper Peninsula (White Pine Copper Company, *The White Pine Story* [White Pine, MI, 1968], 17), a fact repeated by the *Milwaukee Journal*, 8 Feb. 1970. At its absolute peak, the company employed about 3,200.

33. Hyde, *Copper for America*, 161–62.

34. CR, *A.R. for 1963*, 4; and *A.R. for 1972*, 4.

35. CR, *A.R. for 1976*, 6.

36. CR, *A.R. for 1972*, 4.

37. Fields, *Range of Opportunity*, 69–70, 86.

38. Ibid., 88.

39. Ewoldt, "Mining and Milling," 26.

40. Ron Whiton to author, 24 Dec. 2008.

41. CR, *A.R. for 1963*, 4; *A.R. for 1970*, 5; *A.R. for 1972*, 6; and *A.R. for 1973*, 5–6.

42. Ron Whiton to author, 24 Dec. 2008.

43. Ibid.

44. CR, *A.R. for 1963*, 5–6; and *A.R. for 1964*, 6–7. Also see Larry Garfield, ms., "Longwall Mining in the

White Pine Mine," paper for the American Mining Congress meeting, 5–8 May 1968, CR Collection, accession no. 62, box 9, folder 17, MTU.

45. CR, *A.R. for 1965,* 6; *A.R. for 1967,* 3; *A.R. for 1968,* 4–5; *A.R. for 1971,* 5–6; and *A.R. for 1972,* 6. The *A.R. for 1973* said that longwall mining continued to be of long-term interest, but that no work was done on the concept that year (p. 6).

46. CR, *A.R. for 1959,* 7; *A.R. for 1960,* 6–7; and *A.R. for 1961,* 5–6.

47. CR, *A.R. for 1961,* 5–6; and *A.R. for 1963,* 7.

48. Ron Whiton to author, 24 Dec. 2008.

49. CR, *A.R. for 1957,* 11. By 1973, according to the *A.R.* for that year (p. 5), the *average* haulage distance underground via conveyors underground from the mine to the mill had increased to 3.3 miles.

50. CR, *A.R. for 1966,* 7; and *A.R. for 1967,* 5.

51. CR, *A.R. for 1966,* 5; *A.R. for 1967,* 4, 6; and *A.R. for 1968,* 4–5.

52. CR, *A.R. for 1969,* 7; *A.R. for 1970,* 6; *A.R. for 1971,* 5–6; and *A.R. for 1972,* 6.

53. CR, *A.R. for 1970,* 6; *A.R. for 1971,* 5–6; and *A.R. for 1972,* 6.

54. CR, *A.R. for 1972,* 6.

55. CR, *A.R. for 1967,* 7; and *Ontonagon Herald,* 21 Aug. 1968.

56. *Ontonagon Herald,* 21 Aug. 1968; and Robert C. Bacon, "New Automated Transportation System for Open Pit Mines," *Skillings' Mining Review* 51 (24 Aug. 1968): 1, 4, 10.

57. CR, *A.R. for 1972,* 6.

58. Johnson et al., "Geology and Mineralization," 17.

59. CR, *A.R. for 1962,* 4; *A.R. for 1968,* 5; and *A.R. for 1973,* 5. According to later annual reports, the mine operated six mobile crushers underground in 1974 and fourteen in 1975.

60. CR, *A.R. for 1959,* 7.

61. CR, *A.R. for 1963,* 6; *A.R. for 1967,* 3; and *A.R. for 1969,* 5; and Fields, *Range of Opportunity,* 91.

62. CR, *A.R. for 1961,* 10.

63. "Cyanide Leaching," in CR record group, box 12, folder 13, MTU.

64. CR, *A.R. for 1970,* 4; *A.R. for 1972,* 5; and *A.R. for 1973,* 7; and Van Pelt interview of Bingham, pp. 33–34.

65. Smith, *Mining America,* 136–37.

66. Van Pelt interview of Bingham, 2, 5, 35–36.

67. Paper, E. R. Bingham, "Reclaiming Mill Tailings Areas," April 16, 1969, 4, found in Van Pelt Collection, box 12, folder 2, MTU.

68. Bingham, "Reclaiming Mill Tailings Areas," 5; and Saffron and Bingham, "Half Dozen Goslings," 23.

69. CR, *A.R. for 1969,* 9; and *A.R. for 1973,* 7.

70. CR, *A.R. for 1971,* 6; and *A.R. for 1974,* 8.

71. Memo, Larry Chabot to J. R. Van Pelt, 10 June 1969, Van Pelt Collection, box 12, folder 2, MTU.

72. CR, *A.R. for 1971,* 6; and *A.R. for 1974,* 8.

73. *CR News,* July 1968, 3, 12; and CR, *A.R. for 1969,* 9.

74. CR, *A.R. for 1971,* 6–7.

75. CR, *A.R. for 1969,* 8.

76. CR, *A.R. for 1970,* 7.

77. CR, *A.R. for 1974,* 8.

78. *DMG,* 23 Nov. 1974.

79. CR, *A.R. for 1953,* 27; and Moe, "White Pine Mine Development," 386.

80. See, for instance, CR, *A.R. for 1953,* 27; *A.R. for 1958,* 12; *A.R. for 1962,* 14; *A.R. for 1963,* 13; and *A.R. for 1973,* 9.

81. CR, *A.R. for 1952,* 20; *A.R. for 1953,* 27; and *A.R. for 1954,* 28.

82. CR, *A.R. for 1959,* 6.

83. CR, *A.R. for 1960,* 4.

84. Ibid., 4, 6. Also see *A.R. for 1961,* 4.

85. CR, *A.R. for 1960,* 6; and *A.R. for 1963,* 13.

86. Cole's letter of 26 June 1963 is in the Van Pelt Collection, box 9, folder 6, MTU.

87. CR, *A.R. for 1962,* 14; *A.R. for 1964,* 3; *A.R. for 1967,* 3, 10; *A.R. for 1971,* 3; and *A.R. for 1974,* 10.
88. *DMG,* 21 Sept. 1981.
89. Ron Whiton to author, 24 Dec. 2008.
90. In its *A.R. for 1966,* 12, CR noted that a booming economy and "heavy military requirements" could make them labor short.
91. CR, *A.R. for 1967,* 10; *A.R. for 1971,* 3; *A.R. for 1972,* 4; and *A.R. for 1974,* 6.

CHAPTER 21. SOMETHING OLD, SOMETHING NEW: THE WHITE PINE TOWNSITE

1. The author is indebted to Vanessa McLean, a graduate student in the Industrial Communities class at Michigan Tech, for initial research on this new town, written up in her student paper "White Pine, Michigan: A 'Modern' Industrial Community," 2005.
2. *New York Times,* 20 Sept. 1953; *Detroit Free Press,* "Roto Magazine," 27 Sept. 1953; Honeywell Corporation, press release, "Copper Mining Boom Creates Model Town in a Wilderness," 1953, in CR Collection, box 12, folder 12, MTU; and *DMG,* 29 Aug. 1953.
3. *Marquette Mining Journal,* 1 Dec. 1951.
4. *Detroit Free Press,* 23 Jan. 1952.
5. Arnold R. Alanen, "The Planning of Company Communities in the Lake Superior Mining Region," *Journal of the American Planning Association* 45 (July 1979): 270.
6. Transcript of 1983 Betty J. Blum interview with Charles Booher Genther, Chicago Architects Oral History Project, Art Institute of Chicago, pp. 11, 17, 18, 32. Genther had worked on buildings at three U.S. Army flight-training schools in the early 1940s. With a principal interest in housing, he also worked with the Federal Housing Administration.
7. White Pine Copper Company, "Presentation to Mining Symposium, Jan. 12 and 13, 1954," 12, CR Collection, accession no. 549, box 8, folder 12, MTU.
8. Ibid., 12.
9. Alanen, "Planning of Company Communities," 270; White Pine Copper Company, "Presentation to Mining Symposium," 15–16; and "Company Surveys Family Needs—Then Builds Houses to Fit," *American Builder,* Nov. 1953, 75–76.
10. White Pine Copper Company, "Presentation to Mining Symposium," 15, 25; CR, *A.R. for 1952,* 16; and Ewoldt, "Mining and Milling," 21.
11. Quoted from correspondence from Larry Chabot to author, 26 Dec. 2008.
12. CR, *A.R. for 1953,* 22; and *DMG,* 21 Feb. 1953. Relatively soon, the large cafeteria was converted into a restaurant with seating for about two hundred.
13. For plans and elevations of the dormitories and the Staff House, see MTU, drawer 65, folder D. For the Service Building, see drawer 65, folder C.
14. CR, *A.R. for 1955,* 10.
15. "Trailer Park Site Plan," 10 Jan. 1952, MTU map collection, drawer 64, folder C.
16. "White Pine Bulletin," 1, no. 3 (1953).
17. "White Pine Bulletin," April 1954, includes the "White Pine Directory," CR Collection, accession no. 549, box 4, folder 4, MTU.
18. "Company Surveys Family Needs," 76.
19. Ibid., 76; Ramsey, "White Pine Copper, 86; and CR, "White Pine Copper Co.: Plant and Residential Area," ca. 1955, found in Van Pelt Collection, box 8, file 8, MTU.
20. The descriptions, sizes, and features of the White Pine houses, as given in this chapter, are largely based on their architectural drawings as produced by Pace Associates in 1952 and 1953. The various plans and elevations are located in the White Pine Collection at the MTU, drawer 65, folder E.
21. "Constructing $70 Million Michigan Copper Mine," 23. Gundlach built not only the houses, but the dormitories, cafeteria, and staff house.
22. Two typescripts, "Exterior Finishes" (revised 5 May 1953) and "Interior Color Schedule and Key," are found among the architectural drawings in drawer 65, folder E, MTU. The street with fourteen houses along it was not named yet and was called "D-2."

23. Appended to the "White Pine Bulletin" of April 1954 is a "White Pine Directory," listing the addresses of such managers as Harold Ewoldt (99 Maple Avenue), Richard Moe (91 Maple), and Dr. K. L. Olmsted (141 Maple). Found in CR Collection, accession no. 549, box 8, folder 4, MTU.

24. "Company Surveys Family Needs," 74. Reportedly, 86 percent of those surveyed ate all their meals in the kitchen.

25. The drawings "Revised General Plan, Scheme A," 7 Nov. 1952, and "White Pine Townsite," 15 July 1953, are found in MTU, drawer 64, folder A. Also see *Ironwood Daily Globe,* 24 Aug. 1952; "Constructing $70 Million Michigan Copper Mine," 17; and Ramsey, "White Pine Copper," 86.

26. CR, *A.R. for 1952,* 18; and *A.R. for 1953,* 22.

27. CR, *A.R. for 1953,* 22–24; CR, "First Ten Years," 2–3; and *White Pine Bulletin,* Aug. 1953 and April 1954, CR Collection, accession no. 549, box 8, folder 4, MTU.

28. CR, *A.R. for 1953,* 23; Harold B. Ewoldt to Willam P. Nicholls, 22 Oct. 1953, Van Pelt Collection, box 9, folder 11, MTU; and Robert P. Knight to D. M. Goodwin, 19 Nov. 1953, CR Collection, box 18, folder 23, MTU.

29. CR, *A.R. for 1954,* 23; *A.R. for 1958,* 14; and *A.R. for 1960,* 5; and "Address Given Before the Parent Teachers Association . . . by William P. Nicholls," 15 Sept. 1964, CR Collection, box 12 folder 11, MTU. In 1964 the mining company conveyed to the Carp Lake Township Board of Education the deeds to both the grade school and the high school.

30. CR, *A.R. for 1953,* 23; *A.R. for 1954,* 24; *DMG,* 20 Feb. 1954; "Townsite: Complete Buildings, 1953–1954," CR Collection, box 18, folder 23, MTU; and Van Pelt interview of Neil.

31. Moe, "White Pine Mine Development," 386; CR, *A.R. for 1954,* 24; *A.R. for 1956,* 13; *A.R. for 1957,* 13; *A.R. for 1958,* 14; and *A.R. for 1962,* 9; and CR, "First Ten Years," 3. Also see White Pine Copper Company, "White Pine Story," 18.

32. *DMG,* 18 Nov. 1993, reprinted its White Pine article of Nov. 1951; Ramsey, "White Pine Copper," 87; and CR, *A.R. for 1954,* 23. The John Lally quote is found in Allen Shoenfield, "A Miracle in the Making," *Detroit News,* 30 July 1956. Also see CR, *A.R. for 1956,* 13; and *A.R. for 1957,* 13.

33. *DMG,* 18 Nov. 1993 (reprint of Nov. 1951 article).

34. Shoenfield, "Miracle in the Making."

35. Ibid.

36. *Detroit News,* 11 Sept 1953.

37. *DMG,* 21 Feb. 1953.

38. Shoenfield, "Miracle in the Making."

39. Chris Chabot, *Tales of White Pine,* 64.

40. CR, *A.R. for 1958,* 12; map, 10 July 1960, found in White Pine Copper Company, "Community Resources Workshop," CR Collection, accession no. 549, box 8, folder 4, MTU; and White Pine Copper Company, "White Pine Story," 17–18.

41. *Milwaukee Journal,* Feb. 8, 1970.

42. Correspondence, Larry Chabot to author, 26 Dec. 2008.

43. For a discussion of the White Pine safety department, see Van Pelt interview of Neil, pp. 1, 3, 6. A brief report on "Safety" became a common part of CR's annual reports, starting in the one for 1963, p. 14.

44. Van Pelt interview of Neil, 6.

45. CR, *A.R. for 1970,* 7.

46. "Disabling Injury Statements," White Pine, 1958–70, Van Pelt Collection, box 12, folder 9, MTU.

47. CR, *A.R. for 1965,* 4.

48. *Ontonagon Herald,* 9 Jan. 1969; and *DMG,* 9 Jan. 1969.

49. *DMG,* Oct. 21, 1976.

50. Van Pelt interview of Neil, 1; and CR, *A.R. for 1970,* 11.

51. CR, *A.R. for 1960,* 5.

52. CR, *A.R. for 1961,* 8–9; *A.R. for 1962,* 9; and *A.R. for 1963,* 14.

53. CR, *A.R. for 1969,* 17.

54. *Ontonagon Herald,* 30 May 1968; and CR, *A.R. for 1961,* 9.

55. CR, *A.R. for 1966,* 12; and *A.R. for 1969,* 19.

56. CR, *A.R. for 1969,* 7.

57. *DMG,* 17 Sept. 1971.
58. CR, *A.R. for 1970,* 7.
59. Ibid., 17.
60. *DMG,* 17 Sept. 1971; and Unlimited Developments, promotional flyer, "White Pine Community Development Master Plan," ca. 1970, Ontonagon Country Historical Society Museum, White Pine Collection.
61. Unlimited Developments, "White Pine Development"; and *DMG,* 4 July 1970. Also see "An Ideal Community Becomes Reality in Michigan's U.P.," in *Michigan Challenge,* Michigan State Chamber of Commerce (Lansing, MI, May 1971), 9–11.
62. CR, *A.R. for 1971,* 7; and *DMG,* 26 Aug. and 17 Sept. 1971.
63. James Bekkala to author, Dec. 2008.
64. Larry Chabot to author, 16 Dec. 2008.

CHAPTER 22. WHITE PINE: NO SOLUTION

1. CR, *A.R. for 1975,* 2–3.
2. CR, *A.R. for 1976,* 2–3.
3. *Ontonagon Herald,* 1 Jan. and 19 and 26 Feb. 1976; and *DMG,* 25 April 1982 and 8 March 1983.
4. *Ontonagon Herald,* 24 Nov. 1976.
5. *Ontonagon Herald,* 1 June 1977; and LL&E, *A.R. for 1977,* 3.
6. The average take-home pay of the laid-off men had been $180 per week. Initially, they received state unemployment compensation ranging from $97 per week for a single man up to $136 weekly for a man with a wife and four children. In addition, they received $42.50 weekly in supplemental compensation from CR, plus hospital and medical insurance for from six to twelve months, depending on years of service (see *Ontonagon Herald,* 1 Jan. 1976).
7. *Milwaukee Journal,* 2 April 1978.
8. *DMG,* 29 March and 30 Aug. 1976.
9. *DMG,* 22 Nov. 1986.
10. *DMG,* 8 Jan. and 18 April 1979, and 22 May and 18 July 1980.
11. LL&E, *A.R. for 1978,* 26.
12. *DMG,* 18 Nov. 1980. Also see 13 Aug. 1982.
13. *DMG,* 18 Nov. 1980 and 13 Aug. 1981.
14. *DMG,* 13 Aug. 1981.
15. *DMG,* 4 March 1981.
16. *DMG,* 13 Aug. 1981.
17. *DMG,* 21 and 29 Sept. 1981.
18. *DMG,* 22 Jan. 1982.
19. *DMG,* 22 Jan. and 12 March 1982, and 8 March 1983.
20. *DMG,* 26 and 31 March 1982.
21. *DMG,* 4 April 1983.
22. *DMG,* 29 July and 1 Aug. 1983.
23. *DMG,* 19 Nov. 1983.
24. *DMG,* 22 May and 25 Aug. 1984.
25. *DMG,* 22 Aug. and 15 Nov. 1984.
26. *DMG,* 15 Nov. and 12 Dec. 1984.
27. *Toronto Globe and Mail,* 27 Dec. 1984; and *DMG,* 22 Dec. 1984.
28. *Detroit Free Press,* 31 March 1985.
29. *DMG,* 15 and 27 Feb. 1985.
30. *Detroit Free Press,* 21 and 31 March 1985.
31. *Detroit Free Press,* 31 March 1985; and undated article by Colman McCarthy in the *Washington Post,* found in White Pine, Cities and Towns, vertical files, MTU.
32. *DMG,* 26 March 1985.

33. *DMG,* 17 April 1985.

34. *Detroit Free Press,* 31 March 1985.

35. *Milwaukee Sentinel,* 17 May 1985.

36. *DMG,* 25 and 27 April, 1985; and Dixie Franklin, "White Pine Miracle," *Lake Superior Magazine,* July–Aug. 1987, 34.

37. *Milwaukee Sentinel,* 17 May 1985.

38. "White Pine Mine Reopens," in Michigan Department of Labor, *Labor Register,* Nov. 1985; *DMG,* 16 Aug., 11 and 19 Sept., 15 Oct., and 7, 8, and 25 Nov. 1985; and *Detroit News,* 16 Nov. 1986.

39. Franklin, "White Pine Miracle," 34.

40. CR, "Your Planning Team Report," 28 Jan. and 8 April 1986, MTU.

41. "Copper Range Report," no. 3 (19 Feb. 1986). This source lists all benefits covered in full, such as 365 days of hospital care, and other benefits that could be obtained by paying additional charges, such as surgical and obstetrical services.

42. "Copper Range Report," No. 3 (19 Feb. 1986) and No. 6 (29 May 1986). In 1986, all issues of "Copper Range Report" listed the currently monthly employee figure.

43. *DMG,* 19 Nov. and 24 Dec. 1986; and Pat McCarthy, "Copper Range Employees See Benefits," *LAB-ORegister* 2, no. 11 (Nov. 1987): 164–65.

44. "Copper Range Report," No. 5 (24 April 1986).

45. *DMG,* 24 March 1983, 10 Feb. and 10 March 1984, and 22 Nov. 1986.

46. *DMG,* 4 April and 5 May 1987.

47. *DMG,* 13 Feb., 3 and 24 May, and 8 June 1989.

48. *Wall Street Journal,* 29 May 1989.

49. U.S. Environmental Protection Agency, "Mine Site Visit: Copper Range Company, White Pine Mine" (Washington, DC, 5 and 6 May 1992).

50. *DMG,* 15 Aug., 12 Sept., and 13 Oct. 1992.

51. *DMG,* 1 Feb. 1995

52. *DMG,* 30 March 1995.

53. *DMG,* 7 Feb. 1995.

54. *DMG,* 17 July 1995.

55. EPA, "Mine Site Visit," 4.2; and Shepherd Miller, Inc., "Environmental Aspects of the Proposed Solution Mining Operation at the white Pine Mine, White Pine, Michigan" (June 1995), 1.6, found in White Pine Solution Mining Documents, box 1, MTU.

56. *DMG,* 14 July 1995.

57. *DMG,* 13 July 1995.

58. Shepherd Miller, "Environmental Aspects," xx.

59. *DMG,* 14 July, 17 Aug., 6 Sept., and 7 Dec. 1995. Also see EPA, "Copper Range White Pine Mine: Proposed Acid Solution Mining Project—Environmental Analysis Document (Washington, DC, 2 June 1997), ii–iii, White Pine Solution Mining Documents Collection, MS-021, box 2, MTU.

60. The seismologists recorded the White Pine blast because they were interested in determining how they might discern whether seismic activity around the world was caused by a mine collapse, an earthquake—or the underground detonation of an atomic bomb, which would be a violation of the new Comprehensive Test Ban Treaty (see Los Alamos National Laboratory, *Daily News Bulletin,* 17 Dec. 1996, 3–7).

61. *DMG,* 6 Sept. 1995 and 29 Feb. and 6 March 1996.

62. *DMG,* 7 Dec. 1995 and 29 May 1996.

63. DEQ press release, 28 May 1996.

64. Michigan DEQ, "Permit M00942: Authorization to Operate an Underground Solution Mining Facility in the State of Michigan," May 1996, 1; and *DMG,* 6 March 1996.

65. *DMG,* 19 Aug. and 20 Sept. 1996.

66. *DMG,* 19 Aug. 1996.

67. *DMG,* 2 Nov. 1996.

68. EPA, "Copper Range White Pine Mine, 7–34.

69. *DMG,* 19 Aug., 3 Sept., and 3 Oct. 1996.

70. *DMG,* 20 and 26 Sept. and 22 Oct. 1996.

71. *DMG,* 22 Oct. 1996.

72. *DMG,* 15 Oct. 1996.

73. *DMG,* 25 Nov. 1996.

74. EPA, "Copper Range White Pine Mine," xxv–xxvi, 7–22.

75. Ibid, iv.

76. *DMG,* 5 June 1997.

77. *DMG,* 30 May 1997; and *Mining Engineering,* July 1997, 18.

78. *DMG,* 30 May 1997.

79. Michigan DEQ, *Copper Range Company's Remedial Action Newsletter* 1 (March 1998), 1–3; and *DMG,* 30 Oct. 1997.

80. Michigan DEQ, *White Pine Mine Information Bulletin,* Oct. 1999, 3–4; and *DMG,* 3 Dec. 1998 and 16 Aug. 1999.

81. *DMG,* 12 and 27 May, 26 Aug., 2, 8, 9, and 11 Sept. 1998, and 12 Aug. 1999; Michigan DEQ, *White Pine Mine Information Bulletin,* 2; and "Draft Remedial Action Plan, White Pine Mine, May 2000," 16–19, White Pine Remedial Action Plan Collection, box 1, MTU.

82. *DMG,* 15 May 1998, 21 July 1999, and 23 Feb. 2001; and "Draft Remedial Action Plan, White Pine Mine, May 2000," 6.

83. *DMG,* 2 Nov. 1995, 18 Aug. 1999, and 2 and 5 May 1900; *Skilling's Mining Review,* 5 Sept. 1998, 4–5; and HudBay Minerals press release, 17 Oct. 2005.

84. *DMG,* 22 July 2000, 31 July 2001, and 27 April 2002.

85. For various school cutbacks, see *DMG,* 26 and 31 March 1982 and 6 Sept. 1984.

86. For enrollment and consolidation data, see *DMG,* 27 Oct. 1998, 15 June and 30 Oct. 1999, 4 March 2000, and 30 and 31 May 2003. Also see an undated supplement to the *DMG,* "Copper Country 2000," found in MTU.

87. *DMG,* 3 Feb. 2002 and 30 and 31 May 2003; and *Ontonagon Herald,* 21 May 2003.

88. *DMG,* 30 May and 2 June 2003.

CHAPTER 23. LEGACY

1. David T. Halkola, *Michigan Tech Centennial, 1885–1985* (Houghton, MI, 1985), 3–16, 76, 121.

2. *DMG,* 16 and 18 Aug. 2006. EPA documents covering the Torch Lake Superfund cleanup were deposited for review at the Portage Lake District Library in Houghton.

3. Scott Fisher See, "Industrial Landmarks: Shaft-Rockhouses of the Keweenaw Copper Mines" (unpublished master's thesis in Industrial Archaeology, Michigan Technological University, 2006), 85–87.

4. Lankton, "One Family's Journey," 26.

5. See, "Industrial Landmarks," 85.

6. Painesdale is indeed one of the most historic and best-preserved mining communities in the Copper Country (see Irene Jackson Henry, U.S. Department of the Interior, National Register of Historic Places, "Registration Form for Painesdale, Michigan," 3 Nov. 1992).

7. For historical maps and photos of the smelter and for descriptions and assessments of structures and artifacts, see Patrick Martin and Gianfranco Archimede, "The Quincy Mining Company Smelting Works, 1898: Historic Land Use Survey Project," unpublished report for the Keweenaw National Historical Park (Houghton, MI, June 2002).

8. U.S. Department of the Interior, National Park Service, *Keweenaw National Historical Park, Michigan: Final General Management Plan & Environmental Impact Statement* (Washington, DC, April 1998), 173–79. Maps of the park's two units are also found in this publication.

9. Ibid., 173.

BIBLIOGRAPHY

❖ ❖ ❖

ARCHIVAL/PUBLIC RECORDS

Baker Library, Harvard University
    R. B. Dun and Company Credit Record Collection
Bayliss Public Library, Sault Sainte Marie, Michigan
    Steere Special Collection
Bentley Historical Library, University of Michigan
    Daniel D. Brockway Collection
Bishop Baraga Association Archives, Marquette, Michigan
Burton Historical Collection, Detroit Public Library
    Frederic Baraga Papers, 1828–1933
Clarke Historical Library, Central Michigan University
    J. H. Pitezel Papers
Finnish American Historical Archive, Finlandia University (formerly Suomi College), Hancock, MI
    Oral History Collection
Houghton County, Michigan
    Records of Death and Articles of Association
Keweenaw County, Michigan
    Records of Death
Keweenaw National Historical Park
    Photographic collections
Library of Congress, Washington, DC
    Historic American Engineering Record Collection, Quincy Mining Company reports
Michigan Technological University Archives and Copper Country Historical Collections
    Brockway Family Collection
    Calumet and Hecla Mining Company Collections
    Copper Range Consolidated Mining Company Collections

Houghton County Public Records Collections
J. R. Van Pelt Collection
Keweenaw County Public Records Collections
Keweenaw Historical Society Collections
Photographic Collections and Digital Archives
Quincy Mining Company Collections
Roy Drier Collection
Sanborn Fire Insurance Map Collection
White Pine Solution Mining Documents Collection
Wilfred Erickson Collection
National Museum of American History, Smithsonian Institution, Washington, DC
Leavitt Collection
Ontonagon County, Michigan
Records of Death
Ontonagon County Historical Society Museum
Photographic Collection
White Pine Collection
Western Historical Collection, Norlin Library, University of Colorado
Western Federation of Miners Collection
Western Reserve Historical Society
Charles Whittlesey manuscript, "Pioneers of Lake Superior"
Wisconsin Architectural Archive, Milwaukee
Drawings of Alexander C. Eschweiler

NEWSPAPERS

Boston Sunday Globe
Calumet News
Daily Mining Gazette
Detroit Daily Free Press
Detroit Free Press
Detroit News
Ironwood Daily Globe
Lake Superior Journal
Lake Superior Miner
Marquette Mining Journal
Milwaukee Journal
Milwaukee Sentinel
Miners' Bulletin
New York Times
Northwestern Mining Journal
Ontonagon Herald
Ontonagon Miner
Portage Lake Mining Gazette
Red Jacket and Laurium Weekly News
Torch Lake Times
Toronto Globe and Mail
Työmies

MINING COMPANY ANNUAL REPORTS

American Exploring, Mining, and Manufacturing Company
Baltic Mining Company
Calumet and Hecla Mining Company
Calumet and Hecla Consolidated Mining Company
Central Mining Company
Champion Copper Company
Copper Range Consolidated Mining Company
Franklin Mining Company
Lake Superior Copper Company
Louisiana Land and Exploration Company
Minesota Mining Company
Norwich Mining Company
Ohio Trap Rock Mining Company
Pewabic Mining Company
Phoenix Mining Company
Pittsburgh and Boston Mining Company
Quincy Mining Company
Toltec Mining Company
Trimountain Mining Company
Universal Oil Products

GOVERNMENT DOCUMENTS/PUBLICATIONS

Annual Reports of the Adjutant General of the State of Michigan for the Years 1862, 1863, 1864, 1865–1866. 3 vols. Lansing, MI.

Benedict, C. Harry. "Methods and Costs of Treatment at the Calumet and Hecla Reclamation Plant." U.S. Bureau of Mines Information Circular 6357. Washington, DC. 1930.

Butler, B. S., and W. S. Burbank. *The Copper Deposits of Michigan.* U.S. Geological Survey Professional Paper 144. Washington, DC, 1929.

Crane, W. R. *Mining Methods and Practice in the Michigan Copper Mines.* U.S. Bureau of Mines Bulletin 306. Washington, DC, 1929.

Fay, Albert H., comp. *Metal-Mine Accidents in the United States during the Calendar Year 1911.* U.S. Bureau of Mines Technical Paper 40. Washington, DC, 1913.

Foster, J. W., and J. D. Whitney. *Report on the Geology and Topography of a Portion of the Lake Superior Land District in the State of Michigan, Part 1: Copper Lands.* U.S. House, 31st Cong., 1st sess., 1850. House Executive Document 69.

Henry, Irene Jackson. U.S. Department of the Interior, National Register of Historic Places, "Registration Form for Painesdale, Michigan." 3 November 1992.

Houghton County, Michigan. *Annual Reports of the County Mine Inspector.* Ca. 1888–1931.

Los Alamos National Laboratory. *Daily News Bulletin,* 17 December 1996.

McCarthy, Pat. "Copper Range Employees See Benefits." *LABORegister* 2, no. 11 (November 1987): 164–65.

McElroy, G. E. "Natural Ventilation of Michigan Copper Mines." U.S. Bureau of Mines Technical Paper 516. Washington, DC, 1932.

Michigan. *Annual Reports of the Commissioner of Mineral Statistics* (Lansing, MI)
———. *Annual Reports of the Secretary of State on the Registration of Births and Deaths, Marriages and Divorces in Michigan.*
———. *Annual Reports of the Superintendent of Public Instruction.*
———. *Biennial Reports of the Board of Corrections and Charities.*

———. Department of Environmental Quality. *Copper Range Company's White Pine Mine Remedial Action Newsletter* 1 (March 1998).

———. Department of Environmental Quality. "Permit M00942, Authorization to Operate an Underground Solution Mining Facility in the State of Michigan." May 1996.

———. Department of Environmental Quality. *White Pine Mine Information Bulletin,* October 1999.

———. *Journal of the House* (Lansing, 1887).

———. *Journal of the Senate* (Lansing, 1887).

———. *Michigan Manual (Red Book): Official Directory and Legislative Manual.* Lansing, 1901, 1911, 1921, 1931, 1941, 1951–52, 1959–60, and 1969–70.

———. *Public Acts* (Lansing, 1887, 1897, and 1911).

———. *Report of the Board of State Commissioners for the General Supervision of Charitable, Penal, Pauper, and Reformatory Institutions* (Lansing, 1873).

———. *Statistics of the State of Michigan Collected for the Ninth Census of the United States, June 1, 1870* (Lansing, 1873).

Superintendents of the Poor, Houghton County, MI. *Annual Reports for 1920–1922.*

U.S. Bureau of Labor Statistics. *Michigan Copper District Strike.* Bulletin 139. Washington, DC, 1914.

U.S. Department of the Interior, National Park Service. *Keweenaw National Historical Park: Final General Management Plan & Environmental Impact Statement.* Washington, DC, April 1998.

U.S. Engineers Office. "History of the Keweenaw Waterway, Michigan." Duluth, MN, 1940.

U.S. Environmental Protection Agency. "Copper Range White Pine Mine: Proposed Acid Solution Mining Project—Environmental Analysis Document." Washington, DC, 2 June 1997.

U.S. Environmental Protection Agency. "Mine Site Visit: Copper Range Company, White Pine Mine." Washington, DC, 5 and 6 May 1992.

U.S. House Committee on Mines and Mining. *Conditions in Copper Mines of Michigan.* Hearings before subcommittee pursuant to House Resolution 387. 63rd Cong., 2nd sess. 7 pts., 1914.

U.S. Immigration Commission. *Immigrants in Industries. Part 17: Copper Mining and Smelting.* Washington, DC, 1911.

ARTICLES, BOOKS, AND UNPUBLISHED WORKS

Adams, Julia Hubbard. *Memories of a Copper Country Childhood.* Booklet at the Michigan Technological University Archives. By the author, n.d.

Alanen, Arnold R. "The Planning of Company Communities in the Lake Superior Mining Region." *Journal of the American Planning Association* 45 (July 1979): 256–78.

Alanen, Arnold R., and Lynn Bjorkman. "Plats, Parks, Playgrounds, and Plants: Warren H. Manning's Landscapes Designs for the Mining Districts of Michigan's Upper Peninsula, 1899–1932." *IA: Journal of the Society for Industrial Archaeology* 24, no. 1 (1998): 41–60.

Andre, George G. *Rock Blasting: A Practical Treatise on the Means Employed in Blasting Rocks.* London, 1878.

Andrew, Clarence A. "'Big Annie' and the 1913 Michigan Copper Strike." *Michigan History* 57 (Spring 1973): 53–68.

Armstrong, Ray W. "Compound Steam Hoist Installation of the Quincy Mining Company." *Proceedings of the Lake Superior Mining Institute* 22 (1922): 39–41.

Ayer, Frank A. "White Pine Mine . . . Potential Major Copper Producer." *Mining Congress Journal* 36, no. 12 (1950): 26–30, 55.

Bacon, Robert C. "New Automated Transportation System for Open Pit Mines." *Skillings' Mining Review* 51 (24 August 1968): 1, 4–5, 10.

Barkell, William. "Honest John Stanton." Pamphlet. Houghton County Historical Society, July 1974.

Barnett, Leroy. "Lac La Belle, Keweenaw's First Ship Canal." *Michigan History* 69 (January–February 1985): 41–46.

———. *Mining in Michigan: A Catalog of Company Publications, 1845–1980.* Marquette, MI, 1983.

———. "What the Sam Hill!" *Michigan History,* May–June 2004, 14–18.

Bartlett, Richard A. *The New Country: A Social History of the Frontier, 1776–1890.* New York, 1974.

Benedict, C. Harry. *Lake Superior Milling Practice: A Technical History of a Century of Copper Milling.* Houghton, MI, 1955.

———. "Milling at the Calumet & Hecla Consolidated Copper Company." *Mining Congress Journal* 17 (October 1931): 519–27.

———. *Red Metal: The Calumet and Hecla Story.* Ann Arbor, MI, 1952.

Benedict, C. Harry, and H. C. Kenny. "Ammonia Leaching of Calumet and Hecla Tailings." *Transactions of the American Institute of Mining and Metallurgical Engineers* 70 (1924): 595–610.

Biographical Dictionary of Copper Country Architects. http://www.social.mtu.edu/CopperCountryArchitects/.

Biographical Record of Houghton, Baraga and Marquette Counties. Chicago, 1903.

Bjorkman, Lynn. "Calumet Village, Laurium Village, Calumet Township: Historic and Architectural Survey—Phase I." Report by the Western Upper Peninsula Planning and Development Regional Commission. Houghton, MI, 31 August 1995.

Blake, William P. "The Blake Stone- and Ore-Breaker: Its Invention, Forms and Modifications, and Its Importance in Engineering Industries." *Transactions* of the *American Institute of Mining Engineers* 33 (1902): 988–1031.

———. "The Mass Copper of the Lake Superior Mines, and the Method of Mining It." *Transactions of the American Institute of Mining Engineers* 4 (1875–76): 110–11.

Blandy, John F. "Stamp Mills of Lake Superior." *E&MJ,* 27 June 1874, 401.

———. "Stamp Mills of Lake Superior." *Transactions of the American Institute of Mining Engineers* 2 (1874): 208–15.

Bornhorst, Theodore J., and Larry D. Lankton. "Copper Mining: A Billion Years of Geologic and Human History." Chap. 13 in *Michigan Geography and Geology,* edited by Randall Schaetzl, with Joe Darden and Danita Brandt, 150–73. New York, 2009.

Bornhorst, Theodore J., and William I. Rose. *Self-Guided Geological Field Trip to the Keweenaw Peninsula, Michigan.* Proceedings of the Institute on Lake Superior Geology 40, part 2. Houghton, MI, 1994.

Bornhorst, Theodore J., William I. Rose Jr., and James B. Paces. *Field Guide to the Geology of the Keweenaw Peninsula, Michigan.* Houghton, MI, 1983.

Bourland, P. D. "The Medical Dept. of C&H." *Mining Congress Journal* 17 (October 1931): 555–57.

Broderick, T. M., and C. D. Hohl. "Geology and Exploration in the Michigan Copper District." *Mining Congress Journal* 17 (October 1931): 478–81, 486.

Burleigh Rock-Drill Company. *Burleigh Rock-Drill Company, Manufacturers of Pneumatic Drilling Machines, Air Compressors, and Other Machinery.* Fitchburg, MA, ca. 1873–76.

Burt, Roger, ed. *Cornish Mining: Essay on the Organization of Cornish Mines and the Cornish Mining Economy.* Newton Abbot, UK, 1969.

Calvert, Monte A. *The Mechanical Engineer in America: 1830–1910.* Baltimore, MD, 1967.

Cannon, George M. *A Narrative of One Year in the Wilderness.* Ann Arbor, MI, 1982.

Chabot, Chris. *Tales of White Pine: An Illustrated Oral History of White Pine, Michigan & Environs.* Ontonagon, MI, 1979.

Chaput, Donald. *The Cliff: America's First Great Copper Mine.* Kalamazoo, MI, 1971.

Clarke, Robert E. "Notes from the Copper Region." *Harper's New Monthly Magazine* 4 (March 1853): 433–48; (April 1853): 577–88.

Coggin, Frederick G. "Notes on the Steam Stamp." *Engineering & Mining Journal,* 20 March 1886, 210–11; 27 March, 232–34; 2 April, 248–50.

Collins, Frederick L. "Paine's Career Is a Triumph of Early American Virtues." *American Magazine,* June 1928, 40–41, 139–43.

Collins, Glenville A. "Efficiency-Engineering Applied to Mining." *Transactions of the American Institute of Mining Engineers* 43 (1912): 649–62.

Committee of the Copper Country Commercial Club of Michigan. *Strike Investigation.* Chicago, 1913.

"Company Surveys Family Needs—Then Builds Houses to Fit." *American Builder,* November 1953, 75–76.

Conant, H. D. "Copper Smelting in Michigan." *School of Mines Quarterly* 4, no. 32 (July 1911).

———. "The Historical Development of Smelting and Refining Native Copper." *Mining Congress Journal* 17 (October 1931): 531–32.

"Constructing $70 Million Michigan Copper Mine." *Excavating Engineer* 48 (January 1954): 16–23.

Cooper, James B. "Historical Sketch of Smelting and Refining Lake Copper." *Proceedings of the Lake Superior Mining Institute* 7 (1901): 44–49.

Copper Country Commercial Club. *Strike Investigation.* Chicago, 1913.

Copper Range Company. "Controlling the White Pine Environment." *Copper Range News* 8, no. 7 (July 1968): 2–3, 12.

Cowan, Ruth Schwartz. *More Work for Mother: The Ironies of Household Technology from the Open Hearth to the Microwave.* New York, 1983.

DeSollar, T. C. "Rockhouse Practice of the Quincy Mining Company." *Proceedings of the Lake Superior Mining Institute* 13 (1912): 217–26.

Dickinson, John N. *To Build a Canal: Sault Ste. Marie, 1853–1854 and After.* Miami, OH, 1981.

Dorr, John A., Jr., and Donald F. Eshman. *Geology of Michigan.* Ann Arbor, MI, 1977.

Drinker, Henry S. *A Treatise on Explosive Compounds, Machine Rock Drills and Blasting.* New York, 1883.

———. *Tunneling, Explosive Compounds and Rock Drills.* New York, 1878.

Dubofsky, Melvyn. *Industrialism and the American Worker, 1865–1920.* Arlington Heights, IL, 1975.

Dunbar, Willis. *All Aboard! A History of Railroads in Michigan.* Grand Rapids, MI, 1969.

Dunbar, Willis F., and George S. May. *Michigan: A History of the Wolverine State.* Grand Rapids, MI, 1995.

Eckert, Kathryn Bishop. *Buildings of Michigan.* New York, 1993.

———. *The Sandstone Architecture of the Lake Superior Region.* Detroit, 2000.

Egleston, Thomas. "Copper Dressing in Lake Superior." *Metallurgical Review* 2 (May 1878): 227–36; (June 1878): 285–300; (July 1878): 389–409.

———. "Copper Mining on Lake Superior." *Transactions of the American Institute of Mining Engineers* 6 (1879): 275–312.

———. "Copper Refining in the United States." *Transactions of the American Institute of Mining Engineers* 9 (1880–81): 678–730.

Eissler, Manuel. *A Handbook on Modern Explosives.* London, 1897.

———. *The Modern High Explosives.* New York, 1884.

Ewoldt, Harold B. "Mining and Milling at White Pine." *Mining Congress Journal* 41, no. 3 (March 1955): 24–26.

Fadner, Lawrence Trever. *Fort Wilkins, 1844, and the U.S. Mineral Land Agency, 1843, Copper Harbor, Michigan.* New York, 1966.

Fichtel, Carl L. "Underground Power Distribution and Haulage." *Mining Congress Journal* 17, no. 10 (October 1931): 503–6, 508.

Fields, Richard A. *Range of Opportunity: An Historic Study of the Copper Range Company.* Hancock, MI, 1997.

Fischer, A. F. "Medical Reminiscence." *Michigan History* 7 (January–April 1923): 27–33.

Fisher, Nancy Beth. "Quincy Mining Company Housing, 1840s–1920s." Unpublished master's thesis, Michigan Technological University, 1997.

Forero, George A., Jr. "The Copper Range Railroad and the Copper Country." *Mid-Continent Railway Gazette* 37, no. 4 (December 2004): 4–42.

Forster, John H. "Early Settlement of the Copper Regions of Lake Superior." *Michigan Pioneer Collections* 7 (1884): 181–93.

———. "Lake Superior Country." *Michigan Pioneer and Historical Society Collections* 8 (1886): 136–45.

———. "Life in the Copper Mines of Lake Superior." *Michigan Pioneer and Historical Collections* 11 (1887): 175–86.

———. "Some Incidents in Pioneer Life in the Upper Peninsula of Michigan." *Michigan Pioneer and Historical Society Collection* 17 (1892): 332–45.

———. "War Time in the Copper Mines." *Michigan Pioneer and Historical Collections* 18 (1892): 375–82.

Franklin, Dixie. "White Pine Miracle." *Lake Superior Magazine,* July–August 1987, 28–35.

Friedman, Lawrence M. *A History of American Law.* New York, 1973.

Friggens, Thomas. "Fort Wilkins: Army Life on the Frontier." *Michigan History* 61 (Fall 1977): 221–50.

Fuller, George Newman, ed. *Geological Reports of Douglass Houghton, First State Geologist of Michigan, 1837–1845*. Lansing, MI, 1928.

Gates, William B. *Michigan Copper and Boston Dollars: An Economic History of the Michigan Copper Mining Industry*. Cambridge, MA, 1951.

Gedicks, Al. "Ethnicity, Class Solidarity, and Labor Radicalism among Finnish Immigrants in Michigan Copper Country." *Politics and Society* 7, no. 2 (1977): 127–56.

———. "The Social Origins of Radicalism among Finnish Immigrants in the Midwestern Mining Communities." *Review of Radical Political Economics* 8 (Fall 1976): 1–31.

"Giant Lifting Wheel for the Copper Mines." *Scientific American,* n.s., 50, no. 10 (8 March 1884): cover.

Goodwin, L. Hall. "Shaft-Rockhouse Practice in Copper Country." *Engineering & Mining Journal* 99 (19 June 1915): 1061–66; (27 June): 1107–10; 100 (3 July 1915): 7–12; (10 July): 53–57.

Graebner, William. *Coal-Mining Safety in the Progressive Period*. Lexington, KY, 1976.

Gray, Susan E. *The Yankee West: Community Life on the Michigan Frontier*. Chapel Hill, NC, 1996.

Griebel, John E. "Cultural Landscape Report: Nonesuch Mine & Village." Unpublished master's thesis in Industrial Archaeology, Michigan Technological University, 2008.

Halkola, David T. *Michigan Tech Centennial, 1885–1985*. Houghton, MI, 1985.

Haller, Frank H. "Transportation System and Coal Dock." *Mining Congress Journal* 17, no. 10 (October 1931): 515–18.

Halsey, John R. "'Ancient Diggings': A Review of Nineteenth-Century Observations in the Prehistoric Copper Mining Pits of the Lake Superior Basin." Paper presented at the 54th Annual Meeting, Midwest Archaeological Conference, Milwaukee, WI, 17 October 2008.

———. "Ancient Pits of the Copper Country." *Michigan History,* May–June 2009, 40–47.

———. "Miskwabik–Red Metal: Lake Superior Copper and the Indians of Eastern North America." *Michigan History,* September–October 1983, 32–41.

Harrison, Joseph S. "Labor Law and the Michigan Supreme Court, 1890–1930." Unpublished master's thesis, Wayne State University, 1980.

Henry, Irene Jackson. "Registration Form for Painesdale, Michigan." U.S. Department of the Interior, National Register of Historic Places, 3 November 1992.

Hixon, Hiram W. *Notes on Lead and Copper Smelting and Copper Converting*. New York, 1898.

Hoagland, Alison K. "The Boardinghouse Murders: Housing and American Ideals in Michigan's Copper Country in 1913." *Perspectives in Vernacular Architecture: Journal of the Vernacular Architecture Forum* 11 (2004): 1–19.

Hoagland, Alison K., Erik C. Nordberg, and Terry S. Reynolds, eds. *New Perspectives on Michigan's Copper Country*. Hancock, MI, 2007.

Hollingsworth, Sandra. *The Atlantic: Copper and Community South of Portage Lake*. Hancock, MI, 1978.

Hood, O. P. "Deep Hoisting in the Lake Superior District: Some of the Engines Used to Hoist from a Depth of over 4,900 Feet." *Mines and Minerals,* July 1904, 614–17.

Hoy, Suellen. *Chasing Dirt: The American Pursuit of Cleanliness*. New York, 1995.

Hulbert, Edwin J. *"Calumet Conglomerate," an Exploration and Discovery Made by Edwin J. Hulbert, 1854 to 1864*. Ontonagon, MI, 1893.

Hunter, Louis C. *A History of Industrial Power in the United States, 1780–1930*. Vol. 2, *Steam Power*. Charlottesville, NC, 1985.

Hunter, Louis C., and Lynwood Bryant. *A History of Industrial Power in the United States, 1780–1930*. Vol. 3, *The Transmission of Power*. Cambridge, MA, 1991.

Hybels, Robert James. "The Lake Superior Copper Fever, 1841–47." *Michigan History* 34 (1950): 97–119, 224–44, 309–26.

Hyde, Charles K. *Copper for America: The United States Copper Industry from Colonial Times to the 1990s*. Tucson, AZ, 1998.

———. "An Economic and Business History of the Quincy Mining Company." Unpublished report. Historic American Engineering Record, Library of Congress. Washington, DC, 1978.

———. "From 'Subterranean Lotteries' to Orderly Investment: Michigan Copper and Eastern Dollars, 1841–1865." *Mid-America: An Historical Review* 66 (January 1984): 3–20.

———. *The Northern Lights: Lighthouses of the Upper Great Lakes*. Lansing, MI, 1986.

————. "Undercover and Underground: Labor Spies and Mine Management in the Early Twentieth Century." *Business History Review* 60 (Spring 1986): 1–27.

"An Ideal Community Becomes Reality in Michigan's U.P." In *Michigan Challenge*. Michigan State Chamber of Commerce (Lansing, May 1971), 9–11.

Ihlseng, M. C., and Eugene B. Wilson. *A Manual of Mining*. New York, 1911.

Jackman, Sidney W., and John F. Freeman, eds. *American Voyageur: The Journal of David Bates Douglass*. Marquette, MI, 1969.

Jackson, Charles T. "Reports on the Mines and Minerals Belonging to the Lake Superior Copper Company." In *A Brief Account of the Lake Superior Copper Company*, 8–16. Boston, 1845.

Jackson, J. F. "Copper Mining in Upper Michigan: A Description of the Region, the Mines, and Some of the Methods and Machinery Used." *Mines and Minerals,* July 1903, 535–40.

Jamison, James K. *This Ontonagon Country: The Story of An American Frontier*. Calumet, MI, 1965.

Karni, Michael G., Matti E. Kaups, and Douglas J. Olilla, eds. *The Finnish Experience in the Western Great Lakes Region: New Perspectives*. Turku, Finland, 1975.

Klingbeil, Dorothy J. "Copper Mining and the Sandwich Kids of Painesdale." Unpublished graduate paper, Northern Michigan University, 1973.

Kolehmainen, John I. "The Inimitable Marxist: The Finnish Immigrant Socialists." *Michigan History* 36 (1952): 395–405.

Krause, David J. *The Making of a Mining District: Keweenaw Native Copper, 1500–1870*. Detroit, 1992.

Landon, David B., and Timothy A. Tumberg. "Archeological Perspectives on the Diffusion of Technology: An Example from the Ohio Trap Rock Mine Site." *IA: The Journal of the Society for Industrial Archeology* 22, no. 2 (1996): 40–57.

Lankton, Larry D. *Beyond the Boundaries: Life and Landscape at the Lake Superior Copper Mines, 1840–1875*. New York, 1997.

————. *Cradle to Grave: Life, Work, and Death at the Lake Superior Copper Mines*. New York, 1991.

————. *Keweenaw Copper: Mines, Mills, Smelters, and Communities*. Guidebook for the 26th Annual Conference of the Society for Industrial Archeology. Houghton, MI, 1997.

————. "The Machine *under* the Garden: Rock Drills Arrive at the Lake Superior Copper Mines, 1868–1883." *Technology and Culture* 24 (January 1983): 1–37.

————. "One Family's Journey to 'Earthly Paradise.'" *Michigan History,* November–December 2001, 26–30.

————. "Technological Change at the Quincy Mine, ca. 1846–1931." Unpublished Historic American Engineering Record report, Library of Congress. Washington, DC, 1978.

————. "Victoria Dam." Historic American Engineering Record report (HAER No. MI-49). HAER collection, Library of Congress, Washington, DC.

Lankton, Larry D., and Charles K. Hyde. *Old Reliable: An Illustrated History of the Quincy Mining Company*. Hancock, MI, 1982.

Lankton, Larry D., and Jack K. Martin. "Technological Advance, Organizational Structure, and Underground Fatalities in the Upper Michigan Copper Mines, 1860–1929." *Technology and Culture* 28 (January 1987): 42–66.

Lanman, Charles. *A Summer in the Wilderness Embracing a Canoe Voyage up the Mississippi and around Lake Superior*. New York, 1847.

Leavitt, Erasmus D. "The Superior." *Transactions of the American Society of Mechanical Engineers* 2 (1881): 106–21.

Lehto, Steve. *Death's Door: The Truth behind Michigan's Largest Mass Murder*. Troy, MI, 2006.

"Letter from the Secretary of War, in Response to the Deposit of Silt and Sand in Portage Lake, Michigan." In *Executive Documents of the House of Representatives for the Second Session of the Forty-Seventh Congress, 1881. No. 85*. Washington, DC, 1884.

Link, Arthur S., and Richard L. McCormick. *Progressivism*. Arlington Heights, IL, 1983.

Lovell, Endicott R., and Herman C. Kenny. "Present Smelting Practice." *Mining Congress Journal* 17 (October 1931): 533–38.

MacNaughton, James. "History of the Calumet and Hecla since 1900." *Mining Congress Journal* 17 (October 1931): 474–77.

Magnaghi, Russell M. *Miners, Merchants, and Midwives: Michigan's Upper Peninsula Italians.* Marquette, MI, 1987.

"The Manufacture of Power Drills for Mining, Excavating, Etc." *Scientific American* 43 (25 December 1880): 399, 402.

Martin, Patrick, and Gianfranco Archimede. "The Quincy Mining Company Smelting Works, 1898: Historic Land Use Survey Project." Unpublished report prepared for Keweenaw National Historical Park. Houghton, MI, June 2002.

Martin, Susan R. *Wonderful Power: The Story of Ancient Copper Working in the Lake Superior Basin.* Detroit, 1999.

Mason, Philip P., ed. *Copper Country Journal: The Diary of Schoolmaster Henry Hobart, 1863–1864.* Detroit, 1991.

McNear, Sarah. "Quincy Mining Company: Housing and Community Services, ca. 1860–1931." Unpublished Historic American Engineering Record report, Library of Congress. Washington, DC, 1978.

Mendelsohn, Albert. "Mining Methods and Costs at the Champion Copper Mine, Painesdale, Michigan." U.S. Bureau of Mines, Information Circular 6515. Washington, DC, 1931.

Mercer, Harry T. "Fifty Years on the South Range." Unpublished manuscript, Michigan Technological University Archives. Houghton, MI, n.d., 54 pp.

———. "Rock House Practice of the Copper Range Consolidated Company." *Proceedings of the Lake Superior Mining Institute* 17 (1912): 283–89.

Moe, Richard F. "White Pine Mine Development." *Mining Engineering* 6 (April 1954): 381–86.

Monette, Clarence J. *Baltic Michigan.* Lake Linden, MI, 1996.

———. *The Copper Range Railroad.* Calumet, MI, 1989.

———. *Painesdale, Michigan: Old and New.* Lake Linden, MI, 1983.

———. *Trimountain and Its Copper Mines.* Lake Linden, MI, 1991.

Moore, Charles. "The Ontonagon Copper Boulder in the U.S. National Museum." Report of the U.S. National Museum, 1023–30. Washington, DC, 1895.

Munroe, H. S. "The Losses in Copper Dressing at Lake Superior." *Transactions of the American Institute of Mining Engineers* 8 (1879–80): 409–51.

"Nordberg Compound Steam Hosting Engine, Quincy Mining Company, No. 2 Shaft, Hancock, Michigan." *LSMI Proc.* 22 (1922): 192–94.

O'Connell, Charles F., Jr. "A History of the Quincy and Torch Lake Railroad Company, 1888–1927." Unpublished Historic American Engineering Record report, Library of Congress. Washington, DC, 1978.

———. "Quincy Mining Company: Stamp Mills and Milling Technologies, ca. 1860–1931." Unpublished Historic American Engineering Record report, Library of Congress. Washington, DC, 1978.

Paine, Webber. *Paine, Webber & Company, 1880–1930: A National Institution. Boston.* 1930.

Pappas, Efstathios I. "Swedetown Location: Hope and Failure in a Company Neighborhood." Unpublished master's thesis, Michigan Technological University, 2002.

Peters, Edward Dyer, Jr. *Modern American Methods of Copper Smelting.* New York, 1887.

———. *Modern Copper Smelting.* New York, 1895.

———. *The Principles of Copper Smelting.* New York, 1907.

Polk and Company's Houghton County Directory, 7 (1907–8):1067–86.

Polk's 1910 Calumet, Houghton, Hancock, and Laurium Directory. Detroit, 1909.

Piggot, A. B. *The Porphyry Coppers.* New York, 1933.

Pitezel, Rev. John H. *Lights and Shades of Missionary Life: Containing Travels, Sketches, Incidents and Missionary Efforts during the Nine Years Spent in the Region of Lake Superior.* 1857. Reprint, Cincinnati, 1882.

Pope, Graham. "Some Early Mining Days at Portage Lake." *Proceedings of the Lake Superior Mining Institute* 7 (1901): 17–31.

Portage Lake and River Improvement Company. *Articles of Association and By-laws.* Detroit, 1863.

Portage Lake and Lake Superior Ship Canal Company. *Articles of Incorporation.* Detroit, 1864.

Potter, Ocha, and Samuel Richards. "Osceola Lode Operations." *Mining Congress Journal* 17 (October 1931): 487–90.

Preservation Urban Design Incorporated. *Architectural Analysis: Fort Wilkins Historic Complex, Fort Wilkins State Park, Copper Harbor, Michigan.* Ann Arbor, MI, 1976.

Puotinen, Arthur Edwin. "Finnish Radicals and Religion in Midwestern Mining Towns, 1865–1914." Unpublished PhD diss., University of Chicago, 1973.

"The Quincy Hoist." *E&MJ*, 11 December 1920, 11, 126.

Ramsey, R. H. "White Pine Copper: Mine, Mill and Smelter Construction Stepped up as New Michigan Project Nears Production Stage." *Engineering and Mining Journal* 154 (January 1953): 72–78.

Rawlings, J. W. "Recollections of a Long Life." In *Copper Country Tales.* Vol. 1, edited by Roy Drier, 75–127. Calumet, MI, 1967.

Reed, James. *From Private Vice to Public Virtue: The Birth Control Movement and American Society since 1830.* New York, 1978.

Reynolds, Terry S. "A Narrow Window of Opportunity: The Rise and Fall of the Fixed Steel Dam." *Industrial Archeology* 15 (1989): 1–20.

Rice, Claude T. "Copper Mining at Lake Superior." *Engineering & Mining Journal* 94 (20 and 27 July; 3, 10, 17, 24, and 31 August 1912): 119–24, 171–75, 217–21, 267–70, 307–10, 365–68, 405–7.

———. "Labor Conditions at Calumet & Hecla." *Engineering & Mining Journal* 92 (23 December 1911): 1235–39.

———. "Labor Conditions at Copper Range." *Engineering & Mining Journal* 94, no. 26 (28 December 1912): 1229–32.

Rickard, T. A. *The Copper Mines of Lake Superior.* New York, 1905.

———. "The Use of Native Copper by the Indigenes of North America." *Journal of the Royal Anthropological Institute of Great Britain and Ireland* 64 (1934): 265–87.

Robinson, O. W. "Recollections of Civil War Conditions in the Copper Country." *Michigan History Magazine* 3 (October 1919): 598–609.

Root, Robert L., Jr., ed. *"Time by Moments Steals Away": The 1848 Journal of Ruth Douglass.* Detroit, 1998.

Rosenberg, Nathan. *Technology and American Economic Growth.* New York, 1972.

Rowe, John. *The Hard Rock Men: Cornish Immigrants and the North American Mining Frontier.* New York, 1974.

Rowse, A. L. *The Cousin Jacks: The Cornish in America.* New York, 1969.

Royce, Ward. "Scraping and Loading in Mines with Small Compressed-Air Hoists." *Engineering & Mining Journal* 112 (1921): 925–31, 973–77, 1014–19.

Sawyer, Alvah L. *A History of the Northern Peninsula and Its People.* Chicago, 1911.

Schacht, W. H. "The Mining Methods of the Copper Range Company, Houghton County, Michigan." *Proceedings of the Lake Superior Mining Institute* 22 (1922): 76–99.

Schlereth, Thomas J. *Victorian America: Transformations in Everyday Life.* New York, 1991.

Schoolcraft, Henry Rowe. *Narrative Journal of Travels through the Northwestern Regions of the United States.* Albany, NY, 1821.

Schubert, George P. "Development of Copper Smelting Practice at Michigan Smelter." *LSMI Proc.* 27 (1929): 231–47.

"Scrap Recovery Campaign in Michigan Iron and Copper Country a Model." *Mining and Metallurgy* 24 (July 1943): 317.

See, Scott Fisher. "Industrial Landmarks: Shaft-Rockhouses of the Keweenaw Copper Mines." Unpublished master's thesis in Industrial Archeology, Michigan Technological University, 2006.

Sharpless, F. F. "Ore Dressing on Lake Superior." *Proceedings of the Lake Superior Mining Institute* 2 (1894): 97–104.

Skillings, David N. "White Pine Copper Co.: Huge Construction Program Nears End." *Skillings' Mining Review* 43 (25 December 1954): 1–3

———. "White Pine Copper Company, Ontonagon County, Michigan." *Skillings' Mining Review* 41, no. 35 (6 December 1952): 1–2.

Smith, Duane A. *Mining America: The Industry and the Environment, 1800–1980.* Lawrence, KS, 1987.

Soddy, Thomas P., and Allan Cameron. "Hoisting Equipment and Methods." *Mining Congress Journal* 17, no. 10 (October 1931): 509–14.

St. John, John R. *A True Description of the Lake Superior Country.* New York, 1846.

Starr, Paul. *The Social Transformation of American Medicine.* New York, 1982.

Stevens, Horace J. *The Copper Handbook: A Manual of the Copper Industry of the World.* Vols. 1 (1900), 2 (1902), 3 (1903), 6 (1906), 8 (1908), 10 (1910–11). Houghton, MI.

Strasser, Susan. *Never Done: A History of American Housework.* New York, 1982.

Suffron, John R., and Edward R. Bingham. "Half a Dozen Goslings." *Michigan Natural Resources* 43 (November–December, 1974): 22–25.

Swineford, Alfred P. *History and Review of the Material Resources of the South Shore of Lake Superior.* Marquette, MI, 1877.

Thurner, Arthur W. *Calumet Copper and People: History of a Michigan Mining Community.* By the author, 1974.

———. *Rebels on the Range: The Michigan Copper Miners' Strike of 1913–1914.* Lake Linden, MI, 1984.

———. *Strangers and Sojourners: A History of Michigan's Keweenaw Peninsula.* Detroit, 1994.

———. "The Western Federation of Miners in Two Copper Camps: The Impact of the Michigan Copper Miners' Strike on Butte's Local No. 1." *Montana: The Magazine of Western History,* Spring 1983, 30–45.

"The Upper Peninsula of Michigan." *Harper's New Monthly Magazine,* May 1882, 892–902.

Vago, David A. "An Interpretive Plan for the Calumet and Hecla Mill Site." Unpublished master's thesis in Industrial Archaeology, Michigan Technological University, 2005.

Van Gelder, Arthur Pine, and Hugo Schlatter. *History of the Explosives Industry in America.* 1927. Reprint, New York, 1972.

Vivian, Arthur C. "Cutting Mass Copper." *Engineering & Mining Journal* 98 (7 November 1914): 825.

Vivian, Harry. "Mining Methods Used on the Calumet Conglomerate Lode." *Mining Congress Journal* 17 (October 1931): 496–502, 508.

Walling, Regis M., and Rev. N. Daniel Rupp, eds. *The Diary of Bishop Frederic Baraga.* Detroit, 1990.

Western Historical Society. *History of the Upper Peninsula of Michigan Containing a Full Account of Its Early Settlement: Its Growth, Development and Resources.* Chicago, 1883.

White Pine Copper Company. *A Visit to the White Pine Copper Company, a Division of the Copper Range Company.* White Pine, MI, 1963.

———. *The White Pine Story.* White Pine, MI, 1968.

"White Pine Uses New Methods and Equipment for Mine Development." *Mining World* 16 (April 1954): 34–38.

Whittlesey, Charles C. "Ancient Mining on the Shores of Lake Superior." *Smithsonian Contributions to Knowledge* 13, no. 4 (1863): 1–29.

Wilson, Thomas. "Quincy Hoist—Largest in World." *Power* 53, no. 3 (18 January 1921): 90–95.

Wright, H. G. "The Quincy Amygdaloid Mills." *Engineering & Mining Journal,* 22 July 1911, 166–68.

Wright, James North. "The Development of the Copper Industry of Northern Michigan." *Publications of the Michigan Political Science Association* 3 (January 1899): 127–41.

Young, Otis E., Jr. "The American Copper Frontier, 1640–1893." *The Speculator: A Journal of Butte and Southwest Montana History* 1 (Summer 1984): 4–15.

INDEX

❖ ❖ ❖

Page numbers in italics refer to illustrations.